MAPPING THE MIND

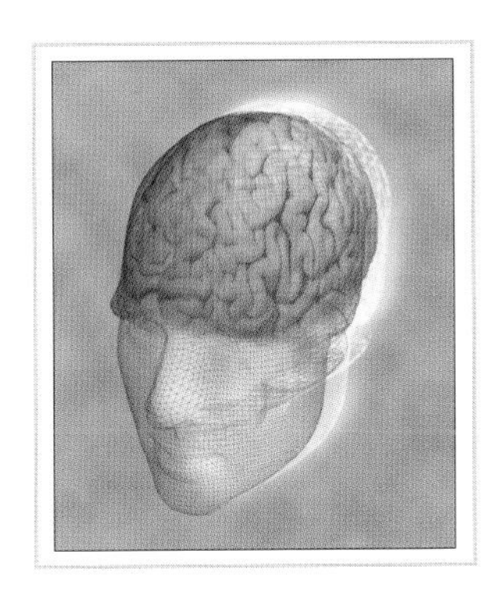

RITA CARTER

Consultant: Professor Christopher Frith

WEIDENFELD & NICOLSON
LONDON

First published in Great Britain in 1998
by Weidenfeld & Nicolson

A CIP catalogue record for this book is available from the British Library

Designed and illustrated by MoonRunner Design Ltd
Printed and bound in Italy

Weidenfeld & Nicolson

The Orion Publishing Group Ltd
5 Upper Saint Martin's Lane
London WC2H 9EA

Contents

Introduction

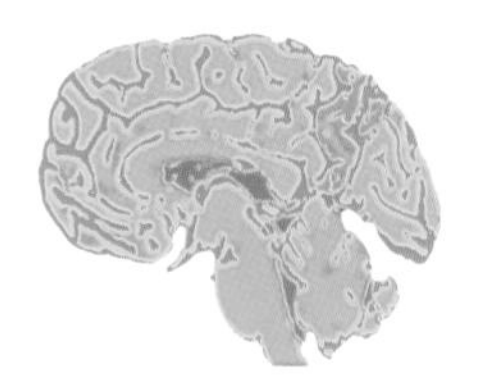

The human brain has been slow to give up its secrets. Until recently the machinations that give rise to our thoughts, memories, feelings and perceptions were impossible to examine directly – their nature could only be inferred by observing their effects. Now, however, new imaging techniques make the internal world of the mind visible, much as X-rays reveal our bones. As we enter the twenty-first century functional brain scanning machines are opening up the territory of the mind just as the first ocean-going ships once opened up the globe.

The challenge of mapping this world – locating the precise brain activity that creates specific experiences and behavioural responses – is currently engaging some of the finest scientists in the world. This book brings news of their discoveries in a way that will make them comprehensible even to those with no knowledge of, or specific interest in, science.

Everyone should be enthralled by this venture because it is giving us greater understanding about one of the oldest and most fundamental of mysteries – the relationship between brain and mind. It is also providing fascinating insights into ourselves and shedding light on aberrant and bizarre behaviour. The biological basis of mental illness, for example, is now demonstrable: no one can reasonably watch the frenzied, localized activity in the brain of a person driven by some obsession, or see the dull glow of a depressed brain, and still doubt that these are physical conditions rather than some ineffable sickness of the soul. Similarly, it is now possible to locate and observe the mechanics of rage, violence and misperception, and even to detect the physical signs of complex qualities of mind like kindness, humour, heartlessness, gregariousness, altruism, mother-love and self-awareness.

The knowledge that brain mapping is delivering is not only enlightening, it is of immense practical and social importance because it paves the way for us to recreate ourselves mentally in a way that has previously been described only in science fiction. Rather as knowledge of the human genome will soon allow us to manipulate the fundamental physical processes that give rise to our bodies, so brain mapping is providing the navigational tool required to control brain activity in a precise and radical way.

Unlike genetic engineering, gaining this control does not depend on the development of tricky new technology – all it will take is a little refinement of existing methods and techniques like drugs, surgery, electrical and magnetic manipulation and psychological intervention. These are limited only in that, at present, they are (literally) hit-and-miss. When our brain maps are complete, however, it will be possible to target psychoactive treatments so finely that an individual's state of mind (and thus behaviour)

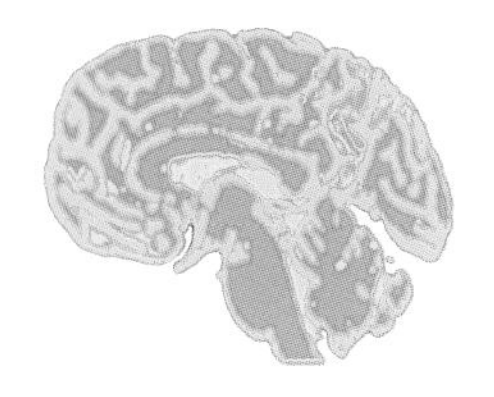

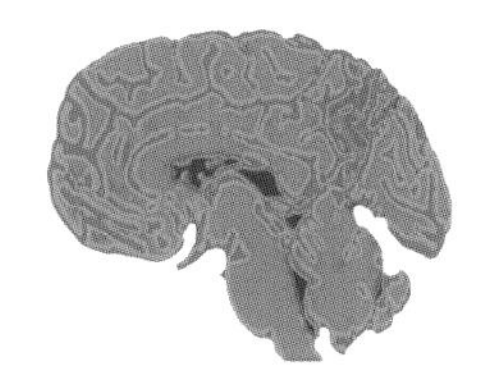

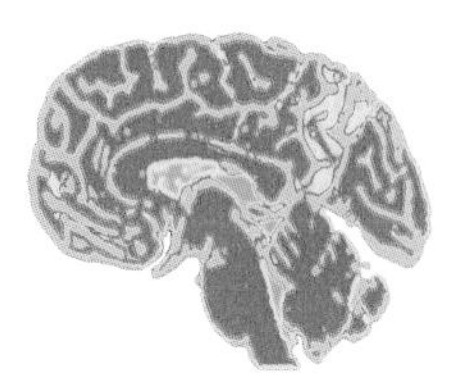

will be almost entirely malleable. It may even be possible to alter individual perception to the extent that we could, if we chose, live in a state of virtual reality, almost entirely unaffected by the external environment.

This is an old ambition, of course, reflected in our perpetual attempts to alter our consciousness through drugs, sensation-seeking and self-entrancement. What is new is that brain mapping may soon make it possible without any of the usual drawbacks. The personal, social and political implications of this are awesome, and one of the most serious ethical questions we will face in the new century is deciding how this powerful new tool should be deployed.

Those who are actually engaged in brain mapping loathe this sort of talk. For people at the leading edge of scientific research, where findings are often hyped in the scramble for funding, they are oddly reticent about the potential uses of their work. One reason for this is that modern behavioural neuroscience is a new discipline and its practitioners have come into it from many different fields: physics, radiology, neurology, molecular biology, psychology and psychiatry – even mathematics and philosophy. They have yet to develop a group mentality or a commonly agreed purpose beyond their immediate task of charting brain function. Many neuroscientists are also terrified of what might happen if their work is ever subjected to the tabloid treatment that has been meted out to their opposite numbers in genetics. The Human Genome Project has led to endless apocalyptic headlines, and as a result the geneticists are now closely scrutinized and controlled. Brain researchers can do without that sort of attention. At a brain mapping conference in 1997 (at which I was the sole reporter) a *Time* magazine cover about neuropsychology was held up by a speaker as a warning of what could result from loose talk to outsiders. The story to which the cover related was not inaccurate or sensational – its fault seemed to be its very existence.

The result of this reticence is that while we all debate and fret about the ethical and practical implications of genetic engineering, brain mapping tends to be regarded as the geeks' corner of psychology – interesting, no doubt, for those who like that sort of thing, but of no practical importance. When news leaks out it tends to be in isolated blips: one tiny piece of brain tissue is found to be the source of fear; the connection between the two hemispheres appears to be denser in women than in men; damage is found in the frontal lobes of a disproportionate number of murderers on Death Row. Each of these stories generates a brief flurry of speculation, but their full significance is rarely elucidated.

One of the purposes of this book is to draw attention to the social implications of what at first sight may seem a purely technological

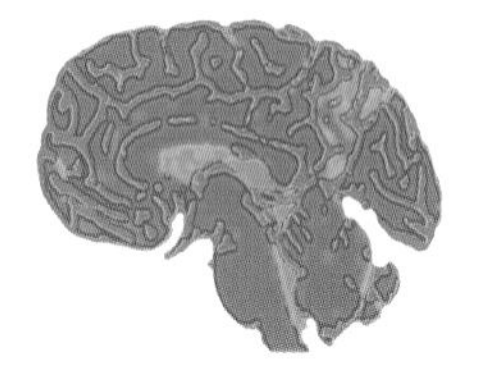
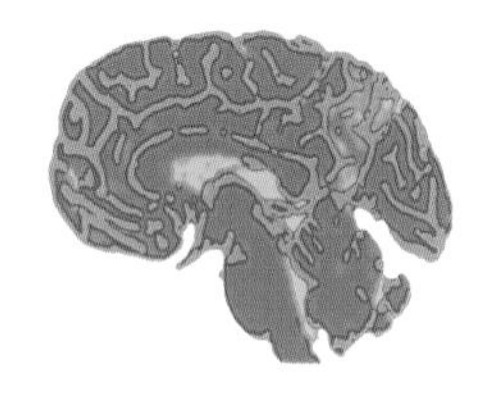
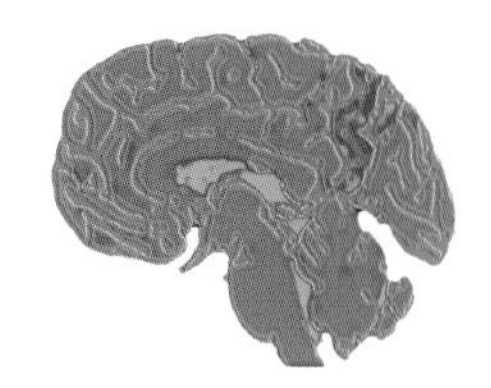

advance. Another is to examine the extent to which behavioural neuroscience is contributing to the age-old brain/mind conundrum and the puzzle of consciousness. Brain mapping is, of course, only one part of the current task. Along with it goes exploration of the workings of individual brain cells, the ebb and flow of neurotransmitters and the phenomenally complex interactions of the brain's various parts – things that are merely touched upon here but are of no less importance.

The more optimistic of today's brain explorers believe that when, or if, all of this is brought together – when each minute brain component has been located, its function identified and its interactions with each other component made clear – the resulting description will contain all there is to know about human nature and experience. Others think this reductionist approach will never fully explain why we feel and behave as we do, let alone yield the secret of the brain's most extraordinary product – consciousness. By their lights, a map of the brain can tell us no more about the mind than a terrestrial globe speaks of Heaven and Hell.

The work described here does not settle the debate about the nature of existence, but it seems to me to provide tantalising clues about it. Please remember, though, that these are the early days of mind exploration and the vision of the brain we have now is probably no more complete or accurate than a sixteenth-century map of the world. Most of what you will read is actually more complicated than I might make it appear, and some of it will almost certainly turn out to be wrong. This is because many of the findings are too new to have been replicated. There are also huge areas where very little is known and – as is the nature of leading-edge science – everyone is essentially guessing. Some of the foremost scientists in this venture have been generous enough to offer their thoughts and theories for inclusion here, and the diversity of their opinions demonstrates how far we still are from consensus in this field.

The cartographers who produced those early world maps filled the gaps in their knowledge with the medieval equivalent of scientific bluster. 'Here,' wrote one map-maker confidently, 'be Dragons.' I have tried to keep the dragons out of this chart, but others are bound to spot them, along with misleading signs and dubious landmarks. Such things are, I suspect, unavoidable on virgin territory, so those who prefer to travel only on well-worn paths should wait for the tourist guides that will be along later. Those who want to explore, read on and I will show you some strange and wonderful things.

Acknowledgements

My thanks to those explorers of the human mind who have been kind enough to enhance this book either by their written contributions, by finding time to answer my questions, or by granting me permission to reproduce passages from previously published work. Particular thanks are due to Chris Frith, whose unerring ability to spot a fragile speculation has (I hope) ensured that everything here is grounded in good science; to Malcolm Godwin whose right-brain talents have made this book beautiful; to Ravi Mirchandani for recognizing what I wanted to do and making it possible; and to my editor Judith Flanders for making it a pleasure. My family and friends Graham Campbell, Alex Laird and David Carlisle have been particularly tolerant and deserve acknowledgement for putting up with my near-obsessive brain talk over the last few years. Their special mix of challenge, query and teasing helped me shape my ideas more than they know. And I shall always be grateful to my late colleague Tricia Ingrams, who took time to encourage me in this when she had too little time to spare.

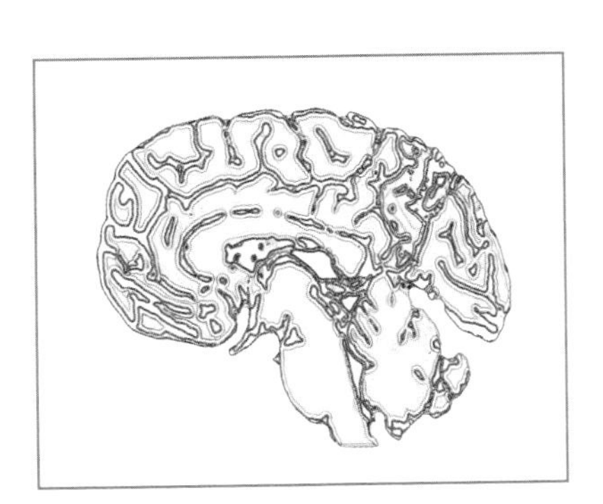

CHAPTER ONE
THE EMERGING LANDSCAPE

The human brain is made of many parts. Each has a specific function: to turn sounds into speech; to process colour; to register fear; to recognize a face or distinguish a fish from a fruit. But this is no static collection of components: each brain is unique, ever-changing and exquisitely sensitive to its environment. Its modules are interdependent and interactive and their functions are not rigidly fixed: sometimes one bit will take over the job of another, or fail, owing to some genetic or environmental hiccup, to work at all. Brain activity is controlled by currents and chemicals and mysterious oscillations; it may even be subject to quantum effects that distort time. The whole is bound together in a dynamic system of systems that does millions of different things in parallel. It is probably so complex that it will never succeed in comprehending itself. Yet it never ceases to try.

If you place a finger on the nape of your neck then edge it up and outwards, you will come across a bump formed by the base of your skull. Have a little feel. According to Franz Gall, the founder of phrenology, this particular protuberance marks the location of the Organ of Amativeness, 'the faculty that gives rise to the sexual feeling'. Now slide your finger an inch or so further up towards your crown. You are now passing over the Organ of Combativeness.

Those of a warm and peaceful disposition should, in theory, find the second area of their skull flatter than the first. But don't be too concerned if your bumps do not match your self-perceptions – Gall arrived at the Amativeness Organ by seeking out the area of greatest heat

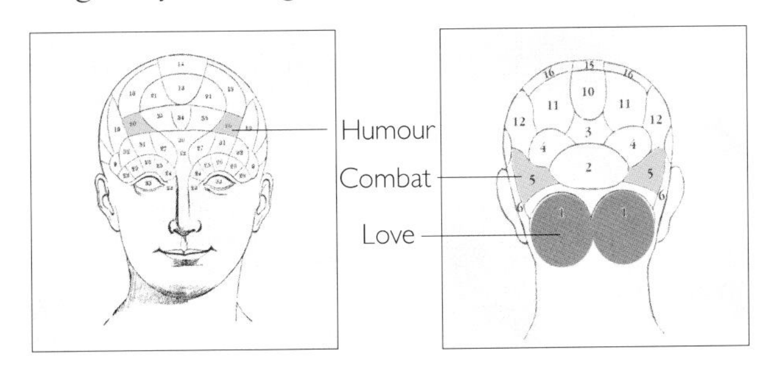

in the skulls of two recently widowed and 'emotional' young women, and he fixed the Organ of Combativeness by observing that area's smallness in 'most Hindoos and Ceylonese'.[1] It was dubious methodology even for the early nineteenth century.

Bump-reading was nonsense, anyway, because the soft tissue of an individual brain generally has no effect on the shape of the skull. Yet it wasn't all wrong. Feel your skull again, on top this time, just a little forward and to the left of the crown. This is where Gall placed his Organ of Mirthfulness. Surgeons at the University of California Medical School recently applied a tiny electric current near to this part of the left brain of a sixteen-year-old girl.

Stimulus here produced laughter

Gall's organ of mirthfulness

The patient had intractable epilepsy and the stimulus was part of an established procedure designed to locate the foci of her seizures prior to their removal. She was conscious, and as the stimulus passed through this particular part of her cortex she started to laugh.[2] It was no empty rictus but a genuine, humour-filled chortle, and when the surgeons asked her what was so amusing she had an answer: 'You guys are just so funny – standing around.' The doctors applied the current again and this time the girl suddenly saw something comic in a picture, which happened to be in the room, of a perfectly ordinary horse. A third time she was struck by something else. The surgeons seemed to have come across a part of the brain that creates amusement, however unpromising the circumstances. Gall's pinpointing of it nearly two hundred years ago was no doubt accidental. Yet his basic idea – of a brain made of functionally discrete modules – was prescient.

Phrenology's downfall came, ironically, from the discovery of *real* brain modules. Towards the end of the nineteenth century a craze for biological psychiatry took root in European universities, and neurologists started to use localized electrical brain stimulation and animal lesion experiments to find out which bits of the brain did what. Others observed the association between certain behaviour and certain brain

Language area discovered by Broca

Language area discovered by Wernicke

Phrenologists' language area

injuries. Many important landmarks were identified during this first era of brain mapping, including the language areas discovered by the neurologists Pierre Broca and Carl Wernicke. Embarrassingly for the phrenologists, these areas were found in the side of the brain, above and around the ear, while Gall's Organ of Language was firmly located near the eye.

The language areas identified by Broca and Wernicke still bear those names today. If early twentieth-century scientists had continued the search for functional brain areas, today's charts might now be crowded with the names of other long-dead individuals instead of the dull labels – primary auditory cortex, SMA , V1 – now attached to each newly discovered region. Instead scientific brain mapping fell out of fashion along with phrenology, and the modular brain theory was largely abandoned by scientists in favour of the theory of 'mass action', which held that complex behaviour arose from the action of all the brain cells working together.

On the face of it, the mid-twentieth century should have been a bad time for anyone with ambitions to use physical methods to treat mental illness or alter behaviour. Yet psychosurgery flourished. In 1935 the Lisbon neurologist Egas Moniz heard about some experiments in which aggressive, anxious chimps had certain fibres in the front of their brains lesioned.[3] The operation – leucotomy – made them quiet and friendly. Moniz promptly tried it on similarly afflicted humans and found it worked. Frontal leucotomy (which later evolved into the more radical frontal lobotomy) rapidly became a standard treatment in mental hospitals, and during the 1940s at least twenty thousand such operations were carried out in America alone.[4]

Looking back, the approach to brain surgery at that time seems shockingly cavalier. It was used for almost any mental disorder – depression, schizophrenia, mani – even though no one then had a clue what was causing the symptoms or why making cuts in the brain should relieve them. Travelling surgeons went from hospital to hospital with their implements in the back of their cars, knocking off a dozen or so operations in a morning. One surgeon described his technique like this:

'[There's] nothing to it. I take a sort of medical icepick...bop it through the bones just above the eyeball, push it up into the brain, swiggle it around, cut the brain fibers and that's it. The patient doesn't feel a thing.' [5]

Unfortunately, the lack of feeling extended, in some patients, to a long-term flattening of emotion and a curious insensitivity that left them seemingly only half-alive. It didn't always cure aggression either: Moniz's career was brought to an end when he was shot by one of his own lobotomized patients.

On balance, the mid-century craze for cutting up the brain probably relieved more suffering than it caused, but it cre-

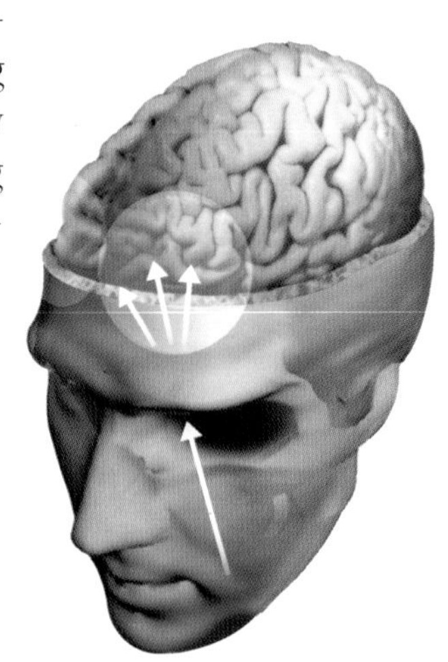

Frontal leucotomy involved cutting through fibres which connect the unconscious brain, where emotions are generated, to the cortical area where they are consciously registered.

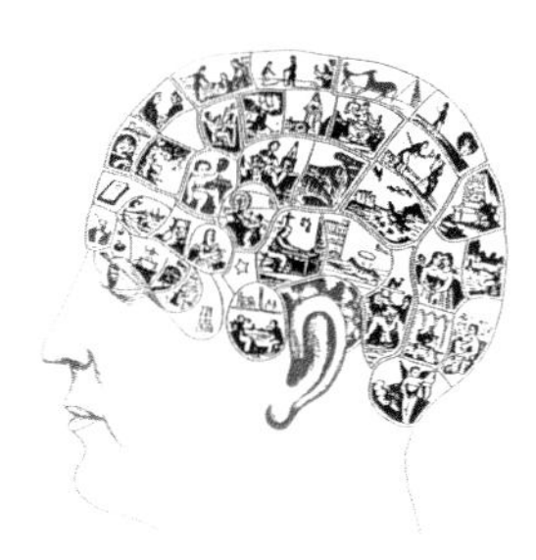

ated a deep sense of unease in the medical profession and a profound and still-lingering suspicion of psychosurgery among lay people. When effective psychotropic drugs came along in the 1960s, surgery for mental disorders was almost entirely abandoned.

As we enter the twenty-first century the idea of modifying behaviour and relieving mental anguish by manipulating brain tissue is creeping back. This time, though, any tinkering that is less to be done will be based on a much greater understanding of how the brain works. Recently developed techniques – functional brain scanning in particular – allow researchers to explore the living, working brain. Watching the brain at work gives new insight into both mental illness and the nature of our everyday experiences.

Take, for example, pain. Common sense might predict a specialized pain centre in the brain, connecting, perhaps, with another bit that registers sensation in the affected part of the body. In fact, scans show that there is no such thing as a pain centre. Pain springs mainly from the activation of areas associated with attention and emotion. Seeing what pain is, in terms of neurological activity, it becomes clear why we feel it so much more when we are emotionally stressed and why we often don't notice it – even when our bodies are quite badly injured – when more pressing things have captured our attention.

While apparently simple functions like pain are showing themselves to be more complex than might be expected, some seemingly imponderable qualities of the mind are looking to be surprisingly mechanistic. Religious belief and experience are usually regarded as beyond scientific exploration, yet neurologists at the

University of California San Diego have located an area in the temporal lobe of the brain that appears to produce intense feelings of spiritual transcendence, combined with a sense of some mystical presence. Canadian neuroscientist Michael Persinger, of Laurentian University, has even managed to reproduce such feelings in otherwise unreligious people by stimulating this area. According to Persinger:

'Typically people report a presence. One time we had a strobe light going and this individual actually saw Christ in the strobe...[another] individual experienced God visiting her. Afterwards we looked at her EEG [electroencephalogram] and there was this classic spike and slow-wave seizure over the temporal lobe at the precise time of the experience – the other parts of the brain were normal.[6]

The fact that we seem to have a religious hot-spot wired into our brains does not necessarily prove

Stimulating parts of the temporal lobes can produce feelings of spiritual transcendence.

A TOUR OF THE TERRITORY

Zoom in close enough on a section of brain and you will see a dense network of cells. Most of them are glial cells – relatively simple-looking structures whose main known purpose is to glue the whole construction together and maintain its physical integrity. It is possible that glial cells also play a role in amplifying or synchronizing electrical activity within the brain, but that, to date, is speculation.

The cells that actually create brain activity – about one in ten of the total – are neurons, cells which are adapted to carry an electrical signal from one to another. There are long thin ones that send single snaking tendrils to the far reaches of the body; star-shaped ones that reach out in all directions; and ones that bear a dense branching crown like absurdly overgrown antlers. Each neuron connects with up to ten thousand neighbours. The bits that join up are the branches, of which there are two kinds: axons, which conduct signals away from the cell nucleus, and dendrites, which receive incoming information.

If you zoom in even closer, you will see that there is a tiny gap where each axon meets a dendrite. This gap is called the synapse. In order for the current to cross the synapse each axon secretes chemicals, called neurotransmitters, that are released into the space when the cell is suitably fired up. These chemicals trigger the neighbouring cell to fire, too, and the resultant chain effect produces simultaneous activity in millions of connected cells.

The things that happen down here among cells and molecules create the foundations of our mental lives and it is by manipulating them that most existing physical psychiatric therapies work. Antidepressants, for example, work on neurotransmitters – mainly

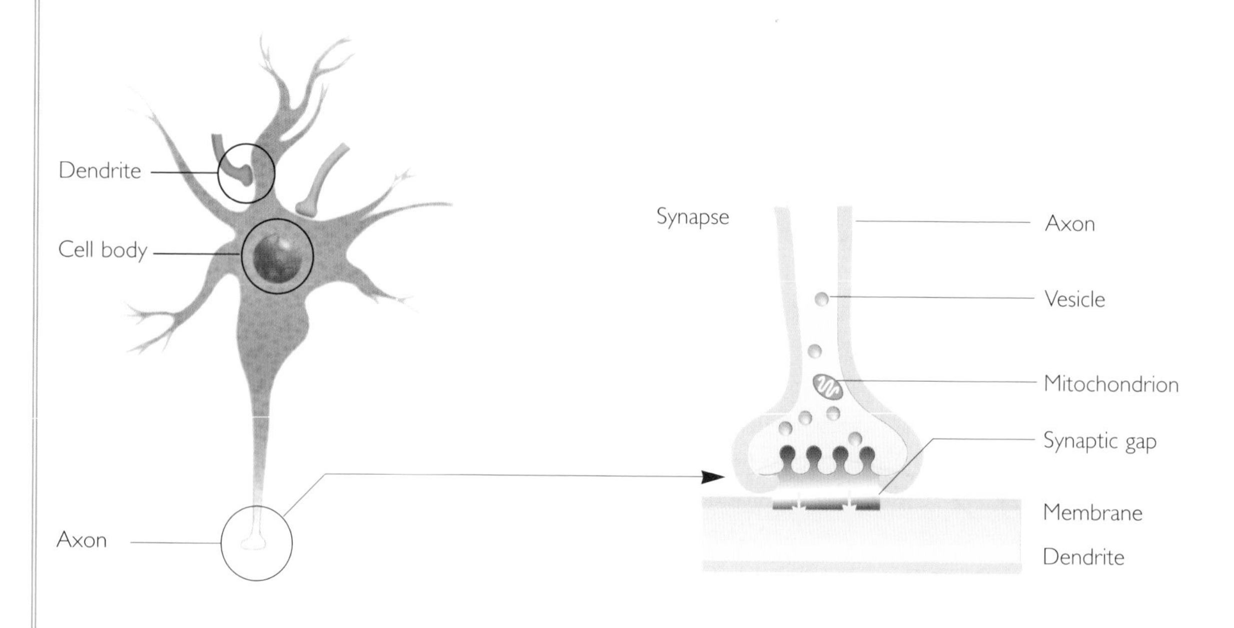

enhancing the action of serotonin. Research into microscopic brain processes is currently helping to develop drugs to relieve dementia, Parkinson's disease and stroke damage. Some scientists believe the secret of consciousness is to be found here or even at a more fundamental level in the quantum processes thought to occur in the tiniest recesses of brain cells.

For our purposes, though, we need to zoom out again. The human brain is as big as a coconut, the shape of a walnut, the colour of uncooked liver and the consistency of chilled butter. It has two hemispheres, which are covered in a thin skin of deeply wrinkled grey tissue called the cerebral cortex. Each infold on this surface is known as a sulcus, and each bulge is known as a gyrus. The surface landscape of each individual's brain is slightly different, but the main wrinkles – like nose–mouth grooves and crow's feet on an ageing face – are common to all and are used as landmarks. At the very back of the main mass of brain, tucked under its tail and partly fused to it, lies the cerebellum – the 'little brain'. Aeons ago, this was our mammalian ancestors' main brain but now it has been superseded by the larger area, the cerebrum.

Each half of the cerebrum is split into four lobes, their divisions marked by various folds. At the very back lies the occipital lobe; the lower side, around the ears, is the temporal lobe; the top section is the parietal lobe; in front of that is the frontal lobe. Each lobe processes its own clutch of things: the occipital lobe is made up almost entirely of visual processing areas; the parietal lobe

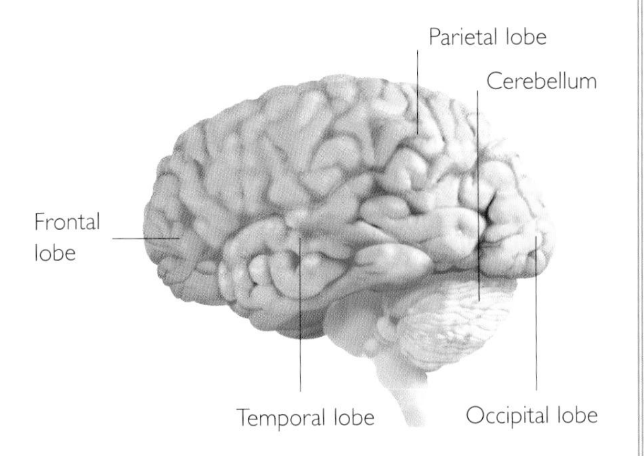

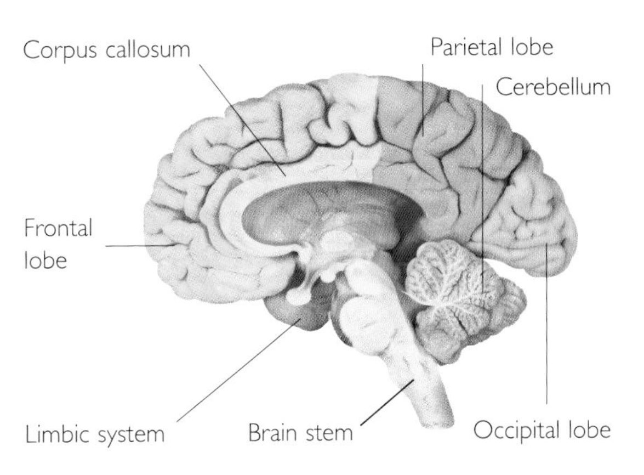

deals mainly with functions connected with movement, orientation, calculation and certain types of recognition; the temporal lobes deal with sound, speech comprehension (usually on the left only) and some aspects of memory; and the frontal lobes deal with the most integrated brain functions: thinking, conceptualizing and planning. They also play a major part in the conscious appreciation of emotion.

If you slice the brain in half down the centre line so that the two hemispheres fall apart, you see that beneath the cortex lies a complex conglomeration of modules: lumps, tubes and chambers. Some of them can be likened in size and shape to nuts, grapes and insects, but many of them look like nothing you have ever seen. Each one of these modules has its own function or functions, and they are all interconnected by criss-crossing ropes of axons. Most modules are greyish – a colour lent by the densely packed neuron bodies. The bands that connect them, however, are lighter because they are covered in a sheath of white substance called myelin, which acts as insulation, allowing electricity to flow swiftly and directly along them.

Apart from a single little bit – the pineal gland in the centre base of the brain – every brain module is duplicated in each hemisphere. Throughout this book each module is referred to in the singular but in fact there are always two. When it is necessary to distinguish between the two it is made clear which one is indicated.

The Limbic System
The main modules

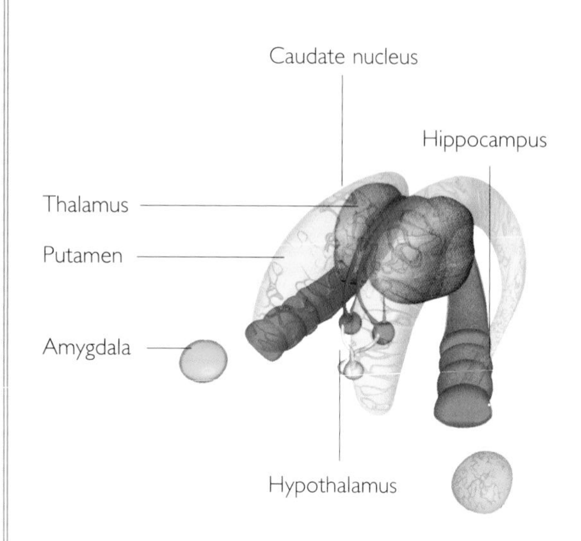

Looking sideways on at the sliced brain, the most noticeable thing is a curved band of white tissue that forms a curvaceous division between the folded cortex and the mass of modules beneath. The corpus callosum itself joins the two hemispheres and acts as a bridge between them, constantly shunting information back and forth so that most of the time they are effectively one. The modules that nestle beneath the corpus callosum are generally known as the limbic system. This area is older than the cortex in evolutionary terms and is also known as the mammalian brain because it is thought to have first emerged in mammals. This part of the brain – and the even older areas below it – is unconscious, but it has a profound effect on our experience because it is densely connected to the conscious cortex above it and constantly feeds information upwards.

Emotions – our most basic cerebral reactions – are generated in the limbic system, along with most of the many appetites and urges that direct us to behave in a way that (usually) helps us to survive. But the limbic modules have many other functions besides: the thalamus is a sort of relay station, directing incoming information to the appropriate part of the brain for further processing. Beneath it the hypothalamus, together with the pituitary gland, constantly adjusts the body to keep it optimally adapted to the environment. The hippocampus (named after a seahorse but actually looking more like a giant

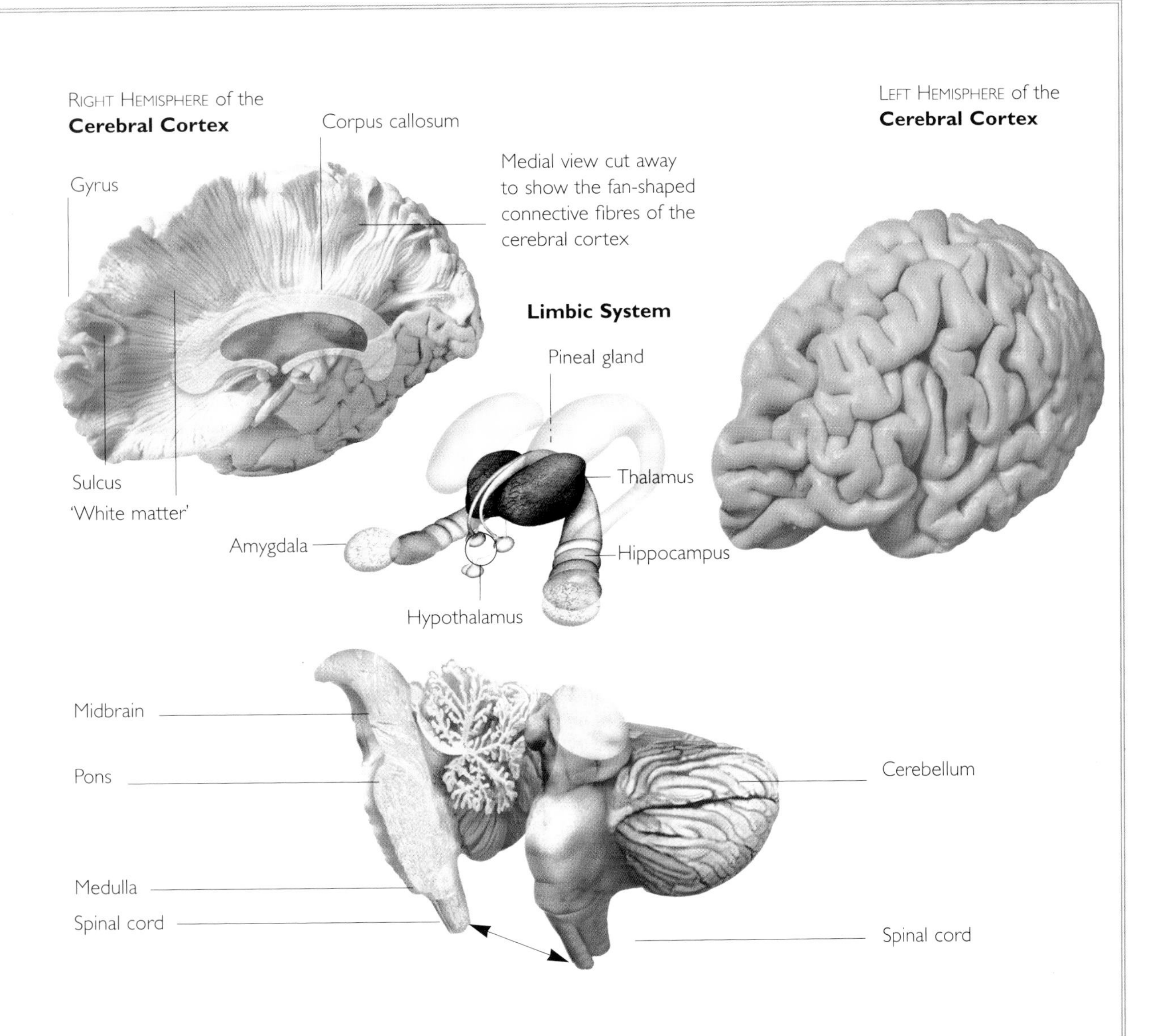

paw) is essential for the laying down of long-term memory. The amygdala, in front of the hippocampus, is the place where fear is registered and generated.

Going down even further you get to the brainstem. This is the most ancient part of the brain – it evolved more than 500 million years ago and it is rather like the entire brain of present-day reptiles. For that reason it is often called the reptilian brain. The brainstem is formed from the nerves that run up from the body via the spinal column and it carries information from the body into the brain. Various clumps of cells in the brainstem determine the brain's general level of alertness and regulate the vegetative processes of the body, such as breathing, heartbeat and blood pressure.

Breaking down the brain

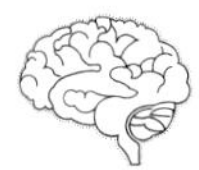

Horace Barlow
Royal Society Research Professor, Physiological Laboratory, University of Cambridge

Can we learn about the mind in the same way that we might seek to understand a machine – by taking it apart and examining its parts? Neurophysiologist Horace Barlow believes this approach can bring important insights but can never tell the full story.

The reductionist approach to the brain shows promise of revolutionizing our ideas about what single neurons can do, but reductionism is limited because its drive is to look for explanations at lower levels in the organizational tree. The isolated preparations that have been so important for the success of the reductionist approach can tell us about extracellular and intracellular processes but not about subjective experience or the survival value of a gene. Obviously, you cannot find out about what you have thrown away or deliberately chosen to ignore, and reductionism ignores a lot. Isolated preparations, for instance, will never lead us to understand interactions between humans, yet these interactions are crucial for understanding the human mind and its role in moulding society. So reductionism is not only limited but it is also unlikely ever to give us the whole answer. To obtain more knowledge one may have to look up from the simpler, more basic levels of organization towards the complex higher levels. All the same, the lower levels may provide the best way forward because they give us lasting knowledge about the very complex system that is the brain.

I think of the brain as a set of networks nested like Russian dolls.

At the innermost level are the billions and billions of neurons – ten to the power of ten – each of which contains its own computational network of interacting molecules. These neurons are linked to each other in the kind of network that artificial neural nets model. The outermost network is made up of the communicating, interacting set of brains that compose a human society.

The links in the outer network result from the fact that our brains can report on some aspects of their own working and can also interpret similar reports received from others. It is this network of conscious communication that mediates social organization. The advantages conferred by social organization could explain the survival value of the conscious communication that makes it possible. Although reductionism will never lead us to understand human relationships, I am firmly convinced that the reductionist approach is heuristically correct. When you find out, for example, how one molecule communicates with another, you have a building block for the next stage up. In contrast, questions like 'Why do we converse?', though fascinating, don't give answers, just more questions.

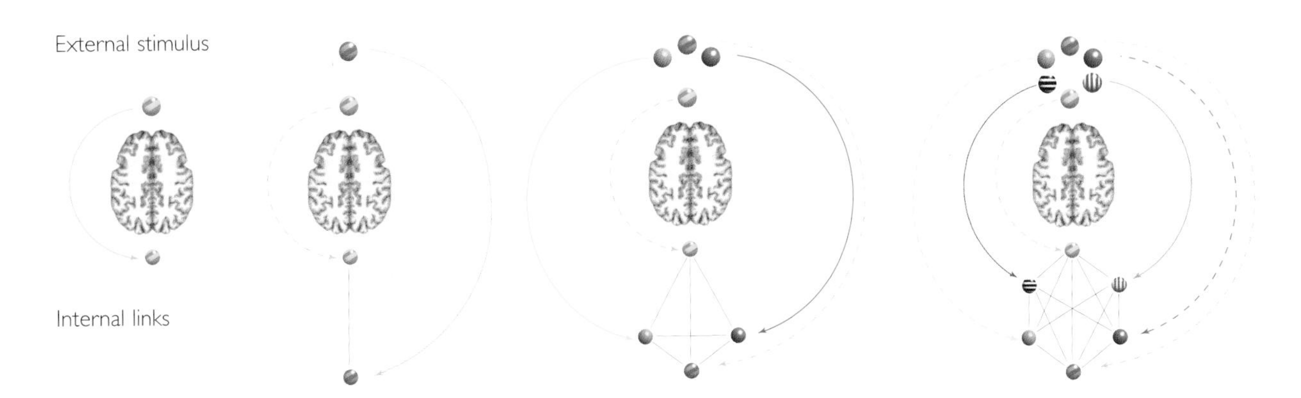

The sensory impact of something in the outside world alters our subsequent perception of it, which in turn creates an altered impact, which further alters our perception …

that the spiritual dimension is merely the product of a particular flurry of electrical activity. After all, if God exists, it figures He must have created us with some biological mechanism with which to apprehend Him. Nevertheless, it is easy to see that being able to get your God Experience from a well-placed electrode could – at the very least – undermine the precious status such states are accorded by many religions. How believers will cope with what many might see as a threat to their faith is one of many interesting challenges that brain science will throw up in the coming millennium.

Firedance

How does the conglomeration of neuronal clumps and cat's cradle wiring actually do what brains do? Essentially the neurons get fired up, joined up and dancing – on a huge scale.

The firing of a single neuron is not enough to create the twitch of an eyelid in sleep, let alone a conscious impression. It is when one neuron excites its neighbours, and they in turn fire up others, that patterns of activity arise that are complex and integrated enough to create thoughts, feelings and perceptions.

Millions of neurons must fire in unison to produce the most trifling thought. Even when a brain seems to be at its most idle a scan of it shows a kaleidoscope of constantly changing activity. Sometimes, when a person undertakes a complex mental task or feels an intense emotion the entire cerebrum lights up.

New neural connections are made with every incoming sensation and old ones disappear as memories fade. Each fleeting impression is recorded for a while in some new configuration, but if it is not laid down in memory the pattern degenerates and the impression disappears like the buttocks-shaped hollow in a foam rubber cushion as you stand up.

Patterns that linger may in turn connect with, and spark off, activity in other groups, forming associations (memories) or combining to create new concepts. In theory, each time a particular interconnected group of neurons fires together it gives rise to the same fragment of thought, feeling or unconscious brain function; but in fact the brain is too fluid for an identical pattern of activity to arise – what really happens is that similar but subtly mutated firing patterns occur. We never experience exactly the same thing twice.

Little explosions and waves of new activity, each with a characteristic pattern, are produced moment by moment as the brain reacts to outside stimuli. This activity in turn creates a constantly changing internal environment, which the brain then reacts to as well. This creates a feed-

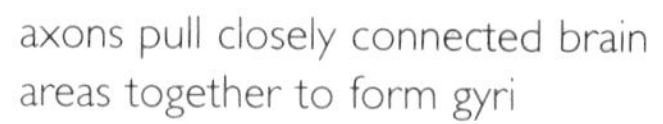

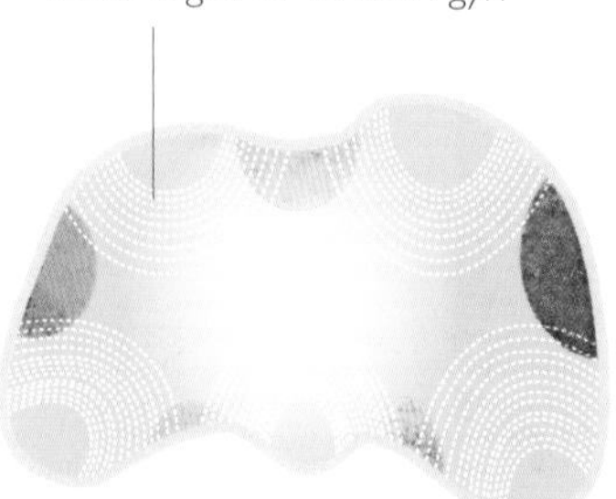

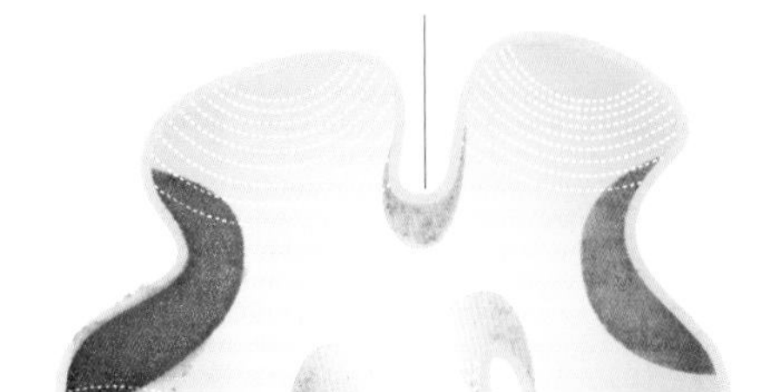

Right: *The surface of the brain develops numerous furrows and bulges as it grows. The pattern they make is different in each of us, and it affects how we perceive and think about the world.* Below: *The increasing complexity of the cortex of an infant as shown from 25 weeks to 40 weeks*

back loop that ensures constant change.

Part of the brain's internal environment is a ceaseless pressure to seek out new stimuli. This greed for information is one of the fundamental properties of the brain and it is reflected in our most basic reactions. People can have their conscious mind totally destroyed, yet their eyes will still scan the room and lock on to and track a moving object. The eye movements are triggered by the brainstem and are no more significant of consciousness than the turning of a flower to the sun. Yet even when you know this, it is deeply disturbing to have your movements followed by the eyes of a person you know is for all intents and purposes dead.

The loop-backs between brain and environment are a bootstrap operation *par excellence*. Computer simulations of neural networks show that the simplest network can develop phenomenal complexity in a short time if it is programmed to replicate patterns that are beneficial to its survival and junk those that aren't. Similarly, brain activity evolves in the individual.

This process – sometimes referred to as neural Darwinism – ensures that patterns that produce thoughts (and thus behaviour) that help the organism to thrive are laid down permanently while those that are useless fade. It is not a rigid system – the vast majority of brain patterns we produce are quite irrelevant to our survival – but overall this seems to be the way that the human brain has come to be furnished with its essential reactions.

Some of this furniture has been built in at genetic level. Certain patterns of brain activation – even quite complex ones like speech production – are so strongly inherited that only an extraordinarily abnormal environment can distort them. The pattern of brain activation during, say, a word retrieval task is usually similar enough among the dozen or so participants who

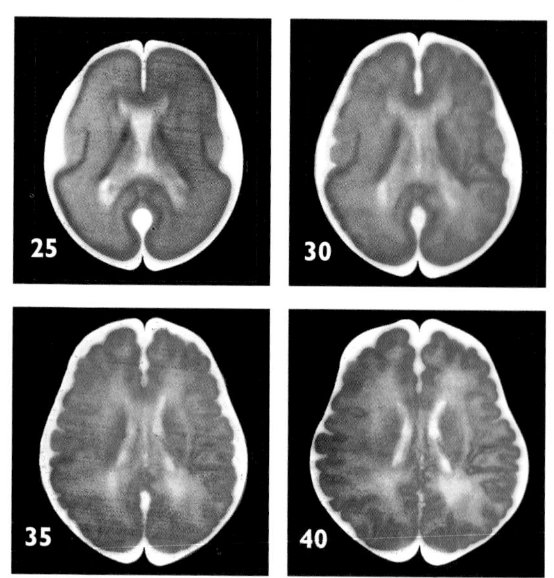

typically take part in such studies for their scans to be overlaid and still show a clear pattern. This is how brain mappers can confidently talk of a chart of 'the' brain, rather than 'a' brain.

This is not to say that everyone thinks alike. Thanks to the infinitely complex interplay of nature and nurture no two brains are ever exactly the same. Even genetically identical twins – clones – have different brains by the time they are born because the tiny divergence in the foetal environment of each is enough to affect their development. The cortex of human twin babies is visibly different at birth, and structural variations inevitably produce differences in the way brains function.[7]

During foetal development the brain develops, bulb-like, at the upper end of the neural tube that forms the spine. The main sections of the brain, including the cerebral cortex, are visible within seven weeks of conception and by the time the child is born the brain contains as many neurons – about 100 billion – as it will have as an adult.

These neurons are not mature, however. Many of their axons are as yet unsheathed by myelin – the insulation that allows signals to pass along them – and the connections between them are sparse. Hence large areas of the brain, and particularly the cerebral cortex, are not functioning. PET studies of newborn brains show the active areas are those associated with bodily regulation (the brainstem), sensation (thalamus) and movement (deep cerebellum).[8]

The uterine environment has a profound effect on the wiring of the infant brain. Babies born to drug addicts are frequently addicted at birth, and those born to mothers who eat curries while they are pregnant take to spicy food more readily than others,[9] suggesting that their tastes are primed by exposure to the residue of such foods in their mothers' blood.

Life in the womb provides a good example of how genes and environment are inextricably combined. A male foetus, for example, has genes that provoke the mother's body to produce a cascade of hormones, including testosterone, at certain times in its development. This hormonal deluge physically alters the male foetus's brain, slowing the development of certain parts and speeding the development of others. The effect of this is to masculinize the foetal brain, priming

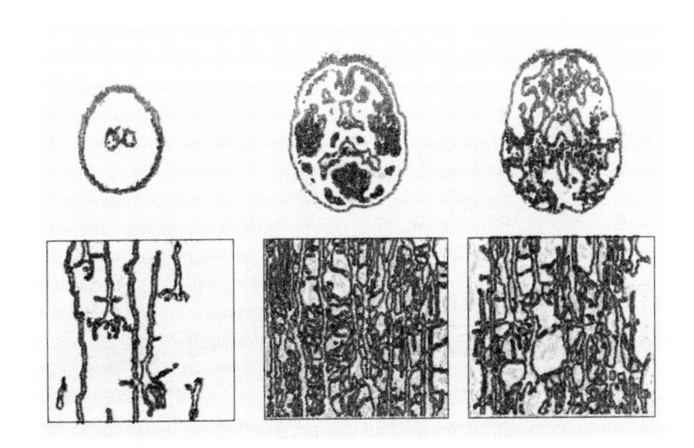

Neural connections are sparse at birth (left), *but new connections are made at a terrific rate during infancy and by the age of six* (middle) *they are at maximum density. Thereafter they decrease again as unwanted connections die back.* (right.) *Adults can increase neural connections throughout their life by learning new things. But if the brain is not used the connections will become further depleted.*

it to produce male sexual behaviour. It also creates many of the typical differences seen between the sexes, like girls' superiority at speech and boys' at spatial tasks. If a male foetus does not get the appropriate pre-birth hormone treatment, his brain is likely to remain more typically female; if a female foetus gets exposed to a male-pattern hormonal sequence, she is likely to be more typically masculine.

Inside the developing brain individual neurons race about looking for a linked team of other neurons to join as though in some frantic party game. Every cell has to find its place in the general scheme, and if it fails it dies in the ruthless pruning process known as apoptosis or programmed cell death. The purpose of apoptosis in the immature brain is to strengthen and rationalize the connections between those that are left, and to prevent the brain becoming literally overstuffed with its own cells. This 'sculpting' process, though essential, may also have a price. The connections that get killed off during it include some that may otherwise confer the sort of intuitive skills we label gifts. Eidetic (photograph) memory, for example, is quite common among young children but usually disappears during the years of cerebral pruning. Incomplete apoptosis may account for the astonishing abilities of so-called *idiot savants*, as well as being one causal factor of their deficits. Conversely, apoptosis that runs wild and strips out far too many connections is thought to be one of the causes of impaired intelligence in Down's syndrome. It is also probably the reason why Down's people are more likely than others to develop Alzheimer's disease

Climbing to consciousness

A baby's brain contains some things that an adult's does not. There are, for example, connections between the auditory and visual cortices, and others between the retina and the part of the thalamus that takes in sound. These connections probably give the infant the experience of 'seeing' sounds and 'hearing' colours – a condition that occasionally continues into adulthood and is known as synaesthesia. Babies show emotion dramatically, but the areas of the brain that in adults are linked to the conscious experience of emotions are not active in a newborn baby. Such emotions may therefore be unconscious.

'Unconscious emotion' sounds like a contradiction in terms – what is emotion if not a conscious feeling? In fact, the conscious appreciation of emotion is looking more and more like one quite small, and sometimes inessential, element of a system of survival mechanisms that mainly operate – even in adults – at an unconscious level.[10]

This does not necessarily mean that early traumas do not matter. Unconscious emotion may not be, strictly speaking, experienced, but it may lodge in the brain just the same. We cannot remember things before the age of about

FOURTEEN WEEKS

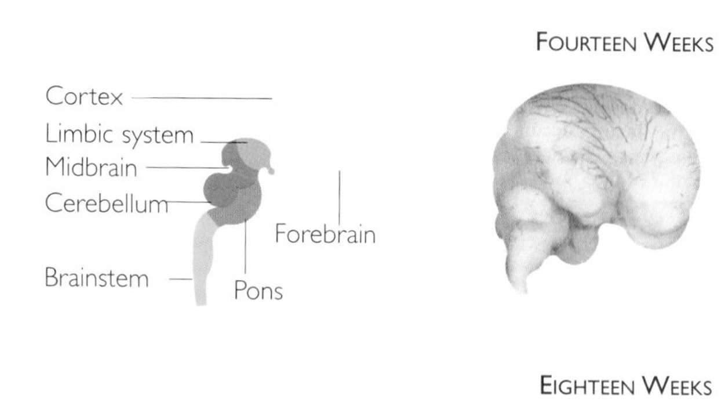

EIGHTEEN WEEKS

Cortex
Limbic system
Midbrain
Cerebellum
Forebrain
Brainstem
Pons

Parietal lobe
Frontal lobe
Occipital lobe
Limbic area

SIX MONTHS

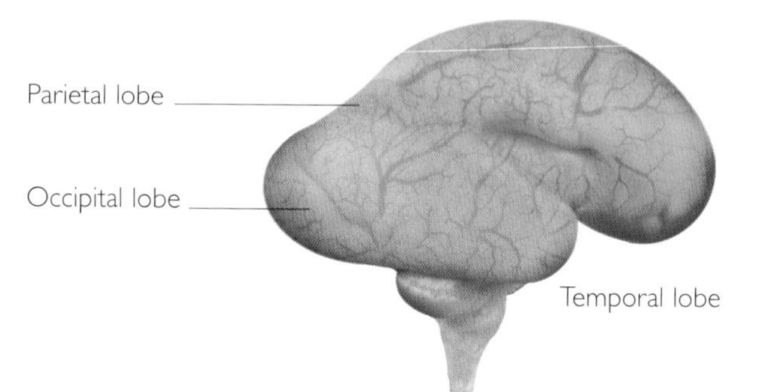

three because until then the hippocampus – the brain nucleus that lays down conscious long-term memories – is not mature. Emotional memories, however, may be stored in the amygdala, a tiny nugget of deeply buried tissue that is probably functioning at birth.[11]

As the baby gets older myelinization creeps outward and brings increasing numbers of brain areas 'on- line'.[12] The parietal cortex starts to work fairly soon, making babies intuitively aware of the fundamental spatial qualities of the world. Peek-a-boo games are endlessly intriguing once this part of the brain is working because babies then know that faces cannot really disappear behind hands – yet the brain modules that will one day allow them to know why have not yet matured.

The frontal lobes first kick in at about six months, bringing the first glimmerings of cognition. By the age of one they are gaining control over the drives of the limbic system – if you offer two toys to a child of this age, they will make a choice rather than try to grab both. Up until the age of about a year babies are, in the words of one developmental psychologist, 'robotic looking machines' – their attention can be caught by more or less any visual stimulus. After that age they get an agenda of their own – not always one that fits in with other people's.

The language areas become active about eighteen months after birth. The one that confers understanding (Wernicke's area) matures before the one that produces speech (Broca's area), so there is a short time when toddlers understand more than they can say – a frustrating condition that probably does much to fuel the tantrums that typify the 'Terrible Twos'.

Around the same time as the language areas become active myelinization gets under way in the prefrontal lobes. Now children develop self-consciousness – they no longer point at their reflection in the mirror as though they see another child and if a dab of coloured powder is put on their face while they look at their reflection, they rub it off – they don't rub the mirror as younger children do. This self-consciousness suggests the emergence of an internal executor – the 'I' that most people say they feel exists inside their heads.

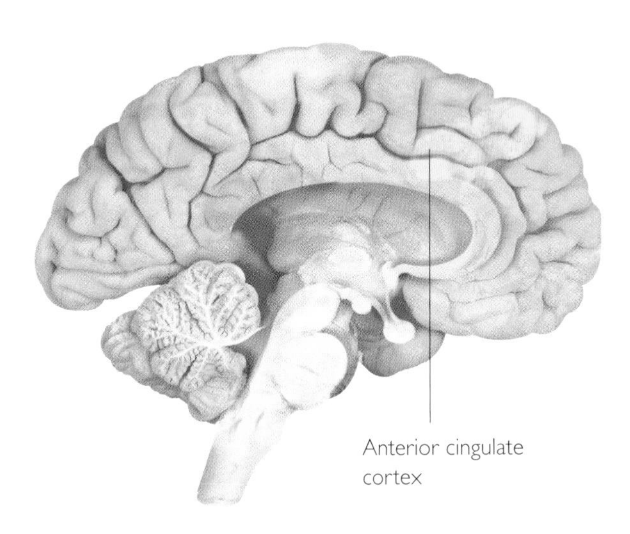

This part of the brain lights up when it does something of its own volition – it is one of the areas which seems to contain the 'I' we all feel we have inside us.

Certain brain areas take many years to mature. A nucleus called the reticular formation, for example, which plays a major role in maintaining attention, usually only becomes fully myelinated at or after puberty, which is why prepubescent children have a short attention span. The frontal lobes do not become fully myelinated until full adulthood. This is one reason, perhaps, why younger adults are more emotional and impulsive than those who are older.

Human brains are at their most plastic during infancy. You can take away an entire hemisphere from a child's brain and the other will rewire itself to take on the tasks of both. It will even manage to develop functions that are usually exclusive to its other half. As we age, however, brain functions become more rigid and more distinctive. By the time we are adults our mental landscapes are so individual that no two of us will see anything in quite the same way. A couple watching the same film, for example, would probably have entirely different patterns of neural activity because each would be cogitating on different aspects of the show and associating what they see with personal thoughts and memories. She might be wondering when the tiresome couple on screen will finally reach their feel-good ending so she can get some dinner; he might be thinking how the heroine's cute upper lip reminds him of his ex-girlfriend.

That is why experiments designed to reveal which brain areas are responsible for what have to involve artificially rigid and narrow tasks. The subjects who lay for more than two hours in a PET scanner doing nothing more than lifting a finger in response to a given signal, for example, must at times have wondered what possible insight could be gained from such a tedious manoeuvre.

In fact, wonderful discoveries are emerging from these unpromising exercises. The finger-lifting experiment, carried out by Chris Frith and colleagues at the Wellcome Department of Cognitive Neurology in London, revealed something that, not so long ago, would have been expected to remain one of life's eternal mysteries: the source of self-determination. They arrived at it by designing a procedure that first narrowed down what was going on in the participants' brains to a few things that they already knew from previous work would show up as particular patterns of activity in certain brain localities. In this case they got the subjects to move a specified finger on cue – a task that duly provoked activity in the auditory cortex (when the cue was a noise) and the motor cortex (the area that controls movement). They then added the element of the task for which they wanted to find a brain location: self-willed activity. Instead of telling the subjects which finger to lift

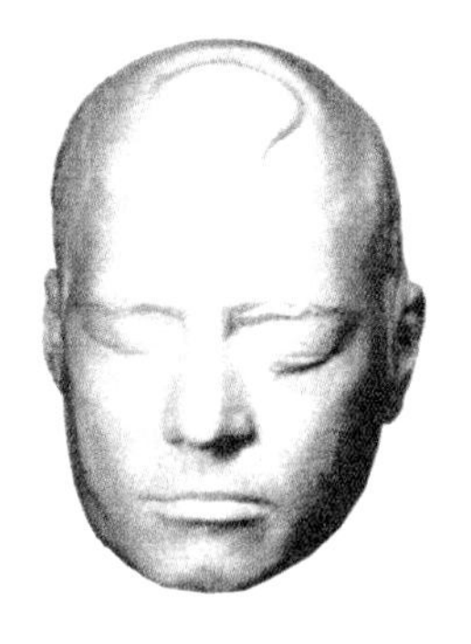

The death mask of Phineas Gage showing the massive injury to the skull.

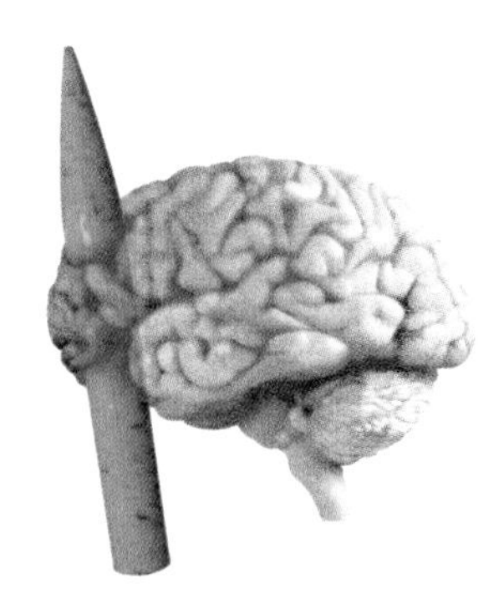

Reconstruction of the position of the rod which passed through the frontal lobe of Gage's brain.

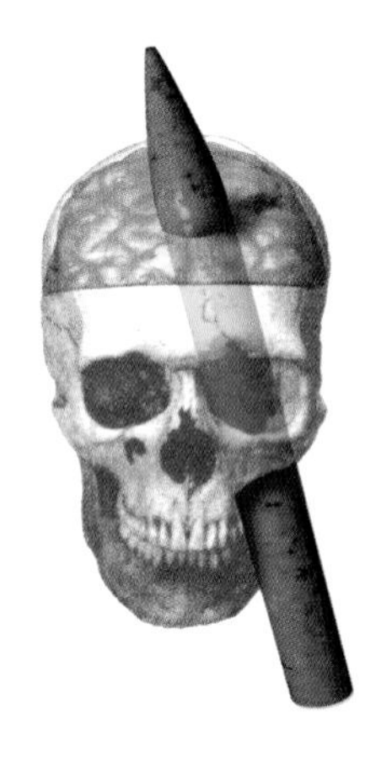

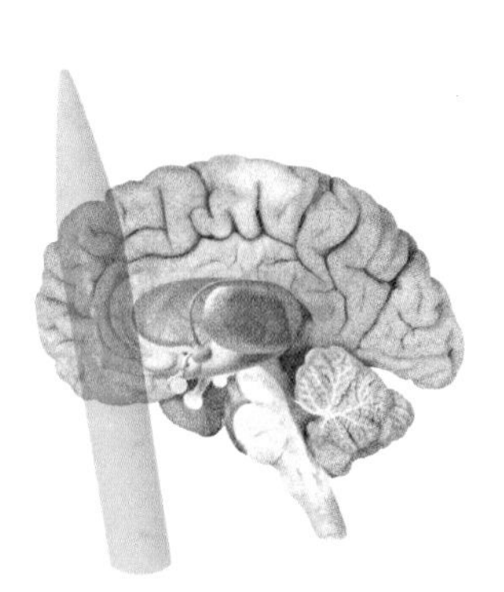

they left it to them to decide which one to move. Then they watched to see how the brain activity involved in doing this differed from the brain activity involved in lifting an externally specified finger.

The difference was clear: as soon as the participants started to make their own decisions a previously 'dead' area of their brains sprang into life. Various controls were in place to ensure that this new activity did not just represent the extra effort required to think about the task once it was not a simple matter of obeying an order. The elegant and careful design of the experiment made sure that the bit of brain identified is almost certainly that which, quite specifically, allows people to do things of their own volition.

Can identifying the brain activity involved in deciding which of two fingers to lift possibly shed any light on decision-making in the messy and infinitely more complicated world outside a brain research laboratory?

Indirectly, yes. The region of brain in which the self-will area was found is the prefrontal cortex, a region of the frontal cortex which lies mainly behind the forehead. Injuries to this area often produce a characteristic change in behaviour that includes loss of self-determination on a grand scale. The classic case is that of Phineas Gage, a nineteenth-century rail-worker who lost a large chunk of forebrain when a steel rod was blown through his skull by a mistimed explosion. Gage survived, but from the time of the accident he changed from a purposeful, industrious worker into a drunken drifter. John Harlow, the doctor who treated him, described the new Gage as 'at times pertinaciously obstinate, yet capricious and vacillating, devising many plans of future operations which are no sooner arranged than they are abandoned... a child in his intellectual capacity and manifestations yet with the animal passions of a strong man.' Ladies were

THE SEARCH FOR THE MIND

The first known brain map is found on an Egyptian papyrus thought to date from 3,000 to 2,500 BC.[17] The idea of modularity cropped up again in medieval times with 'cell theory' that placed various human attributes – spirit, thinking and so on – in the ventricles, the brain cavities where cerebro-spinal fluid is secreted. Then, in the early seventeenth century the French philosopher René Descartes conceived the notion that mind existed in a separate sphere from the material universe, a concept that lingers still. In his scheme the brain was a sort of radio receiver that tapped into the dimension of mind via the pineal gland – the only brain component Descartes could find that was not replicated in each hemisphere.

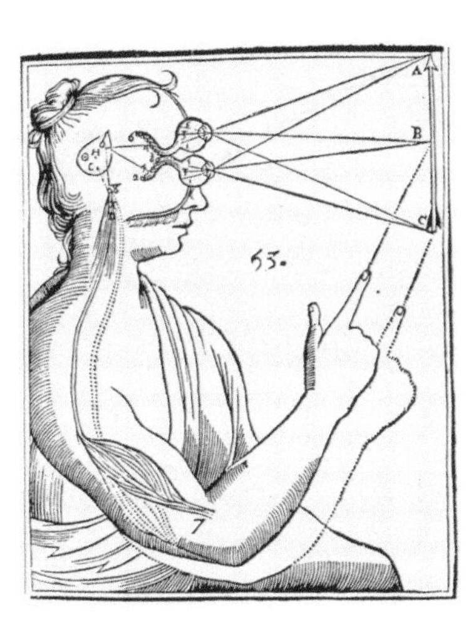

Illustration from Traité de l'Homme *by René Descartes, 1664 in which the pear-shaped pineal gland was thought to be responsible for consciousness and the soul.*

Cartesian dualism was dominant for centuries and still infiltrates our thinking today. But there were always scientists who held that mind and brain function were one and the same, and during the nineteenth and early twentieth centuries many of them worked feverishly hard to produce coherent brain maps. They were helped by several historical events: the French Revolution provided fresh heads to dissect, and the First World War produced countless brain-injured soldiers to observe. Brain mapping went out of favour, though, when the American neurologist Karl Lashley persuaded most of his colleagues that higher cognitive functions were the result of 'mass action' of neurons, and were therefore not susceptible to localization. Psychosurgery suggested that this was not necessarily so, and now brain scanning techniques are showing just how precisely it is possible to pin down even the most sophisticated and complex machinations of the human brain.

SCANNING THE BRAIN

Magnetic Resonance Imaging (MRI, sometimes called nuclear magnetic resonance imaging – NMR) works by aligning atomic particles in the body tissues by magnetism, then bombarding them with radio waves. This causes the particles to give off radio signals that differ according to what sort of tissue is present. A sophisticated software system called Computerized Tomography (CT) converts this information into a three-dimensional picture of any part of the body. A brain scan taken this way looks like a greyish X-ray, with different, clearly delineated types of tissue.

Functional MRI (fMRI) elaborates this basic anatomical picture by adding to it the areas of greatest brain activity. Neuronal firing is fuelled by glucose and oxygen, which are carried in blood. When an area of the brain is fired up, these substances flow towards it, and fMRI shows up the areas where there is most oxygen. The latest scanners can produce four images every second. The brain takes about half a second to react to a stimulus, so this rapid scanning technique can clearly show the ebb and flow of activity in different parts of the brain as it reacts to various stimuli or undertakes different tasks. fMRI is proving to be the most rewarding of scanning techniques, but it is phenomenally expensive and brain mappers often have to share a machine with clinicians who have more pressing claims to it. For this reason a lot of experimental work is still done by the older technique of:

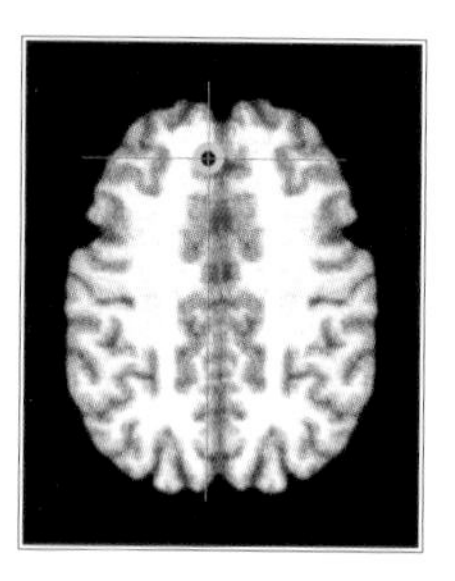

Positron Emission Topography (PET). PET achieves a similar end result to fMRI – it identifies the brain areas that are working hardest by measuring their fuel intake. The pictures produced by PET are very clear (and strikingly pretty) but they cannot achieve the same fine resolution as fMRI. The technique also has a serious drawback in that it requires an injection into the bloodstream of a radioactive marker. The dose of radioactivity given in each scan is tiny but, for safety, no one person is generally allowed to have more than one scanning session (usually twelve scans) a year.

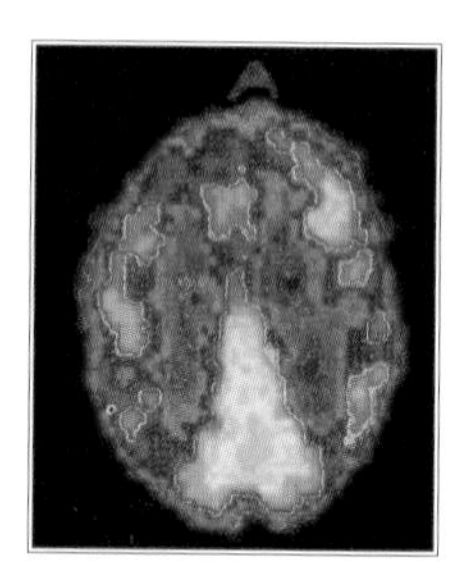

Near-Infra-red Spectroscopy (NIRS) also produces an image based on the amount of fuel being gobbled at any moment by each part of the brain. It works by beaming low-level light waves into the brain and measuring the varying amount that is reflected from each area. NIRS is cheaper than fMRI and does not use radioactivity but it cannot (yet) give a clear picture of what happens in the deepest regions of the brain.

Electroencephalography (EEG) measures brainwaves – the electrical patterns created by the rhythmic oscillations of neurons. These waves show characteristic changes according to the type of brain activity that is going on. EEG measures these waves by picking up signals via electrodes placed in the skull. The latest version of EEG takes readings from dozens of different spots and compares them, building up a picture of varying activity across the brain. Brain mapping with EEG often uses Event-Related Potentials (ERPs), which simply means that an electrical peak (potential) is related to a particular stimulus like a word or a touch.

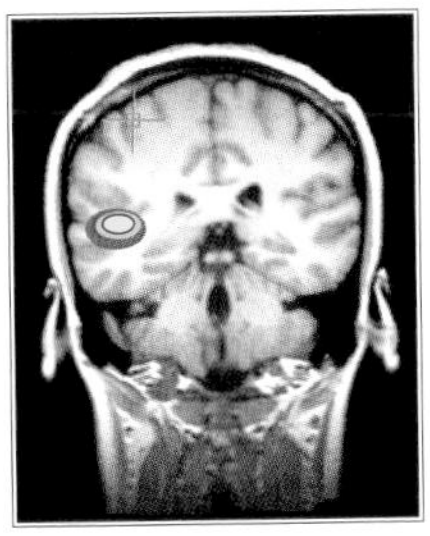

Magnetoencephalography (MEG) is similar to EEG in that it picks up signals from neuronal oscillation, but it does it by homing in on the tiny magnetic pulse they give off rather than the electric field. It still has teething problems: the signals, for example, are weak and easily masked by interference. Yet it has enormous potential because it is faster than other scanning techniques and can therefore chart changes in brain activity more accurately than fMRI or PET.

Multi-modal Imaging, which is becoming increasingly popular, combines two or more of these techniques to give a more complete picture.

advised not to stay in his presence. The hallmark of Gage's new condition was his complete inability to direct or control himself.[13]

If self-determination lies in a specific bit of tissue, it follows that those who appear not to have it may simply be unlucky – victims of a sluggish brain module. So is it reasonable to blame the Phineas Gages of today for their ways? Should we be unsympathetic to addicts who fail to conquer their habit, or punish recidivist criminals?

The current discoveries about the brain should breathe new life into this old debate. If antisocial behaviour can be linked to malfunctioning brain modules, perhaps we should start investigating how to turn the modules on or off. This is not science fiction – it is already being tried. A technique called Transcranial Magnetic Stimulation (TMS) uses a powerful magnetic field to stimulate or inhibit precise areas of the brain. Trials in several countries have shown that it can help relieve depression, and doctors at the US National Institute of Neurological Disorders and Strokes are currently trying it on people with obsessive–compulsive disorder (OCD), post-traumatic stress disorder (PTSD) and mania.[14] Current research into rage, psychopathy and the sort of behaviour typified by Phineas Gage suggests that these might be treated in the same way. If the idea sends shivers down your back, think of what we do to such people now. Is an artificially induced change of mind worse than a stretch in prison?

Windows on the Mind

The safety video that comes with one brand of MRI scanner shows a man walking up to the machine with a metal wrench in his hand. When he gets within a few feet the hand with the wrench in it suddenly shoots up and his arm locks into a rigid horizontal salute with the wrench pointing straight at the scanner. The next few seconds have a cartoon quality as the man engages

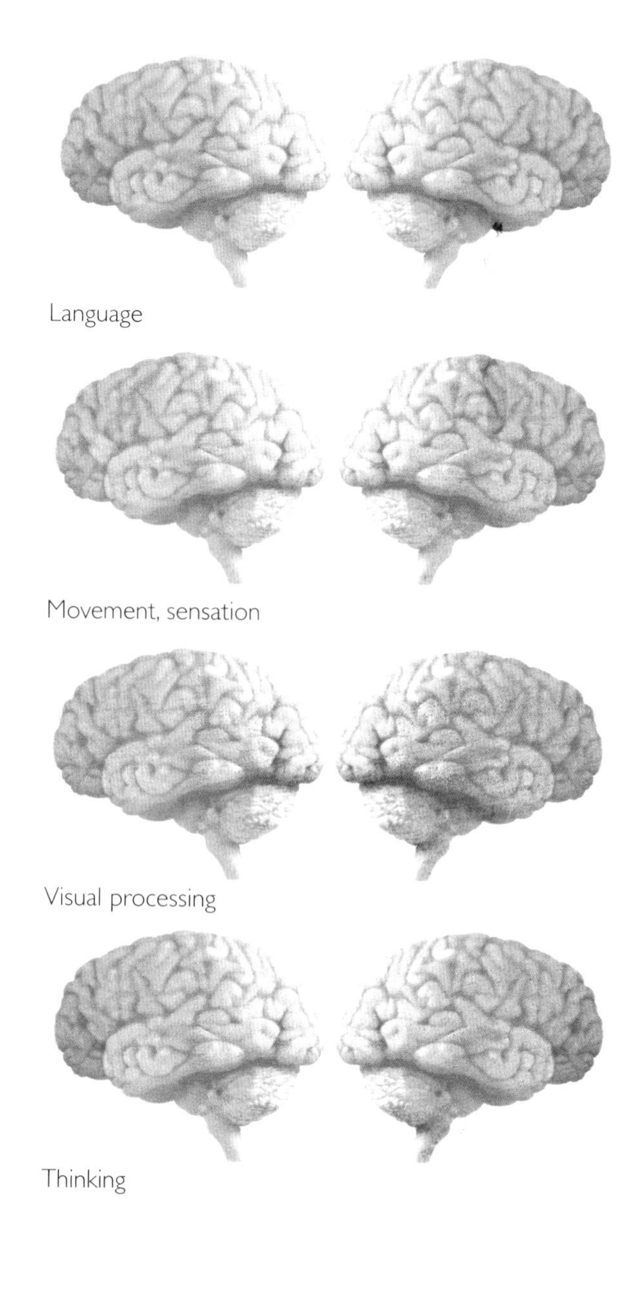

Functional MRI has helped identify which bits of the brain are involved in specific mental tasks.

in a tug-of-war with an unseen rival. As he inches nearer to the machine the wrench starts to quiver like a banner in a wind-tunnel, then slips through his clenched fingers towards the mouth of the scanner. The man clutches it with both hands and leans back, but he obviously cannot keep hold. Suddenly the tool jettisons forward into the tube, where it smashes into a strategically placed brick. The force of the impact is so great the brick is pulverized.

The scene is meant to show the dangers of taking metal near an MRI scanner. These machines are basically massive, circular magnets. Their gravitational pull is some 40,000 times greater than that of the earth – it is easy to see that the consequences of entering one with, say, a heart pacemaker in place, would be catastrophic. If you do not happen to have any metal about your person, however, MRI scanning seems to be perfectly safe – no one has yet reported any detrimental biological effect from it.

Powerful scanning techniques like fMRI are opening the brain to scrutiny in a way undreamt of until a couple of decades ago. But brain mapping began long before fancy scanning machines were invented.

The two main language areas, which are still among the most important cortical landmarks on the map, were identified by Broca and Wernicke more than a hundred years ago. They did it by looking at the brains of people with speech disorders and noting that those with the same problems all had damage in the same place. Broca located the area that allows us to articulate speech by post-mortem examination of the brains of people who were unable (usually as the result of a stroke) to get words out. His classic case was a man named Tan.

Tan was so-called because that was what he said when he was asked his name. It was also what he said when he was asked his date of birth, his address and what he wanted for dinner. 'Tan' was all that Tan could say, even though he understood speech perfectly well.

Broca had to wait until Tan was dead before he could look at his brain and see which bit was injured. Today, scanners allow neuroscientists to

Rivers of the Mind

Different types of cells secrete different neurotransmitters. Each brain chemical works in widely spread but fairly specific brain locations and may have a different effect according to where it is activated. Some fifty different neurotransmitters have been identified, but the most important seem to be:

Dopamine: controls arousal levels in many parts of the brain and is vital for giving physical motivation. When levels are severely depleted – as in Parkinson's disease – people may find it impossible to move forward voluntarily. Low dopamine may also be implicated in mental stasis. Overly high levels seem to be implicated in schizophrenia and may give rise to hallucinations. Hallucinogenic drugs are thought to work on the dopamine system.

Serotonin: the neurotransmitter that is enhanced by Prozac, and has thus become known as the 'feel-good' chemical. Serotonin certainly has a profound effect on mood and anxiety – high levels of it (or sensitivity to it) are associated with serenity and optimism. However, it also has effects in many other areas, including sleep, pain, appetite and blood pressure.

Acetylcholine (ACh): controls activity in brain areas connected with attention, learning and memory. People with Alzheimer's disease typically have low levels of ACh in the cerebral cortex, and drugs that boost its action may improve memory in such patients.

Noradrenaline: mainly an excitatory chemical that induces physical and mental arousal and heightens mood. Production is centred in an area of the brain called the locus coeruleus, which is one of several putative candidates for the brain's 'pleasure' centre.

Glutamate: the brain's major excitatory neurotransmitter, vital for forging the links between neurons that are the basis of learning and long-term memory.

Enkephalins and Endorphins: endogenous opioids that – like the drugs – modulate pain, reduce stress and promote a sensation of floaty, oceanic calm. They also depress physical functions like breathing and may produce physical dependence.

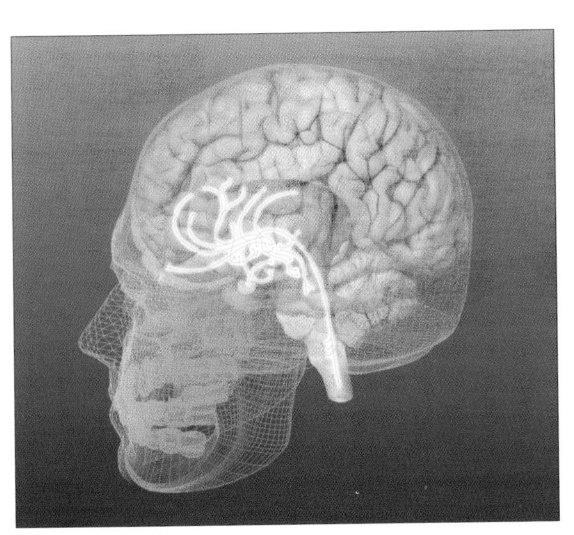

Dopamine pathways

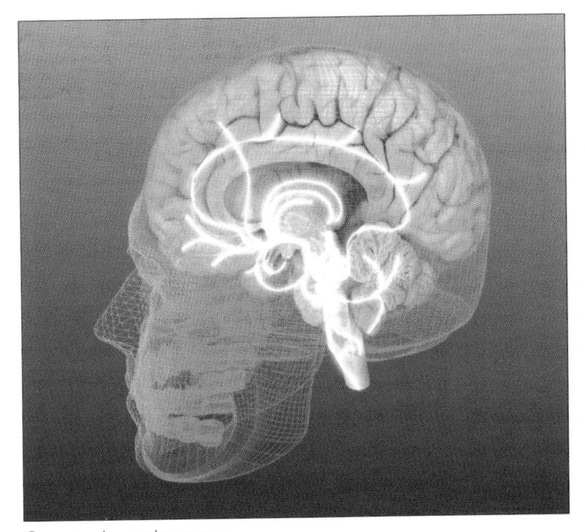

Serotonin pathways

COGNITIVE FLUIDITY: THE ROOT OF RACISM?

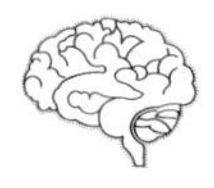

STEVEN MITHEN
Lecturer in Archaeology, University of Reading

Steven Mithen is an archeologist who has studied the mind rather as though it were a living fossil. In* The Pre-History of the Mind *he proposes that the brain evolved through a process of increasing 'cognitive fluidity', which has given humans their unique intelligence. He explains how this might lead to conflict between different races.

At first – prior to six million years ago – the ancestral mind was dominated by a domain of central intelligence, a suite of general-purpose learning and decision-making rules. Behaviour could be modified in light of experience in any behavioural domain but learning would be slow, errors frequent and complex behaviour could not be acquired. The second phase of its evolution gave rise to minds in which general intelligence was supplemented with multiple specialized intelligences where each cognitive domain was devoted to a specific area of behaviour and each worked in isolation from the others. Learning within these domains was rapid and with a minimum of errors, complex behaviour patterns could be acquired and easily modified. But thought at domain interfaces would appear far simpler than that within a single domain. Such domain-specific intelligence was possessed by all hominid species prior to 100,000 years ago, including homo erectus and the Neanderthals. In the final phase, the modern human mind of h. sapiens sapiens, multiple specialized intelligences appear to be working together with a flow of knowledge and ideas between behavioural domains. Experience gained in one domain can now influence that in another resulting in an almost limitless capacity for imagination. I refer to these phase three minds, with their ability simultaneously to manipulate ideas from disparate domains, as having cognitive fluidity. I suggest that it is this very cognitive fluidity, and hence the integration of social and technical intelligence, that creates the possibility that people can be treated as physical objects without rights or emotions – that is, racism.

Early Humans with their Swiss army knife mentality could not think of other humans as either animals or artefacts. For Neanderthals people were people were people. They may have lacked the belief that other groups or individuals had different types of mind from their own – the idea that others are 'less than human' that lies at the heart of racism. Believing that differences exist between human groups is very different from believing that some groups are inherently inferior to others. For this view we seem to be looking at the transfer into the social sphere of concepts about manipulating objects, which indeed do not mind how they are treated because they have no minds at all. My argument is that the cognitive fluidity of the Modern Human mind provides a potential not only to believe that different races of humans exist but that some of these may be inferior to others owing to the mixing up of thoughts about humans, animals and objects. There is no compulsion to do this, simply the potential for it to happen. Unfortunately, this potential has been repeatedly realized throughout the course of human history.

locate injured tissue in living patients, but the basic technique of deducing a brain area's normal activity by seeing what happens in people in whom it is damaged remains an important one.

Another time-honoured technique is to stimulate different areas of the brain directly and see what effect it has. It was by doing this that the surgeons in California brought about so much amusement in their epilepsy patient and thus identified what seems to be the (or a) humour module.

Direct stimulation was pioneered in the 1950s by the Canadian neurosurgeon Wilder Penfield, who charted large regions of the cerebral cortex by applying electrodes to different areas in the brains of hundreds of epilepsy patients. He demonstrated in this way that the entire body surface is represented on the brain's surface as though it had been drawn on: the bit that affects the arm lies next to the bit that affects the elbow, and the bit that affects the elbow is next to the bit that affects the upper arm and so on. More famously, he found that stimulating points in the temporal lobes produced what seemed to be vivid childhood memories, or snatches of long-forgotten tunes.

Most patients reported these recollections as dream-like, yet crystal clear. 'It was like ... standing in the doorway at [my] high school,' reported a twenty-one-year-old man. 'I heard my mother talking on the phone, telling my aunt to come over that night,' said another, '... my nephew and niece were visiting at my home ... they were getting ready to go home, putting on their coats and hats ... in the dining room ... my mother was talking to them. She was rushed – in a hurry.'[15]

Penfield's findings have been widely interpreted as showing that memories are stored in discreet packets ('engrams') just waiting to be revived, but recent work indicates a more complex arrangement. Research by Steven Rose and colleagues at the Open University in Britain suggests that memories are cloned and that each clone is laid down in a different sensory area of the brain: visual, auditory and so on.[16] Stimulation of one of these clones may in some way trigger the others, to give an integrated, multimedia experience. Penfield was probably stimulating just one sensory facet of the memory but eliciting a response in many.

The area stimulated in the laughing patient may similarly be just one node of a much larger brain module. Indeed, many of the neat little 'spots' currently marking particular functions may turn out to be exposed peaks of mainly buried neural conglomerates – icebergs of the mind.

It is possible, too, that areas that light up when a certain mental task is performed are not themselves responsible for that task, but are simply passing on a stimulus to the bit that is. There is an apocryphal story of a scientist who claimed to have discovered that frogs heard through their legs. When challenged to prove it he produced a frog that he had trained to jump on command. After demonstrating the frog's trick he picked the animal up and cut off its legs. Then he put it back on the table and told it, again, to jump. Naturally, it did not. 'There!' said the scientist, triumphantly. 'You see – it can no longer hear my voice!'

Brain mappers are trying hard to avoid falling into that trap but sometimes, inevitably, they do. According to some there is still too much of the gold rush about this science – too many researchers trying to stake out new claims rather than replicating the findings of others. Yet the ground is firming up. In the last couple of years standard scanning protocols have been adopted that have drastically reduced the number of rogue results, and methodology – particularly the problems of designing experiments that give unambiguous results – is a constant concern. The New Phrenologists are determined that their discoveries – unlike those of Franz Gall – will stand the test of time.

EVOLUTION

The human brain carries the story of its evolution in its anatomy. It began in water when fish developed a tube to carry nerves from the distant parts of their body to a central control point. First there was just a bulge on top of the spine, then the nerves started to sort themselves into specialized modules. Some became sensitive to molecules and formed what is now our smell brain. Others became light sensitive and formed eyes. These were connected to a clump that controlled movement – the cerebellum. This collection formed the reptilian brain, mechanical and unconscious. Its basic parts are still intact and form the lowest of the three-tier system that has developed since.

fish brain

reptilian brain

On top of this, more modules developed: the thalamus, allowing sight, smell and hearing to be used together; the amygdala and hippocampus, creating a crude memory system; and the hypothalamus, making it possible for the organism to react to more stimuli. This is the mammalian brain, known as the limbic system. Emotions are generated here but not experienced here as it is not conscious.

During mammalian evolution the sense modules triggered the development of a thin matrix of cells, whose shape allowed for many neural connections to be formed between them with only a small increase in size. This skin became the cortex and it is from this that consciousness emerged.

mammalian brain

The mammals that were to evolve into humans developed an ever-larger cortex, pushing the cerebellum back to the position it now occupies. *Australopithecus Africanus* had a fairly human-shaped brain 300 million years ago, but it was only a third the size of a modern brain. One and a half million years ago the hominid brain underwent an explosive enlargement. So sudden was it that the bones of the skull were pushed outwards, creating the high, flat forehead and domed head that distinguish us from primates. The areas that expanded most are those concerned with thinking, planning, organizing and communicating.

human brain

The development of language was almost certainly the springboard for the leap from hominid to human. It gave our ancestors lots to think about, and new brain tissue was needed. The frontal lobes of the brain duly expanded by some 40 per cent to create large areas of new grey matter: the neocortex. This spurt was most dramatic at the very front, in what are known as the prefrontal lobes. These jut out from the front of the brain, and their development pushed the forehead and frontal dome of the head forward, reforming it to the shape of a modern skull.

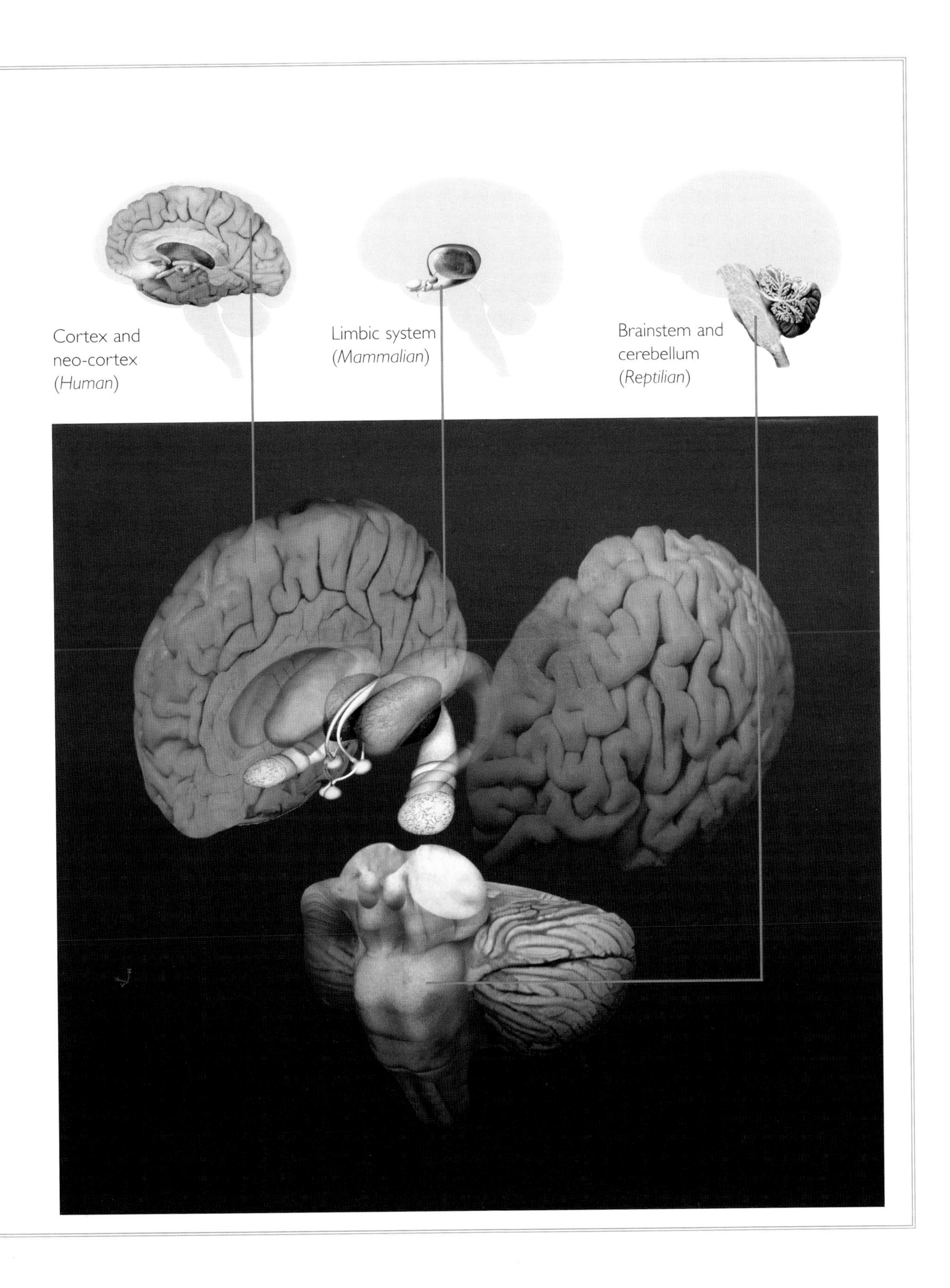

Cortex and
neo-cortex
(*Human*)
Limbic system
(*Mammalian*)
Brainstem and
cerebellum
(*Reptilian*)

CHAPTER TWO
THE GREAT DIVIDE

The human brain is a marriage of two minds. Each of its twin hemispheres is a physical mirror image of the other, and if one hemisphere is lost early in life, the other may take over and fulfil the functions of both. Normally, though, the two are bound together by a band of fibre that conveys a continuous, intimate dialogue between them. Information arriving in one half is almost instantly available to the other and their responses are so closely harmonized that it produces an apparently seamless perception of the world and a single stream of consciousness. Separate these hemispheres, however, and the differences between them become apparent. Each half of a mature brain has its own strengths and weaknesses; its own way of processing information and its own special skills. They might even exist in two distinct realms of consciousness: two individuals, effectively, in one skull.

sinister *adj.* **1.** Threatening or suggesting evil or harm; ominous. **2.** Evil or treacherous. [C15: from Latin **sinister** – on left-hand side] – *Collins English Dictionary, third edition*

THE LEFT BRAIN IS WHAT HAS MADE homo sapiens the spectacularly successful species it is. It is calculating, communicative and capable of conceiving and executing complicated plans. But it has always had a bad press. It is frequently held to represent the worst of the Western world: materialistic, controlling and unfeeling; while the right brain is seen as gentle, emotional and more at one with the natural world – a frame of mind commonly associated with the East.

This notion has spawned a small industry of specialized self-help books and training courses that claim to encourage right-brain thinking. There are books that show you how to draw with your right brain, ride horses with your right brain, even make love with your right brain. You can take all manner of courses to put you back in touch with your right half, and big businesses hire consultants to test their employees for left/right dominance and slot them into appropriate jobs.

Is it all nonsense? Brain scientists will tell you the idea of a rigid divide is a popular myth. They even have a word for the public's enthusiasm for the subject: 'dichotomania'. Like 'modern phrenology' the word is a put-down, intended to imply that the real situation is far too complex for simple conclusions to be drawn.

It is true that the brain is marvellously complicated, and the constant interaction of its two hemispheres makes it extremely difficult to pinpoint what is happening where. Even the most obviously lateralized of skills – language – is atypically organized in about 5 per cent of people. The brain is also very malleable and its wiring can be influenced by all sorts of environmental factors. Given extraordinary circumstances a genetically typical brain may end up organized in a very odd way indeed. Nevertheless, brain imaging studies confirm that the two hemispheres really do have quite specific skills that are 'hard-wired' to the extent that, in normal circumstances, certain skills will always develop on a particular side.

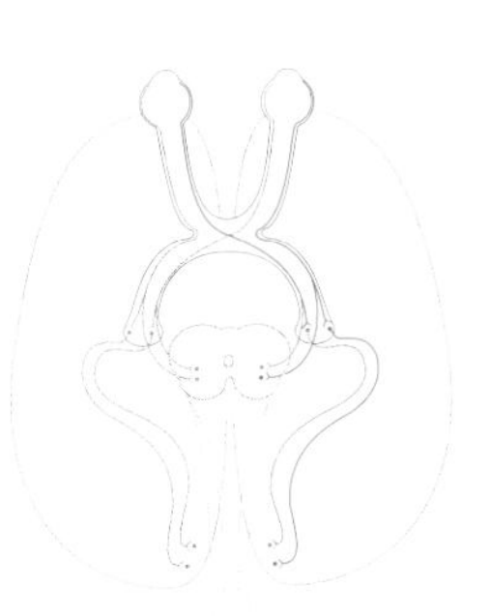

A – vision

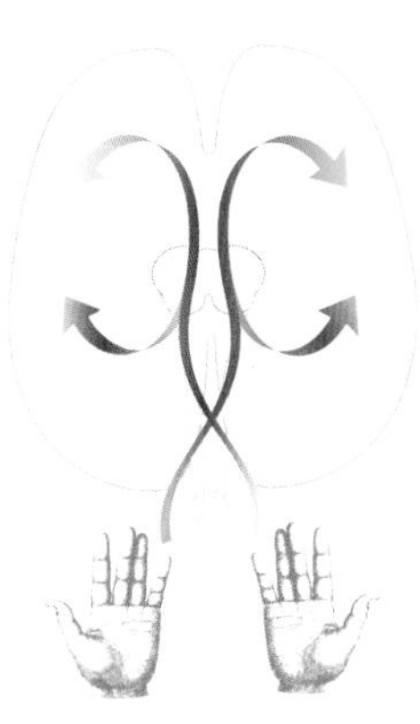

B – touch

C – hearing

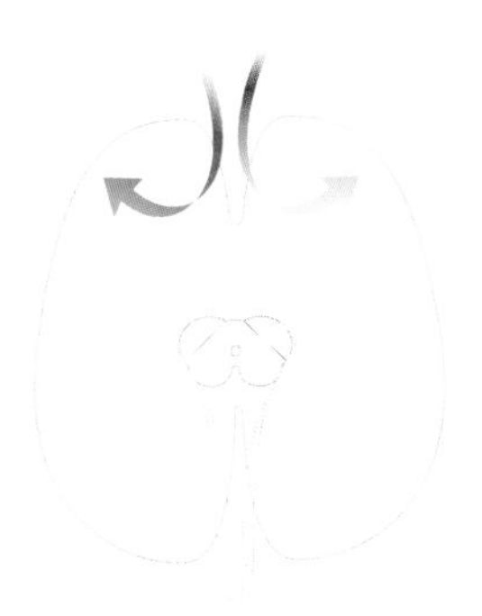

D – smell

Most sensory input to the brain crosses over from the incoming side to the opposite hemisphere for processing. Once the information enters one hemisphere it is swiftly sent on to the other via the corpus callosum. A. Visual input from the left half of each eye goes to the right hemisphere and vice versa. B. Apart from certain facial nerves, the neural pathways from the body also end up in the opposite side of the brain. C. Most auditory input is processed on the opposite side of the brain to the ear through which it enters. D. Smell is the exception to the cross-over rule – odours are processed on the same side as the nostril that senses them.

The layout, furthermore, is more or less as is popularly conceived. The left brain is analytical, logical, precise and time-sensitive. The right brain is dreamier, it processes things in a holistic way rather than breaking them down and it is more involved with sensory perception than abstract cognition.[1]

There is truth, too, in the idea that the right brain is more emotional than the left. In particular it is responsible for fearful and mournful feelings and for general pessimism. This is why people who suffer severe left-brain strokes quite commonly behave as though what has happened is a catastrophe, even if the disabilities they suffer as a result are relatively mild. What seems to happen in these cases is that the damaged left brain can no longer keep the right brain subjugated, so the right hemisphere floods consciousness with its own miserable view of life.

Patients with bad right-brain damage, by contrast, sometimes appear to be entirely unmoved by it, maintaining an optimistic, what-the-hell sanguinity in the face of what would otherwise be quite dreadful suffering. In the most extreme cases they refuse to recognize their shortcomings at all. A very senior American judge is said to have caused huge embarrassment after a severe right-brain stroke because he insisted on continuing at the bench despite having lost his ability to weigh evidence in anything like a sensible way. He maintained an exceptionally jolly courtroom, happily allowing serious criminals to go free while occasionally dispatching minor offenders to lifelong prison sentences. He resisted his colleagues' attempts to persuade him to retire and was finally sacked. Thanks to his right-brain damage he seemed perfectly content – if puzzled – by this turn of events, and subsequently enjoyed a long and happy retirement.

Sometimes this sublime disregard for their condition is taken to such extremes that people with right-hemisphere injuries fail to notice frankly disabling conditions like paralysis or even blindness – a condition known as anosognosia.

Despite the irrepressible jollity of an unopposed left brain it takes the combination of both hemispheres to produce a fully rounded sense of humour. Consider a fairly standard joke: a kangaroo walks into a bar, sits down and asks for a pint of beer. The barman, astonished, gives the kangaroo his drink. 'How much?' asks the animal. The barman, recovering his composure, decides to see if the kangaroo is really as clever as it seems. Winking knowingly at his human customers, he names an exorbitantly high price. The kangaroo duly pays it, and the barman, reassured of the continuing superiority of his own species, relaxes. 'Don't get many kangaroos in here,' he remarks casually ...

Now, here are three possible punchlines: (a) Then the kangaroo gets out a gun and shoots the barman. (b) Then the man on the next stool admits that he is a ventriloquist who has trained the kangaroo to drink beer. (c) 'I'm not surprised,' replies the kangaroo, 'with beer at that price.'

Punchline (c) might seem the obvious choice – but a person with a right-brain lesion would be quite likely to pick the literal and – to most people humourless – option of (b). Left-brain-damaged patients, on the other hand, may choose (a) – a seemingly arbitrary surprise ending.

The difference is probably because the left brain creates the feeling of amusement and so is quite happy to laugh at more or less anything when prompted – rather like the epileptic girl who laughed at the surgeons when part of her left hemisphere was stimulated on the operating table. Hence ordinary situations, with no 'twist', may be seen as jokes if presented as such. The right hemisphere is the one that 'gets' the joke, in that it registers the dislocation in

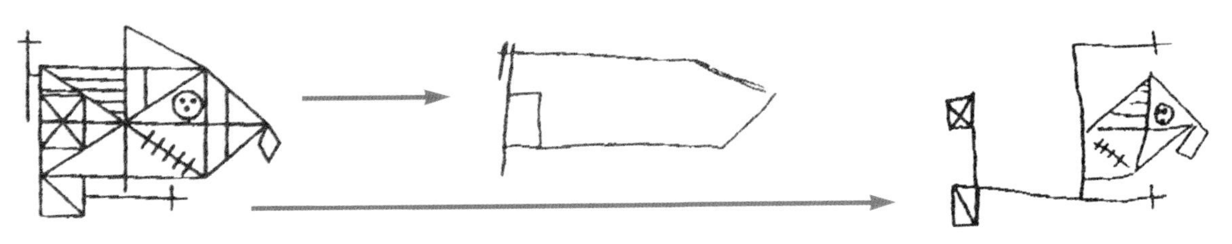

(A) *Patients with single hemisphere brain damage were asked to look at this figure, then copy it.*

(B) *This is typical of an attempt by a patient with left hemisphere damage: the outline is fine but details are neglected.*

(C) *While a patient with right hemisphere damage draws only the details.*

logic that is a hallmark of most formal humour. This is a mild form of alert – a 'something not quite right here!' effect that, on its own, is not funny at all. In fact, it is a mild form of fear. Someone equipped only with this right-brain ability might well think that any surprise ending makes a joke.

The combination of (right-brain) alert and (left-brain) jollity is still not enough to make something funny – humour also needs meaning in order to work. This is why seeing a nice-looking person slipping on a banana skin is not funny but watching a pompous bully come a cropper is. Meaning emerges from the pulling together of all the threads of the joke, including context, assumptions and knowledge of our own prejudices.

Humour is a diffuse, fuzzy sort of thing – a matter of taste, often, and something you don't expect to find in even the most sophisticated computer unless it has been programmed in by a human. In this it is typical of the type of functions that engage both hemispheres. Specialized functions tend, by contrast, to be lateralized. Finding your way around in space, for example, is very much a right-brained activity, and if the parts of the brain that do it (right hippocampus and right parietal cortex) are damaged, a person may find they get lost in what were once the most familiar places. One such patient could not even find his own front door. Whenever he wanted to go out he would spend up to five minutes blundering around his (relatively small) bungalow searching for the way out.[2]

In time most people who suffer one-sided brain damage find that their condition improves. Sometimes this is because the undamaged hemisphere learns to do tasks usually left to the other. It probably does not do them in the same way, though, nor so efficiently. If the ability to negotiate from place to place is lost, for example, the left brain may take on the job, using its own special skills of sequential memory and deduction. So, instead of just 'knowing' (as the right brain did) that the front door is the third door along from the kitchen, the left hemisphere remembers that as a fact and arrives there by counting the intervening doors. Similarly, recognizing familiar faces is a right-hemisphere activity that is sometimes lost after brain injury.

Face recognition does not require thought – again, like all right-brain activity, it just happens. If the right-brain area responsible for this skill is lost, the only way a person can recognize those they know is by deliberately taking note of their distinguishing features, then matching their mental notes to each face they see. This imperfect technique often fails, making social gatherings a nightmare. One man, after a particularly embarrassing incident at a dinner party, insisted his wife wore a red ribbon in her hair when they went out to prevent him again trying to take home the wrong woman.

The right hemisphere is also good at grasping wholes, while the left brain likes detail. Other right-brain strengths include the ability to make out camouflaged images against a complex background and to see patterns at a glance. This could well have had important survival benefits when our ancestors needed to watch for predators. The left, by contrast, is good at breaking down complicated patterns into their component parts. This might have survival value in a bureaucracy, but a left brain alone in the wild would be unable to see the wood for the trees, let alone a half-hidden bear in search of dinner.

Almost every mental function you can think of is fully or partly lateralized. Precisely how it happens is not known, but it seems that incoming information is split into several parallel paths within the brain, each of which is given a slightly different treatment according to the route it takes. Information that is of particular 'interest' to one side will activate that side more strongly than the other. You can see this happen in a brain scan – the side that is 'in charge' of a particular task will light up while the matching area on the other side will glow far more dully.

The tasks that each hemisphere takes on are those that fit its style of working: holistic or analytical. The key to the differences in left- and right-brain style may in part be found in a curious physical difference that exists between them. If you were to cut the hemispheres open you would find they are made up of a mixture of grey and white matter. The grey matter consists of the central bodies of brain cells and is found mainly in the cortex, which is a couple of millimetres thick. The white matter lies beneath this. It is made up of dense bundles of axons – the threads that project from the cell bodies and carry messages between them.

The distribution of white matter to grey is not even throughout the brain – the right hemisphere has relatively more white matter, while the left has more grey.[3] This microscopic distinction is significant because it means that the axons in the right brain are longer than in the left and this means they connect neurons that are, on average, further away from one another. Given that neurons that do similar things or process particular types of input tend to be clustered together, this suggests that the right brain is better equipped than the left to draw on several different brain modules at the same time. The long-range neural wiring might explain why that hemisphere is inclined to come up with broad, many-faceted, but rather vague concepts. It might also help the right brain to integrate sensory and emotional stimuli (as is required to apprehend art) and to make the sort of unlikely connections that provide the basis of much humour. 'Lateral thinking' would be helped, too, by the neural arrangement in the right brain – the sideways extension of axons even makes the phrase literal rather than figurative. The left brain, by contrast, is more densely woven. The close-packed, tightly connected neurons are better equipped to do intense, detailed work that depends on close and quick co-operation between similarly dedicated brain cells.

To draw a fanciful (and very right-brain) analogy:

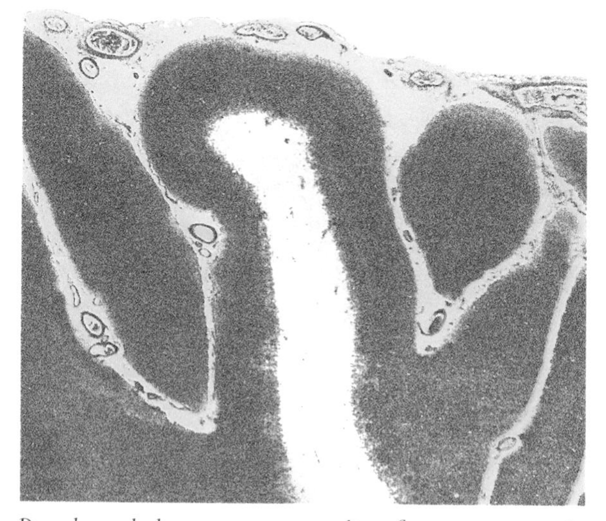

Densely packed neurons create a skin of grey matter in the brain while their axons – tentacles which reach out to other cells – show up as white tissue.

D
D
D
D
D
DDDDD

Concentrating on the 'L' (left) *creates activity in the right hemisphere* (A) *while attending to the 'Ds' causes activity in the left* (B). *These scans demonstrate how the two sides of the brain deal with different aspects of a single stimulus.*

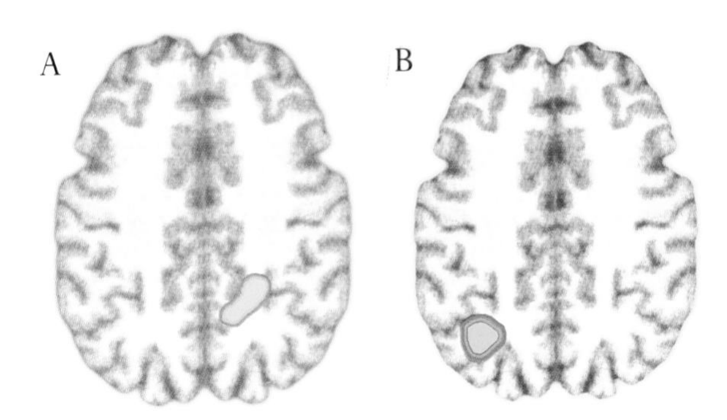

imagine the two hemispheres as two halves of a flat black screen. A film is to be projected, in duplicate, on both halves of the screen concurrently, and you need to extract as much information from this film as you can. The screen needs to be white to reflect the picture, and you have one pot of white paint with which to transform it. Unfortunately, you need double that amount to cover the whole area, and you can't just whiten one half and forget the other – you have to use half the paint on one half of the screen and half on the other. What to do? You cannot afford to risk missing either the outline of the film you are to watch, nor certain details.

One solution would be to spread your paint thinly but completely over one surface, and on the other to create islands of dense white where you think the main action of the film is likely to take place, leaving less important areas uncovered. Now, when the film is projected the right half of the screen will show a shadowy but complete outline, while the left half will show patches of sharp, detailed imagery but no overall shape.

With this arrangement you can get the full picture providing you constantly scan both halves of the screen, looking to the left for details and to the right for an overall impression of what is going on.

This seems to be how the two hemispheres work. They each process their 'halves' of the big picture, and then pool their information by sending signals back and forth via the corpus callosum.

To use another simile: overall the two sides of the brain are like an old married couple who long ago fell into doing things according to a 'his 'n' hers' division of labour. The partner that can communicate takes the dominant role, speaking and acting, most of the time, for both and doing much of the day-to-day business of thinking, calculating, and dealing with the outside world. The other stays largely in the background, quietly doing its moment-by-moment chores and constantly using its singular talents to sniff and taste the social environment for signs of anything that may be a threat or of benefit. They keep each other perfectly informed of what is going on in their own spheres through a continuous, intimate conversation. So natural to them is this pattern of working that they can carry out the most complex tasks together in a perfectly integrated way.

Most of the time the marriage of our two brains is completely harmonious. Conscious decisions, although they may seem to be made by the dominant partner alone, are in fact fully informed by the findings of both hemispheres. Sometimes, though, the conversation between them falters. The dominant hemisphere may ignore the information supplied by its partner and make a decision based purely on what it thinks. The result may be an emotional disquiet that is difficult to explain. Conversely, the non-dominant partner sometimes bypasses the executive control of the other side and triggers an action based purely on instinct. These are the sort of things we look back on (usually with

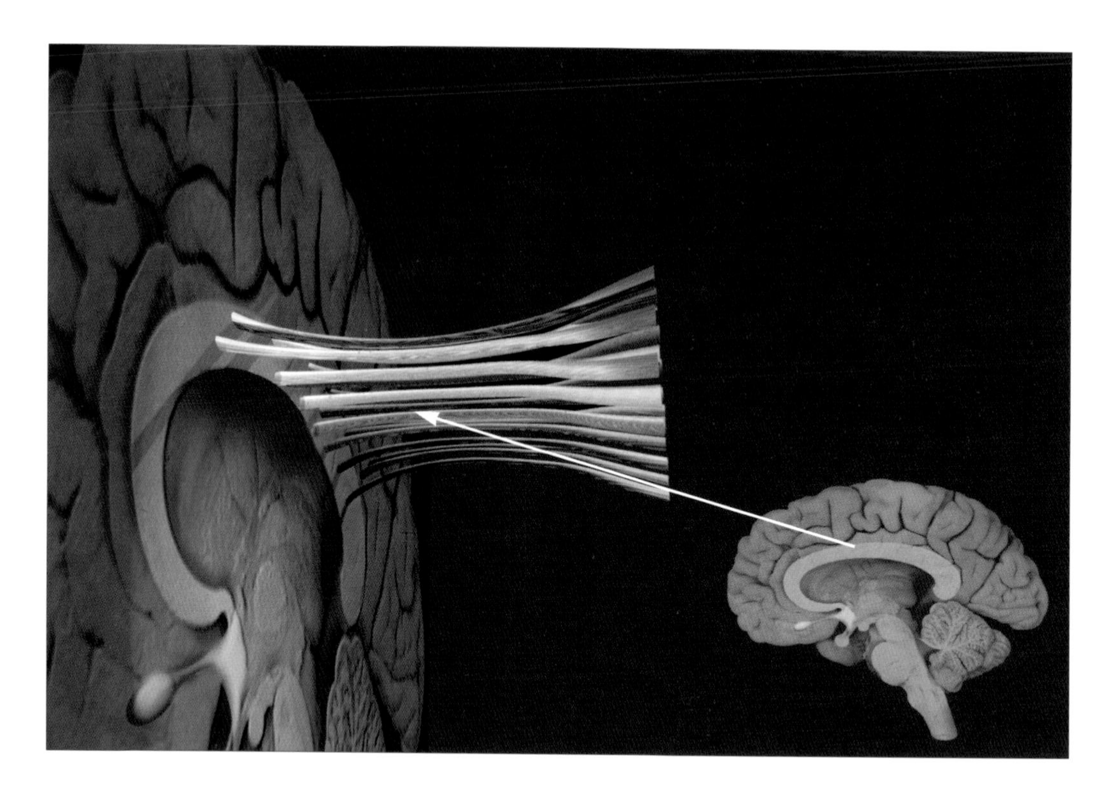

The corpus callosum is a thick band of axons – 80 million or so – which connect the brain cells in one hemisphere to those in the other. The two sides keep up a continuous conversation via this neural bridge.

embarrassment) and say: 'I didn't mean to do it – I just couldn't help myself.'

Sometimes, too, a hemisphere acts unilaterally because it does not receive all the incoming information. The corpus callosum can carry enormous amounts of information from one side of the brain to the other in milliseconds, but sometimes there is a split second during which incoming data lingers in one hemisphere only. And some information – the sort for which one or the other hemisphere has a very strong preference – may be registered only dimly by the other.

We all experience this half-knowledge in subtle ways. Odd remarks that just slip out, feelings we can't explain, silly errors like mistaking one object for another, are traditionally seen as evidence of deep inner conflicts. In fact, many of these may be caused by faulty or incomplete inter-hemispheric communication. The giveaway signs are comments like: 'There's something about her/him/the décor I don't like but I can't put my finger on it', or 'I know something awful has happened but it hasn't hit me yet.' In the first situation the person's right brain may have grasped something of which their left brain is only faintly aware, and in the second the left brain has acknowledged something but the right brain has not yet taken it in.

Not understanding a feeling does not stop us from acting on it. Much human behaviour is based on right-brain hunches. We see millions of things happening around us from minute to minute and only a tiny percentage of them is consciously registered. The rest enter the brain fleetingly as momentary blips of energy, leaving no impression. Some of them are probably just noticeable enough to create a momentary emotional response in the right brain, but not significant enough to create conscious awareness in the left. Such half-seen stimuli may account for the odd, uncalled-for flicker of irritation or the little cloud of inexplicable melancholy that most people experience from time to time.

These subtle shifts of feeling are more likely when the left brain is relatively idle and therefore sending fewer than normal inhibitory signals to its moody twin. This may be one reason why throwing yourself into some left-brain activity like reading, talking, or even doing your tax return often brings relief from mild anxiety or depression. Similarly, feelings of grief may be lessened by frenetic left-brain activity, which inhibits the emotional response of the right. Conventional wisdom has it that the brisk 'get on with it' school of therapy is ultimately damaging – rather like putting a tight lid on a fermenting pot. Instead we are encouraged – not least by the ever-expanding counselling and psychotherapy industry – to talk about our emotions, to 'get it out'.

Psychotherapy does, in some circumstances, work. But this is probably not because it allows emotions free rein, rather that it helps us to elevate them to cortical level where they can be consciously processed. One of the most successful types of psychological treatment is cognitive therapy, which, by definition, involves left-brain activity. Talking and thinking about emotions gives us control over them, so they cease to overwhelm us. Simply allowing emotions to well up until we are engulfed by them is, on the other hand, likely to make them more painful, if they are painful to begin with. Counselling after trauma, for example, can make people feel worse – especially if the therapy involves simply reliving the original experience.[4]

The left/right hemisphere split often shows up in our reactions to art: 'I like it but I don't know why' is not necessarily a philistine reaction – it demonstrates only that the work is being appreciated by the right brain rather than analysed by the left. Much advertising is designed to exploit the gap between the impressionable right brain and the critical left. Those adverts that use visual images rather than words to convey messages are particularly likely to impinge on the right hemisphere without necessarily being registered by the left. The purpose of such communication is, of course, to make us buy whatever is being advertised. We may like to think of that action as the result of a rational decision, but it is really just the fulfilment of an impulse.

Not that we like to admit this. The idea that our actions may be irrational is peculiarly unacceptable to the left hemisphere. A series of famous experiments[5] showed that people hardly ever admit to making arbitrary decisions. In one of the experiments, for example, a selection of nylon stockings was laid out and a group of women were invited to choose a pair. When they were asked why they had made their particular choice all the women were able to give detailed and sensible reasons, citing slight differences in colour, texture or quality. In fact, all the stockings were identical – the women's 'reasons' for choosing them were actually rationalizations constructed to explain an essentially inexplicable piece of behaviour.

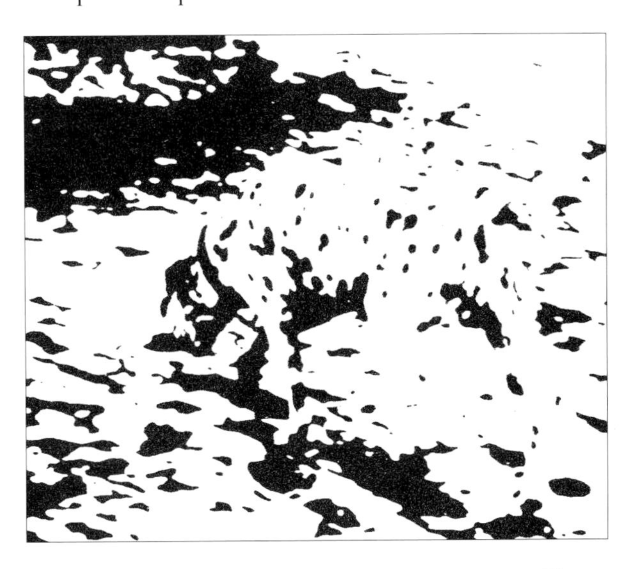

A meaningless collection of blobs, or a Dalmatian sniffing at something on the ground? The left hemisphere sees only the lines – the right sees the dog.

You do not have to use much imagination to realize how such a process may be used to dignify emotional or arbitrary acts in ordinary life: choosing to employ a person of one colour, for example, over that of another. It is also easy to see why so many of us are continuously and compulsively drawn to analyse and explain our own and others' behaviour – spinning elaborate explanations for the way we behave is a built-in hobby. It may also go some way towards explaining why Freudian psychoanalysis – despite an almost complete lack of evidence for its efficacy as therapy – became a near century-long obsession in those societies that could afford it.

Our urge to rationalize behaviour probably has considerable survival value. The human species got where it is largely by forming complex social constructs – from the hunting party to the political party – and making them work. To work they require that we have confidence in them and to have confidence we need to believe that the actions of these organizations are based on sound, rational judgements. At one level, of course, we know we are kidding ourselves. For example, all governments, in all societies, have some policies that are demonstrably irrational. However, no government member, anywhere, ever admits this – not at the time, anyway. Instead they rationalize their policy-making. We may see through it, but basically we like things this way – it makes us feel safe.

Similarly, rationalizing our own actions gives us confidence in our sanity. You can see it happening all around you: watch some harassed mother trying to cope with a lively toddler, for example. Sooner or later she is likely to snap at the child, or slap it, or just withdraw her attention. She does this because she feels irritated, worried or tired. The feelings may be generated by circumstances that are, at that moment, quite out of mind, and nothing to do with the child. Yet the child feels the brunt of it. Can that mother admit this? Later on, perhaps, when she is at home, relaxed, and the child is sleeping she may reflect that she was rather unfair. But at the time, when the child wails: 'Why did you do tha-a-at?' she will rationalize: 'Because you were rude/naughty/disobedient.' If she didn't do this, her confidence in her ability to cope would be dented. Better have a reason for doing things – any reason – than no reason at all.

Splitting the brain

'I don't know why, but I feel scared,' said the woman. 'I feel kind of jumpy ... I know I like Dr Gazzaniga, but right now I'm kind of scared of him' – *split-brain patient to researcher.*[6]

The patient, V.P., had good reason to feel jumpy. She had just seen a particularly violent murder and it had clearly affected her mood. The murder was merely on film, and the neurologist who was investigating her, Michael Gazzaniga, was not the culprit. But his patient did not know these things. Although she had been fully awake and alert throughout the film-show she was now apparently unconscious of having seen anything but a flash of light. She was therefore ascribing the emotions roused by the images to anything in the vicinity – including the man in charge of the experiment.

V.P.'s curious half-knowledge of the trauma she had just been through was caused by a weird and rare condition: her brain had effectively been sawn in half. This drastic operation was carried out to control what had been very severe epilepsy. Cutting the neural connections between one hemisphere and the other stopped the flurries of random electrical activity – her fits – from travelling along them and engulfing her whole brain. Once it had been done the seizures remained localized in the side in which they arose and were far less troublesome. It left her – and similar patients – with some of the most bizarre neurological side-effects ever reported.

V.P. was part of the second cohort of split-brain patients to offer themselves for psychological testing. The first were investigated by

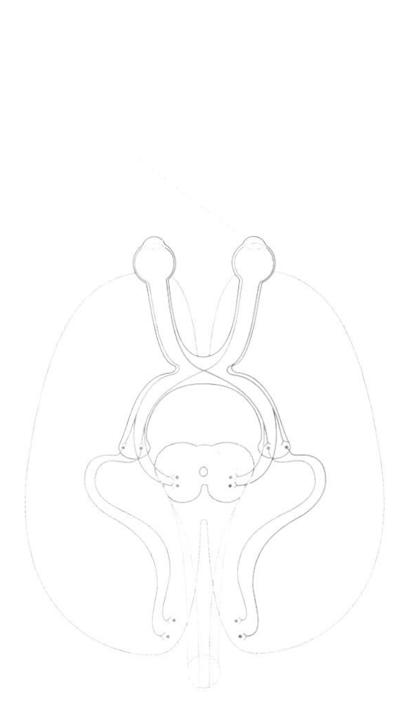

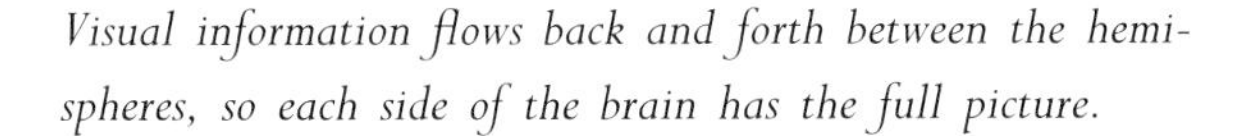

Visual information flows back and forth between the hemispheres, so each side of the brain has the full picture.

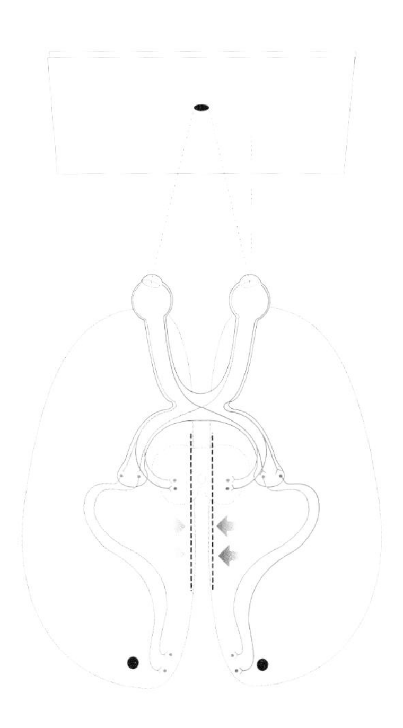

In a split-brain visual information is trapped in each hemisphere. Providing the eyes are kept still, this means that each side of the brain receives information from the opposite half of the visual field only. Below: *The brain is split by severing part of the corpus callosum.*

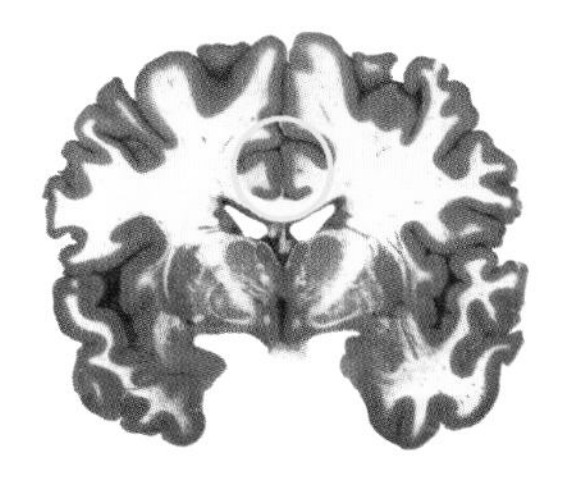

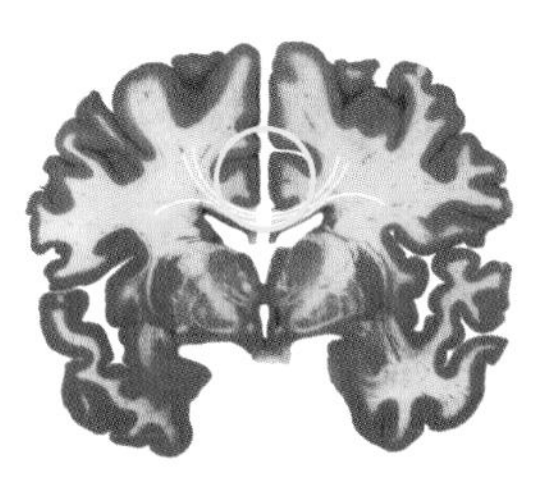

the psychobiologist Roger Sperry, who, in 1981, won the Nobel Prize for his work. Sperry had already done much to demonstrate that the brain was a modular system rather than some homogenous black box. Much of his work was based on cutting connections between various parts of animals' brains and seeing what effect it had. The split-brain technique had already proved particularly revealing. Together with Ronald Myers he had shown, for example, that if you divide a cat's brain you can teach one hemisphere to do a task – pressing a lever for food, say – but leave the other ignorant.

Would human consciousness prove to be similarly divisible? The human split-brain patients provided the perfect means to find out because their hemispheres were already separated. They also held out the wonderful promise of being able to report directly, from inside, on the curious state of half-knowledge that had been observed in the split-brain cat. Sperry duly designed a series of experiments to show how each brain hemisphere worked in isolation, and to tease out the functional differences between them.

A typical experiment involved N.G., a Californian housewife.

The woman was placed in front of a screen that had a small black dot in the centre. She was asked to fix her eyes on this dot, thus ensuring

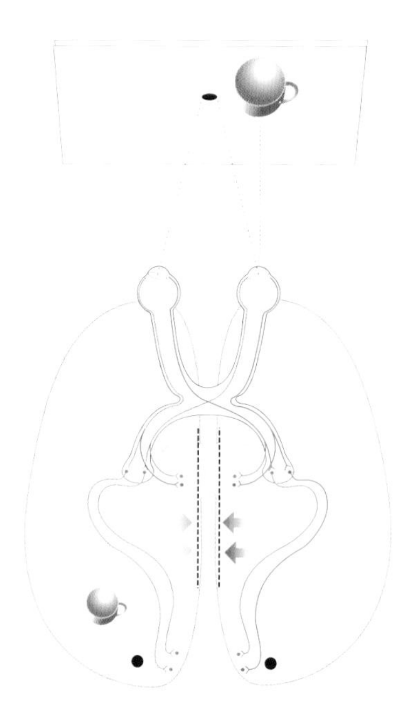

The image of the cup goes to the left hemisphere only and – because the brain is split – it cannot travel to the right hemisphere.

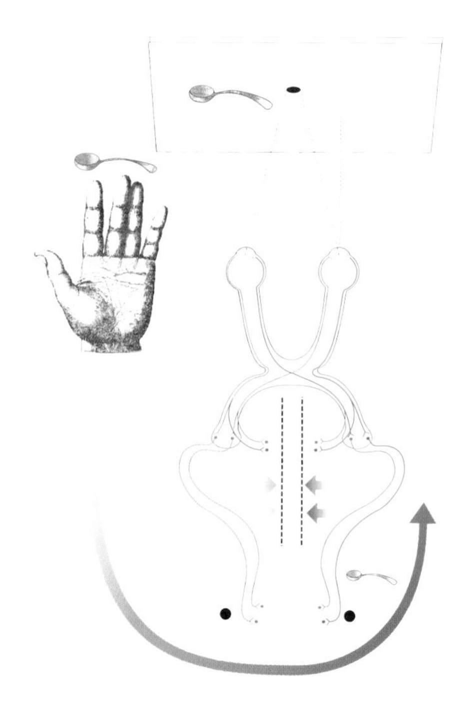

The spoon image went only to the right hemisphere – and because this side does not have speech the subject could not report seeing it. Her left hand 'knew' what was on the screen, however, because it, too, is wired to the right hemisphere.

that images entering her eyes from the side were sent to one hemisphere only. The experimenter then briefly flashed a picture of a cup to the right of the dot. The image stayed on the screen for a very short while – about a twentieth of a second. This is just long enough for an image to be registered, but not long enough for a person to shift their eyes to bring it into focus and thus send it to both hemispheres. So the image on the screen went to N.G.'s left hemisphere only and there it stopped, because the normal route to her other hemisphere – the corpus callosum – was cut. Asked what she saw, N.G. replied, quite normally: 'A cup.'

Next, a picture of a spoon was flashed on the left side of the screen so that it entered N.G.'s right hemisphere. This time, when she was asked what she had seen, she said: 'Nothing.' The experimenter then asked N.G. to reach under the screen with her left hand and select, by touch only, from among a group of concealed items the one that was the same as the one she had just seen. She felt around, pausing briefly at a cup, a knife, a pen, a comb – then settled firmly on a spoon. While her hand continued to hold it behind the screen, the experimenter asked her what she was holding. 'A pencil,' she said. These responses, though inexplicable at first sight, actually gave Sperry and his colleagues a uniquely clear picture of what was going on in N.G.'s brain. N.G. was right-handed, and it was already known that speech, in most right-handed people, is located in the left hemisphere. Thus, when the cup image was sent to the left hemisphere she saw it and named it in the usual way. When the spoon image was fed to her right hemisphere, however, she could not tell the experimenter about it because her right hemisphere was unable to speak. The words that came out of her mouth, 'I see nothing', were

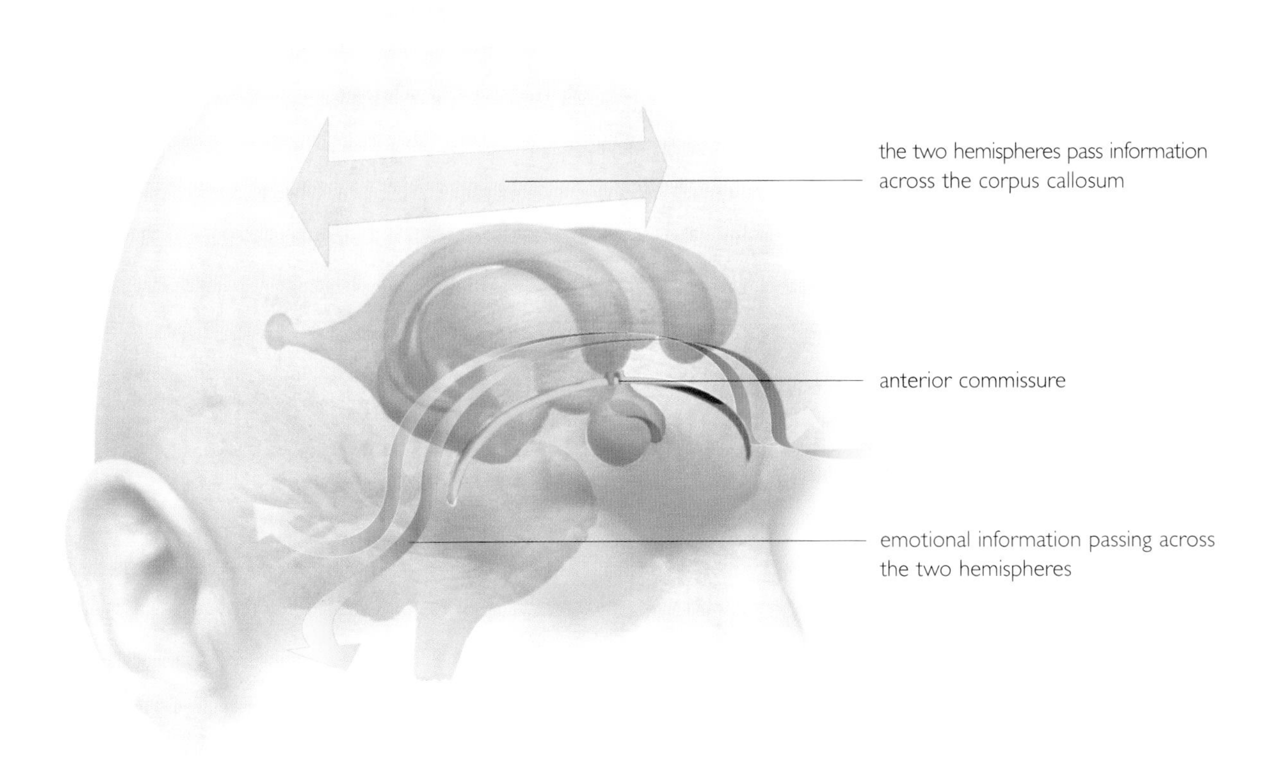

The anterior commissure lies below the corpus callosum. It connects the unconscious limbic structures in both hemispheres and carries emotional information between them. It does not connect the conscious areas of the brain, though, so words and thoughts cannot be exchanged.

uttered by the left hemisphere – the only one that could reply. And, because the left hemisphere is isolated from the right, it was speaking quite truthfully – it had no knowledge of the spoon picture because, without the usual passage of information across the corpus callosum, the image never reached it.[7]

That does not mean that the spoon image did not go in, however. When the subject was asked to use her left hand to select an item it was the knowledge in the right hemisphere (which is wired, remember, to the left hand) that was brought to bear on the task. Thus she selected the spoon. But when the experimenter asked her to name it she was up against the same problem she encountered when asked to name the image of the spoon – the right hemisphere could not tell of it. Instead the left hemisphere kicked in and did the logical thing. Because it was unaware of the spoon image, it had no way of knowing that the left hand had selected a spoon rather than any other item. It could not see the spoon because the hand that grasped it was under the screen. And it couldn't feel it because the sensory stimuli from the left hand were going, as normal, to the right hemisphere, where – in the absence of the corpus callosum – it stayed. The left hemisphere knew that something was in the left hand, but it had to identify it by guesswork or deduction. As it happens the left hemisphere is pretty good at deduction, and it calculated that of all the objects that might be tucked away behind the screen a pencil seemed like a good bet. So 'pencil' it said.

The spoon and cup exercises demonstrated that straightforward factual knowledge does not

The puzzle of the left-hander

More than 90 per cent of people in the world are right-handed. It has been the same throughout the history of man – studies of tools used since the Stone Age, pictures drawn by ancient man on cave walls and analysis of the fractures on baboons' heads presumed to have been made by human hunters all show that the vast majority of people use their right hands in preference to their left for skilled single-handed tasks.

Right-handedness is strongly associated with left-brain dominance. But what of the 5 to 8 per cent of people who use their left hands? Are they cerebral mirror-images of the norm? Not quite, is the answer. Whereas 95 per cent of right-handers have language firmly lodged only in their left hemispheres, left-handed people vary much more in the way their brains are organized.

About 70 per cent of them have language in the left hemisphere only. Of the other 30 per cent, most of them seem to have language in both hemispheres.[14]

Is left-handedness pathological? Culturally, it has always been treated as such and almost every language has some derogatory term derived from the word for left. 'Gauche' – French for left – is used in English to mean awkward, and 'Amancino' in Italian means deceitful as well as left. The Bible offers one of the most striking indictments: in Matthew's vision of the Judgement it states that God will set the sheep on the right, the goats on the left, and then dispatch the former to life eternal while consigning the latter to everlasting fire.[15]

Not surprisingly, given this authoritative prejudice, parents of left-handed children often went to great lengths to encourage their offspring to use their right hands. The fact that many succeeded – and now appear to be right-handed despite having cerebral arrangements that predispose them otherwise – has confounded subsequent research on the subject.

Handedness is well established by the time a baby is born – in fact, the first signs of it can be seen at just fifteen weeks' gestation[16] when most babies start to show a distinct preference for sucking their right thumbs.

The current consensus about left-handedness is that some of it is simply genetically determined and of no particular significance, while in others it is because of some pre- or perinatal disturbance that arrests the normal development of left-brain/right-hand dominance. It might be something like mild brain injury, which affects left hemisphere growth at a crucial time; or it might result from the failure of neuronal apoptosis, or of some neurons to navigate to their proper place. About 20 per cent of twins are left-handed – a far greater proportion than among the general population – and some researchers have speculated that some singleton left-handers – perhaps all of them – are the sole survivors of what started out as a twin pair.[17] The shift in cerebral dominance may have been caused by mild damage sustained when competing with the twin for limited resources in the uterus, or the result of whatever trauma or uterine deficit it was that killed the other twin.

One intriguing theory holds that left-handedness is caused by a particular developmental abnormality that has consequences way beyond the simple business of which hand you use to sign your name. Stanley Coren, Professor of Psychology at the University of British Columbia, claims to have shown that left-handers die, on average, some nine years before right-handers.[18] This finding, if correct, ties in with findings that link left-handedness to a variety of physical abnormalities, most of which can be traced to developmental or immune system dysfunction. They include asthma, bowel and thyroid complaints, myopia, dyslexia, migraine, stuttering and allergies.

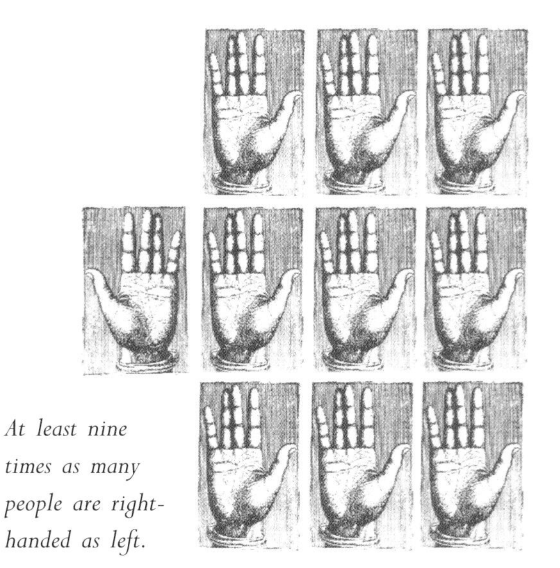

At least nine times as many people are right-handed as left.

travel between hemispheres in a split-brain patient. But what about emotional stimuli? The corpus callosum is the only information conduit to link the cortical areas – the thinking part – of the two hemispheres. But beneath the corpus callosum there is an older pathway between the hemispheres, the anterior commissure. It connects the deep, subcortical regions of the brain commonly known as the limbic system. Here, in this cerebral underworld, raw emotion is generated: alarm bells are set off in response to threat; false smiles are registered; and lust first twitches at the sight of an attractive other.

All of this happens unconsciously, but the limbic system is firmly and broadly connected by millions of two-way neural connections to the conscious, cortical areas of the hemispheres. Everything that the conscious brain takes in is sent down from the cortex to the limbic system where – if it is emotionally significant – a basic response to it is made. That response is then fed back to the cerebral hemispheres where it is processed into the complex, context-sensitive feelings we think of when we speak of fear, anger, embarrassment or love.

These sophisticated expressions of emotion cannot pass from one hemisphere to another in a split-brain patient, but the basic emotional responses generated at subcortical level can, via the anterior commissure. The arrangement is rather like a pair of lifts in a twin-towered building that is connected only by a walk-through basement.

Michael Gazzaniga, of the University of California, together with Joseph LeDoux of New York University, demonstrated this in a series of experiments carried out on a second set of split-brain patients. They were helped by the development by another split-brain researcher, Eran Zaidel, of a special contact lens that refracted the light falling on the eye in such a way that any image presented to it entered only one side of the retina. With the lens in place the subjects could no longer see both fields of vision just by turning their eyes, so much lengthier and more detailed information could be presented to each hemisphere.

In one experiment Gazzaniga and LeDoux showed the right hemisphere of one of their female subjects a series of film vignettes that included a person throwing another into a fire. The patient, V.P., did not consciously know what she had seen, just as N.G. had not been conscious, when asked, of seeing the spoon in Sperry's earlier experiment. 'I think I just saw a white flash – maybe some trees, red trees like in the fall,' she told the researcher who was with her. 'I don't know why, but I feel scared ... I feel kind of jumpy. I don't like this room ... or maybe it's you who's making me nervous.' Dropping her voice, she then said to the researcher: 'I know I like Dr Gazzaniga, but right now I'm scared of him for some reason.' The same sort of unconscious emotional reaction was found when the images presented to the right brain were pleasant: Gazzaniga and LeDoux found that showing scenes of ocean surf, leafy woodland and so on produced feelings of calmness and serenity in their subjects.

Clearly, the right brain, for all its muteness, is capable of making itself felt by the conscious mind. But what actually happens within that hemisphere? Is it possible that it has its own way of seeing things, its own opinions, its own personality, even? What would it say if it could speak?

Sperry's split-brain patients appeared to be almost entirely bereft of language in their right hemispheres, but among the patients who worked with Gazzaniga and LeDoux were two who, unusually, had language ability in the right as well as the left sides of their brains. These patients helped illuminate the wide range of differences between the two hemispheres, and one of them provided the first verbal message known to emanate exclusively from the right side of the brain. It consisted of just two words – but – as we shall see – they demonstrate, beyond doubt, that being 'in two minds' can be more than a figure of speech.

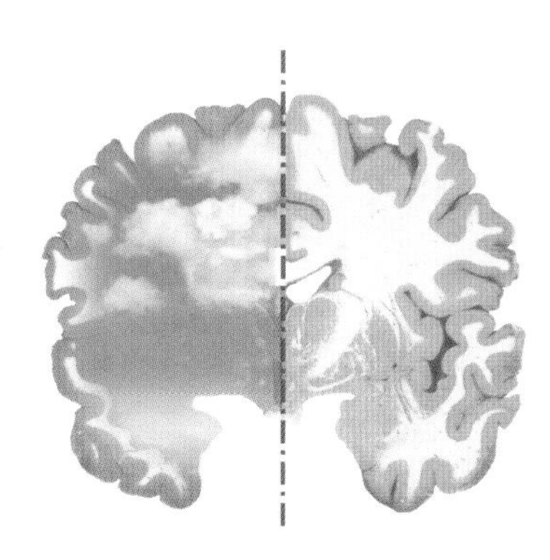

If serene images are fed to the right hemisphere of a split brain they will produce a conscious emotional reaction even though the subject will be unconscious of the actual images.

In two minds

'In living skills M.P. was making good progress with an omelette when the left hand "helped out" by throwing in, first, a couple of additional, uncracked eggs, then an unpeeled onion and a salt cellar into the frying pan. There were also times when the left hand deliberately stopped the right hand carrying out a task. In one instance I asked her to put her right hand through a small hole. "I can't – the other one's holding it," she said. I looked over and saw her left hand firmly gripping the right at the wrist' – *Patient study*[8]

Can you imagine how it would feel to be out of control of one of your hands? To watch, helplessly, as it undid your shirt buttons seconds after your other hand had done them up, or pluck goods you didn't want from supermarket shelves and place them in your pocket? Worse – think of reaching out with one hand to give your lover a gentle caress, only to see the other hand come up and deliver a right hook instead. All of these things have happened to sane and otherwise quite normal-seeming people. Such events are described in the dry jargon of medical reportage, as 'intermanual conflict'. Between themselves researchers call it the 'alien hand'.

Alien hands arise in people who have suffered injury to one or both of two brain areas: one is an area called the supplementary motor area

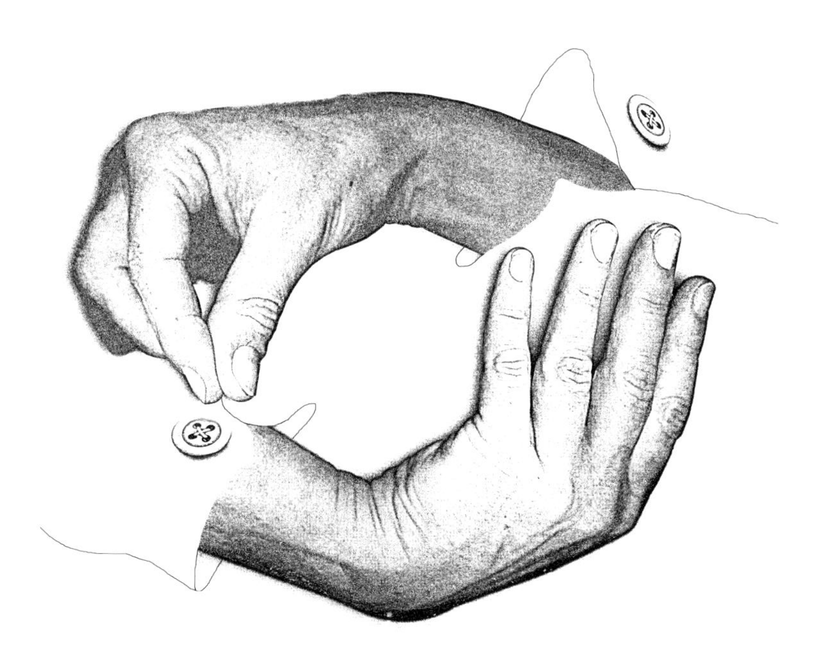

(SMA), a strip of cortex on top of the brain and to the front of the area that controls movement. The other is the corpus callosum. Some people with alien hands, like the woman with the culinary problems above, have had a haemorrhage or stroke. Most of the cases, however, involve people who have undergone split-brain surgery.

Each hemisphere has local control over its own physical realm – mainly the longitudinal half of the body on the opposite side to itself (some facial nerves have a slightly different arrangement). So if the right leg is to be extended, it is the left hemisphere that instigates the movement, and vice versa. Overall control, however, is vested in the dominant (usually left) hemisphere. It is here that the decision to extend the leg is made in the first place. The left brain exercises control by sending commands, mainly inhibitory ones, to the right hemisphere via the corpus callosum. The system makes for smooth running – there is only room for one boss in a single skull.

Sever the connection between the hemispheres, however, and in certain circumstances the command system breaks down. In split-brain patients the inhibitory messages cannot pass from hemisphere to hemisphere, but most of the time this doesn't matter because the two hemispheres are so well versed in their respective roles that things just carry on as normal. Occasionally, though, the non-dominant hemisphere seems to decide that it should be involved in something that is already being handled quite satisfactorily by the left hemisphere, and without its usual line of communication the left brain has no way of stopping it from acting. The two sides can therefore find themselves fighting – literally – for control.

One woman whose brain had been surgically split found, for example, that it often took her hours to get dressed in the morning because her alien hand kept trying to dictate what she should wear. Time and again she would reach out with her right hand and select an item from the wardrobe, only to see her left hand whip up and grab something else. Once her left hand had got hold of something it would not let go, and she, of course, had no way of making it obey her conscious will. Either she had to put on the clothes she was clutching or call someone to help her prise her fingers open. Interestingly, the clothes selected by this woman's alien hand were usually rather more colourful and flamboyant than those the woman had consciously intended to wear.

Another patient had a hand that insisted on pulling down his trousers immediately after his other hand had pulled them up. A third found his alien hand unbuttoned his shirt as fast as his other hand could fasten it. M.P., the woman whose hand chucked uncracked eggs into her omelette, had to put half a day aside to pack when she went away because her alien hand would systematically remove each item from her suitcase just after she had put it in.[9]

Most alien hands are merely irritating or comical. 'It feels like having two naughty children inside my head who are always arguing,' said M.P. Occasionally, though, alien hands seem to be intent on more than mischief. One man reported reaching out with his right hand to give his wife a hug, only to see his left hand fly up and punch her instead. M.P., too, sometimes found her alien hand would prevent her other hand from making affectionate gestures – her husband was often subjected to a tug-of-war as one hand reached out to embrace him while the other pushed him away.[10]

Despite this alien hands rarely do anything seriously violent and the world still awaits the first case of murder to be defended on the grounds that 'it wasn't me who did it – it was my hand'. Some people with alien hands have nevertheless become terrified that they might unwittingly do something catastrophic. One poor man, for instance, was frightened of going to sleep in case his hand strangled him in the night.[11]

Alan Parkin, who has studied the alien hand phenomenon for many years, thinks that it is caused by simple crossed wires in the brain – a neurological fault – that has no psychological significance.

Neat though Parkin's explanation is it remains intuitively rather unsatisfying. The idea of an alien hand as an emissary from some deeper level of the psyche has a compelling attraction that no amount of scientific illumination is likely to dispel entirely.

Perhaps, anyway, there is some truth in the notion of a cantankerous alter ego at work. Alien hands are nearly always on the left, so their actions are triggered by the right hemisphere. This, as we have seen, is usually the mute half of the brain and its inability to communicate has led many researchers to conclude that it must merely be an unconscious servant of its dominant partner, with no ability to form intentions or concepts of its own.

This may not be so. Among the split-brain patients that Gazzaniga worked with at the University of California was one, identified as P.S., who had enough language ability in his right hemisphere for it to be able to understand short phrases as well as individual words. Even more unusually – it could use words to communicate back.

To make contact verbally and exclusively with his right hemisphere nevertheless required an elaborate experimental set-up. Spoken questions, unlike images, cannot be sent to one hemisphere or the other, even in split-brain patients. If a question is spoken in the normal way, the left brain 'grabs' and answers it, and words cannot easily be presented to the right hemisphere only because ear-to-brain wiring is not conveniently separable in the way that eye-to-brain wiring is.

LeDoux and Gazzaniga got around this problem by presenting P.S. with spoken phrases and questions minus the keywords that would make them answerable. This essential information was then sent to the right hemisphere only by presenting the keywords visually. Thus they might say: 'Please would you spell out ...' and then flash the word 'hobby' in his left visual field. This convoluted exercise ensured that the right hemisphere was the only half with all the information required to formulate a reply. P.S.'s right hemisphere could not generate speech, but

The difference between the two brain hemispheres is reflected in the face. The self-portrait of Durer above (centre) has been split and the two halves matched with mirror images of themselves. The resulting two portraits show distinctly different characters.

it was able to write. It therefore spelt out its answers, using P.S.'s left hand (the one under right-brain control) to organize Scrabble letters into words. The resulting conversation was stilted, to say the least. But it was also revealing.

Most of the responses from P.S.'s right hemisphere were similar to those given when the same questions were asked of his left. But the right hemisphere showed distinctive likes and dislikes. When both hemispheres were asked, separately, to evaluate a long string of things – foods, colours and personal things like his and his girlfriend's names – the right hemisphere consistently rated them less highly than the left. More startling than this was the difference that emerged when the investigators asked the two hemispheres about their ambitions.

'What do you want to do when you graduate?' they asked one day. The question was addressed, first, to the boy's left (dominant) hemisphere. 'I want to be a draftsman,' he said. 'I'm already training for it.'

'What do you want to do when you ...?' – this time the word 'graduate' was flashed on a screen to P.S.'s left. The boy's left hand reached for the Scrabble letters. To the amazement of all in the laboratory, including, presumably, P.S. himself, the hand arranged the letters to form the words 'A-U-T-O-M-O-B-I-L-E R-A-C-E-[R]'[12]

These two words constitute about the longest and most complex verbal message ever to emanate from a non-dominant hemisphere. The fact that it was so clearly in conflict with the boy's declared ambition suggests that it was not just conveying a concept that had been constructed by the dominant hemisphere and somehow 'leaked' to it. It was its very own idea – one that had hitherto been hidden from its twin and from the outside world to whom, until that sophisticated laboratory set-up was in place, the right hemisphere had been unable to speak.

The possible implications of this are mind-boggling. It suggests that we might all be carrying around in our skulls a mute prisoner with a personality, ambition and self-awareness quite distinct from the day-to-day entity we believe ourselves to be. Our consciousness may be a single stream because it is the consciousness of our dominant hemisphere only. And if this is so, where is the other entity actually doing its experiencing?

There are other interpretations: the apparent

THE ALIEN HAND

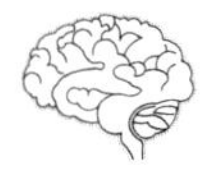

ALAN J. PARKIN
Professor of Experimental Psychology, University of Sussex

Alan Parkin has studied many cases of 'alien hand'. Here he considers a possible explanation for the perverse actions taken by these anarchic limbs.

It is very tempting to think of the antics of an alien hand as the expression of an anarchic unconscious, let loose by the surgeon's knife. This notion chimes in beautifully with popular conceptions of neurosis – the idea that beneath our rational surface lies a naughty, child-like other self that is held in control only by some kind of cerebral police force.

However, florid psychological explanations may not be required to explain why alien hands seem to act in direct opposition to their sensible twins. It may be that, in a maladroit way, these wayward limbs are really just trying to be helpful.

The supplementary motor area (SMA) – the area that, in addition to the corpus callosum, is implicated in alien hand – springs into action when the brain prepares to execute complex volitional bodily action. It does not actually trigger the action itself – instead it acts rather as a motor executive, sending 'move it' signals to the neighbouring motor cortex, which in turn sends the 'get moving' message to the appropriate muscles. As with all other brain parts, the SMA is cross-wired – the left cortex controls the right side of the body and vice versa. Brain scans (Tanji and Kurata, 1982) show that, in a normal brain, the SMA on both sides of the brain is activated even when action is consciously planned for only one side of the body.

The activation on the side that is not actually going to move is pretty weak, but it may be enough to cause movement unless it is stopped. Normally, this inhibition comes from the SMA on the side that is actually meant to move – it sends a message to its opposite number that effectively reads: 'Do not carry through ... leave this one to me.' This message passes through the corpus callosum, so in split-brain patients it does not get through. As a result both SMAs send 'move it' messages to their respective limbs, even though the conscious brain had plans to move only one.

Why, then, do alien hands always seem to work against a person's conscious will, rather than in service of it? It could be that the seemingly mischievous antics of alien hands are not designed that way at all – rather their dogged determination to undo everything the other hand does is because that is all that is left to them.

Say there is some simple task to be done like opening a door. The dominant hand duly does the deed. Then the alien hand – dragging along behind, as it always will – arrives on the scene. The task it came to help with has been done. But the hand 'knows' it was sent to do something in the area and – without the leadership of a conscious, thinking mind – it does the closest thing there is to the open-door manoeuvre it came to do: it closes it.

Alien hands spring into action when the non-dominant side of the brain gains momentary control.

bifurcation of consciousness observed in split-brain patients and occasionally in ourselves may simply reflect a repeated alternation of consciousness from one hemisphere to the other – the meandering of a single stream rather than two separate flows. Or it could be that there are many streams of consciousness in each of us, and that the split personality observed in P.S. was simply the result of his being able to bind them altogether because of his condition.

The notion of some parallel universe peopled by our other halves – each watching helplessly as we hurtle through life oblivious to their cries – is the most sensational of all the explanations, but it was one that Roger Sperry came to believe in after observing split-brain patients at close quarters for months on end. 'Everything we have seen indicates that the surgery has left these people with two separate minds,' he wrote. 'That is, two separate spheres of consciousness.'[13]

CHAPTER THREE
BENEATH THE SURFACE

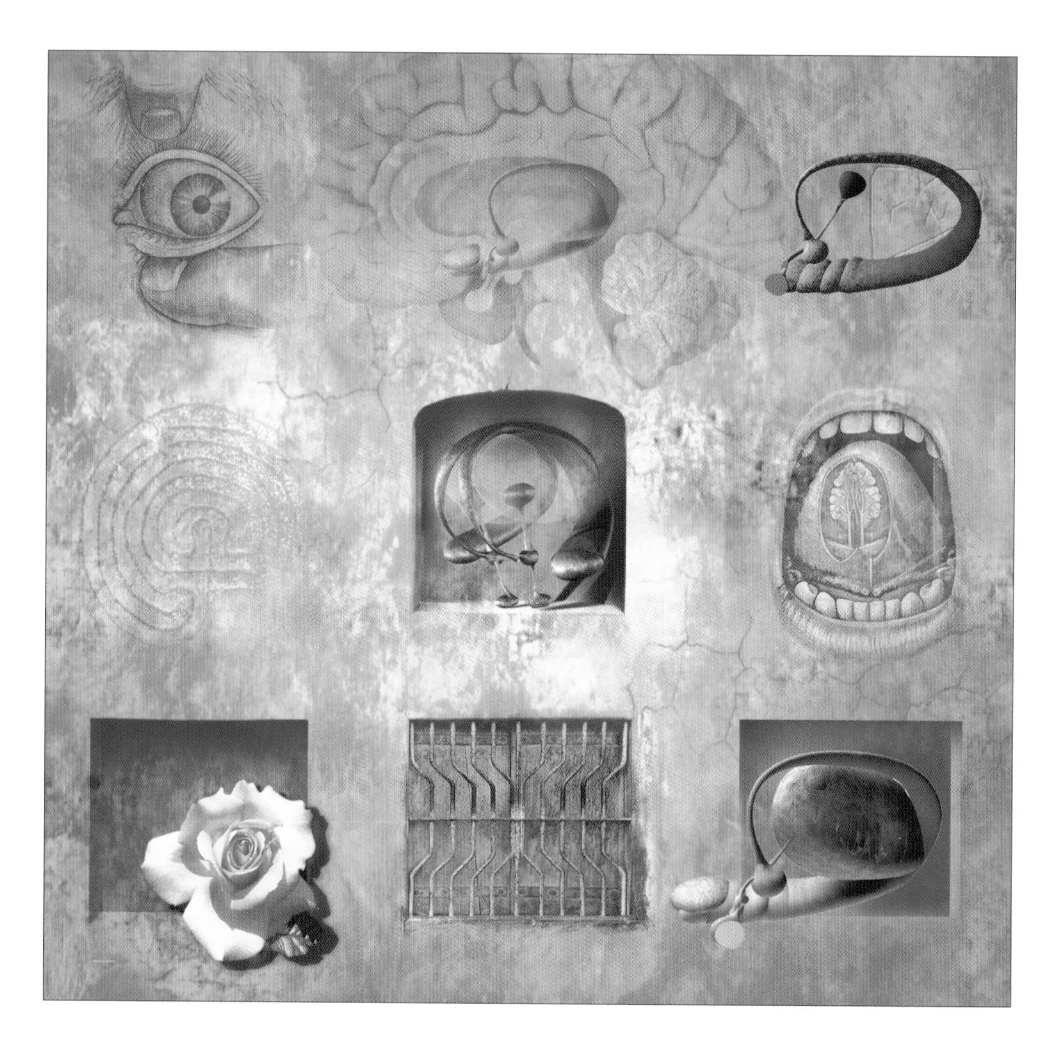

The architecture of the brain is more complex than the familiar wrinkled cortex suggests. At its centre lies a cluster of strange-shaped modules that together are known as the limbic system. This is the powerhouse of the brain – generator of the appetites, urges, emotions and moods that drive our behaviour. Our conscious thoughts are mere moderators of the biologically necessary forces that emerge from this unconscious underworld; where thought conflicts with emotion, the latter is designed by the neural circuitry in our brains to win.

'The heavings...began in my tummy and over the course of a second or two rose up within my body past my chest eventually to my mouth, throat and vocal chords as if I was being "verbally sick". I...suggested [to hospital staff] that the impulse [to shout obscenities] might make partial use of the same circuitry as that in the sneeze mechanism, as the heavings, and even the thoughts accompanying them were totally eliminated when I deliberately held my nose. This was treated as completely irrelevant... what I was suffering from was a motivational problem and all I had to do was get my anger out and the problem would disappear.' – *Peter Chadwick, psychologist and one-time Touretter*[1]

PEOPLE WITH GILLES DE LA TOURETTE'S syndrome can cut a clean swathe through the most crowded pavement. They lurch along, their faces a-twitch and with a stream of weird noises – barks, disjointed phrases or obscenities – issuing from their mouths. Some people glare at them; children giggle; occasionally, someone swears back. Most people adopt a preoccupied expression and scuttle out of their way.

If you find these passing encounters embarrassing, try to imagine what it must be like for the Touretters themselves. Most are of normal or above-average intelligence and they are often painfully aware of how ridiculous or offensive they appear to others. Coprolalia – the uttering of foul language – is particularly distressing because, more than anything else, it causes people to shun them. Some can control their symptoms by concentrating very hard on some cerebral activity (a dozen or so Touretters work, quite safely, as surgeons). When they cease to make an intense, conscious effort, however, or if they become emotionally aroused, their jerks and tics, animal noises and spat curses well up from the unconscious underworld of their brains and emerge with explosive force.

Tourette's syndrome was around long before the French physician Georges Gilles de la Tourette lent it his name in the late nineteenth century. Several medieval reports of people supposedly possessed by devils clearly describe its symptoms. More recently, Freudian psychoanalysts seized on the condition as a perfect demonstration of what happens when anger is repressed: 'Look! The anger must out!' Treatment was therefore directed towards revealing the 'root causes' of the supposed anger, or encouraging the patient to express it more openly. This approach didn't work and often made things worse but that did not lead to its abandonment (in fact, it is still handed out even today – as Peter Chadwick discovered).

The theoretical underpinning of the suppressed anger approach to Tourette's syndrome was irretrievably undermined by the discovery, in the 1960s, of a drug that drastically reduced its symptoms and in some cases took them away altogether. The drug – it transpired – fitted into the receptors designed to take the neurotransmitter dopamine. By filling these chemical 'locks' on the surfaces of cells it prevented dopamine from activating them and once these neurons were quietened the tics stopped. Today Tourette's syndrome is coming to be recognized as one of a wide range of mental disorders associated with malfunctions in the elaborate chemically modified system that ensures that necessary urges – the ones that keep us going – are properly attended to.

The brain's main function is to keep the organism of which it is part alive and reproducing. All its other tricks – the ability to appreciate music, to fall in love, to create a unified theory of the Universe – arise out of that single overriding ambition. So it shouldn't be too surprising to find that a huge part of brain structure and function is given over to making sure its neighbouring body parts do whatever is required to find food, sex, security and other vital requirements.

It does it by means of an elaborate carrot-and-stick control system. There are three basic steps. First, in response to the appropriate stimuli the brain creates an urge that demands to be satisfied. If the stimulus is, say, falling blood glucose, the urge will be hunger, and if the stimulus is sexual, the urge will be lust. More complicated stimuli like social isolation or separation from familiar surroundings may produce less easily identifiable drives – a desire to socialize, or a hankering to return home. Whatever form it takes, urges are often accompanied by a feeling of 'emptiness'. This may resemble a literal emptiness like an empty stomach. Or it may be something more vague – an emptiness of spirit. Whatever the feeling its purpose is the same: to trigger action.

Second, the action brought about by step one – eating, sex, homecoming, socializing – is rewarded with positive feelings of pleasure. Note – the *action* is the thing. Not just the food, or the sex, or the being at home. Having nutrients pumped into the blood keeps you alive but it does not give the same pleasure as a meal that has to be prepared, served, chewed and swallowed. This is why so many essential functions are elaborated with rituals. The preparation of a feast, the courtship that precedes sex, the journey home – these are not just necessary peripherals; they are the very things that make life enjoyable.

Third, when the action is complete the rush of pleasure is replaced by a sense of contentment and – note the word – fulfilment.

Most of the time, this system works away quietly and efficiently, creating cycles of 'desire – action – satisfaction' that mould our behaviour and provide the background rhythm of our daily lives. We feel hungry when our bodies are short of fuel; we eat, which is pleasurable; then we feel satiated – a serene feeling that lasts until our bodies again need fuel. Yet sometimes – indeed, quite often – the system breaks down. Either our urges cease to prompt appropriate action, or normal actions cease to be enough to satisfy them.

The first type of breakdown may be catastrophic. At its most mechanical, the ability to make purposeful movements may be lost, resulting in the physical stasis seen in Parkinson's disease and similar movement disorders. When higher brain areas are affected the result may be more subtle but just as disruptive to normal life. If a person loses the urge for self-protection, for example, or if their natural drives are sublimated to an overriding ambition – to climb a mountain, perhaps, or win a sporting contest – they appear reckless and may well injure themselves. If they lose the urge to keep clean, their health will suffer. If hunger fails to make itself felt or is overruled by conscious self-denial, they may die of starvation.

Conversely, if a person's urges become insatiable, they cease to behave normally. The insistent demands of the body force them to repeat and repeat the actions that once gave them relief: food is stuffed, sex is taken wherever it can be found; self-comforting rituals like hand-washing, door-checking or talk-talk-talking are indulged in to the point of exhaustion. Yet still the nagging hunger, lust or anxiety continues.

Physical Tourette-type tics are an example of this. The twitches are fragments of skills – each a tiny, degraded echo of some once-purposeful movement – that are triggered into existence by bursts of activity in an area of the unconscious brain called the putamen. The putamen is one part of the complex intertwined knot of nuclei that makes up the basal ganglia, deep in the centre of the brain. Its function is to look after automatic movements – those that have been learned by repetition – and to keep them flowing smoothly so the conscious brain can get on with the grander business of deciding how to direct those movements and learning new ones. Pedalling a bike (providing the rider is experienced), for example, is controlled by the putamen, while the movements needed to execute a new and complicated dance, say, would be controlled by different brain areas.

A survey of 135 people with Tourette's syndrome found that 93 per cent could identify 'premonitory urges' and that they could – by making a conscious effort – suppress the movement that would normally follow.[2] Blocking tics like this does not, however, stop them. Until the urge has been translated into movement it continues to beat up against the walls of consciousness like an ever-intensifying itch demanding to be satisfied.

One man, who has an elaborate combined shoulder-jerk and jaw-thrusting routine that he generally does about five times a minute, says:

'I can hold it off for a few minutes or even for an hour or so if I have to. When I meet someone new, or if I am doing something important I can appear quite normal for the duration. But when the pressure is off I have to catch up by ticcing some extra – I usually lock myself in the bathroom for ten minutes to do it.

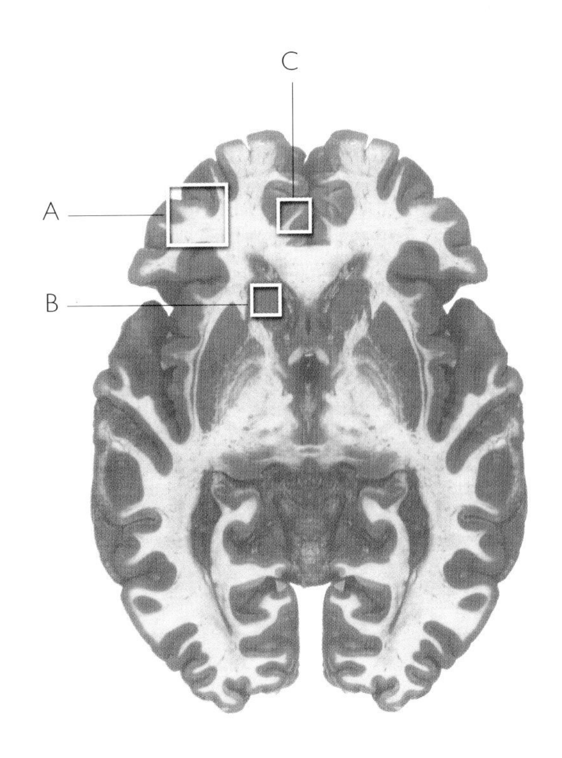

Brain scans of 50 people with Tourette's syndrome showed significant lack of activity in three brain areas, all in the left hemisphere. One was the dorsolateral prefrontal cortex (A) *an area which is concerned with generating appropriate actions. Another was in the left basal ganglia* (B) *concerned with the control of automatic movements. The third was the anterior cingulate cortex* (C) *an area which is concerned with focussing attention on actions. The lack of activity in these areas allows inappropriate fragments of actions – tics – to 'burst through'.* [Ref: Moriarty J. et al. 'Brain Perfusion abnormalities in Gilles de la Tourette's syndrome', British Journal of Psychiatry 167 (2): 249–254 August 1995]

People say to me then: "If you can control it, why do you do it at all?" I explain to them: it's like holding your breath – you can do it for so long, but eventually you have to let go. And when you do, you pant for a while.'[3]

The shouts and curious vocal habits of Tourette's come from overactivity in another part of the dopamine-fuelled pathway that links the unconscious brain to the conscious. Here the

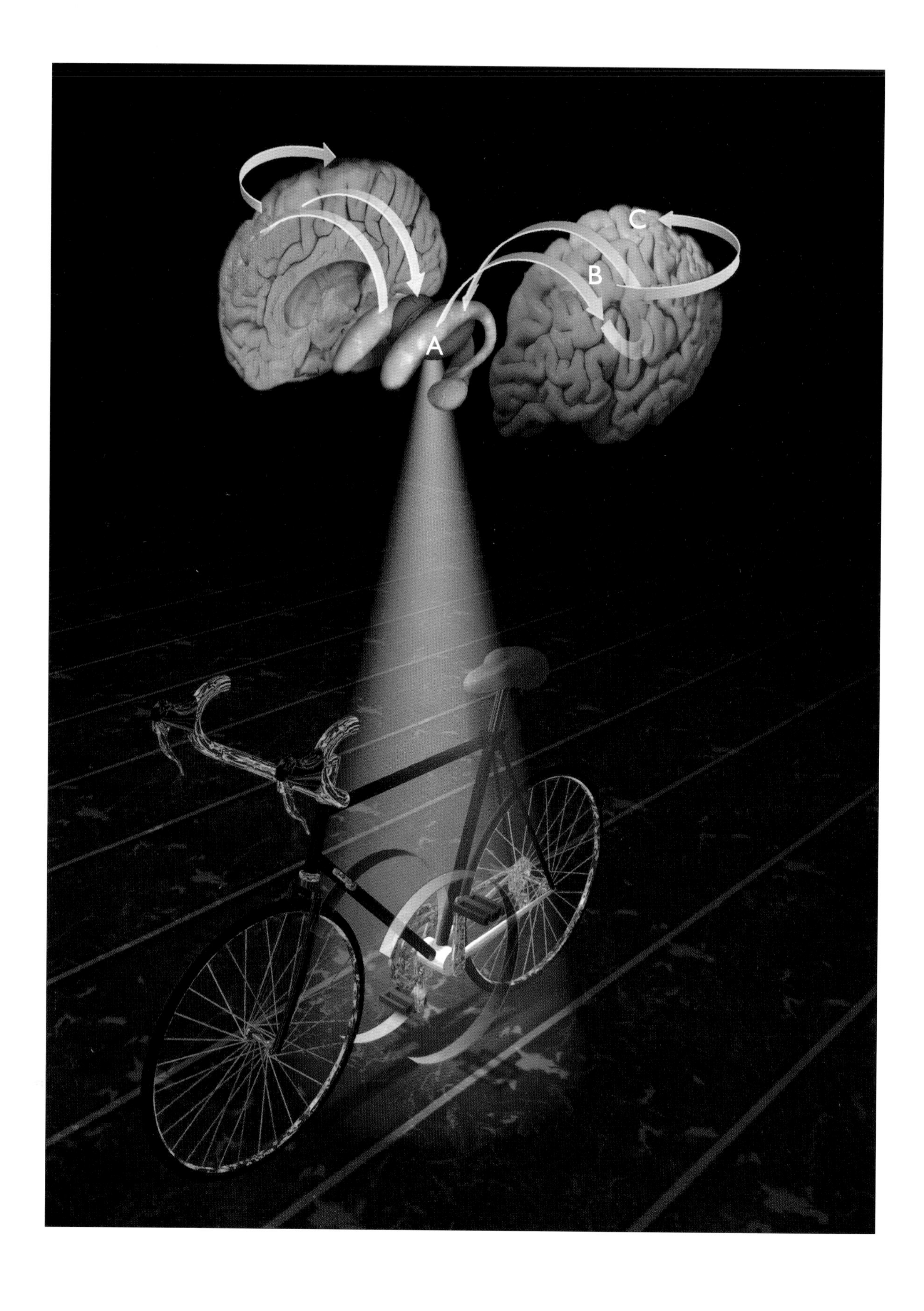
A
B
C

pathways that are affected touch on the language areas in the temporal zone. The words seem to be tiny residual fragments of otherwise long-forgotten phrases: 'Hi Patsy!' was one of the tics constantly uttered by the surgeon with Tourette's observed by Oliver Sacks in his book *An Anthropologist on Mars*. 'Hideous' was another. Patsy, apparently, was an old girlfriend, but the surgeon had no idea why one particular greeting to her should have lodged itself so stubbornly in his brain that he was forced to repeat it for decades after. The origin of 'hideous' was lost to him entirely. Perhaps he had once heard the word uttered in circumstances that encouraged it to be etched into his mind and his forced repetition had kept it as an intact neural trace long after the memories that surrounded it had faded from his consciousness.

Tourette's syndrome, with its obvious physical manifestations and sometimes extravagant behaviour, seems on the face of it to be light years removed from the quiet mental anguish felt by people with obsessive–compulsive disorder (OCD). Recently, however, the two conditions have come to be recognized as different manifestations of the same underlying biological disturbance.

In OCD the urges are more complicated than in Tourette's. Instead of being compelled to, say, shout a word or move a limb in a particular way, people with OCD are driven to carry out complicated routines to still an ever-present feeling of unsettlement or doubt.

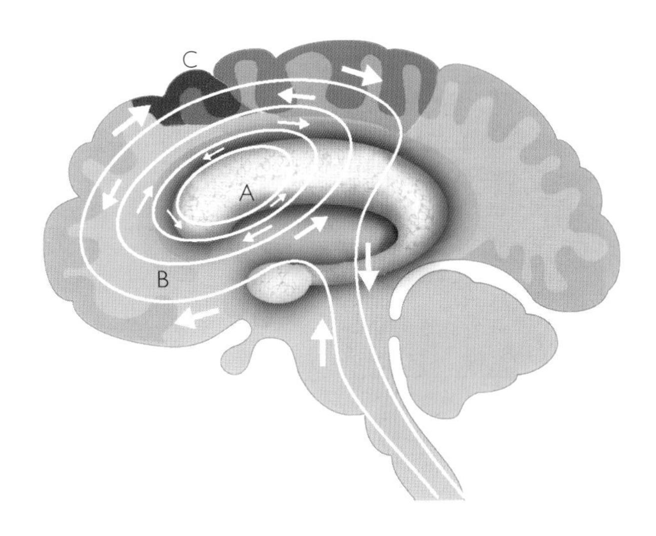

Above: *When people with obsessive-compulsive disorder (OCD) are put in situations where they feel uneasy a loop of neural activity sparks up in their brains. It runs between the caudate* (A), *which triggers the urge to 'do something', through the orbital prefrontal cortex* (B), *which gives the feeling that 'something is wrong', and back through the cingulate cortex* (C) *which keeps attention fixed on the feeling of unease.*

The routines may be purely mental or they may involve elaborate behavioural rituals. Counting is common. One woman says:

'I have to count to seven between each mouthful of food. If someone asks me a question while I have food in my mouth I have to finish the counting before I can swallow to reply. If I try to swallow without completing the sequence, I gag. And if for some reason I lose count, I have to spit the food out, count seven again and take a new mouthful.'[4]

Another OCD patient has a fixation with the number four. Everything has to be done four times: four foldings back of the duvet in the morning before getting out of bed; four steps to

Left: *Familiar motor skills, like riding a bicycle, are controlled by the putamen* (A), *which is part of the unconscious limbic system. The putamen is connected via a complex loop of nerves to the premotor cortex* (B) *– the part of the conscious brain which creates the urge to move. If the putamen is stimulated it passes the message on to the premotor cortex which, in turn, passes a 'move it' message to the adjoining cortex* (C). *The motor cortex then instructs the appropriate muscles to contract. In Tourette's syndrome the putamen is overactive, and triggers the acting out of fragments of old learned skills at inappropriate moments.*

the door, teeth-scrubbing movements in groups of four and so on. He has a particular horror of being left on an uneven number. Once his girlfriend told him that she loved him. He wasn't too sure if he reciprocated this feeling but the words 'sort of hung about in the air...like a big number one', so he told her he loved her, too. His tone of voice was not, perhaps, convincing enough for the girl to feel satisfied with this exchange, and she said it again: 'I love you.' Now, of course, the words hung like a huge number three, so he had to repeat them again, to make it up to four. Satisfied by this, the girl then said she wanted to marry him – a proposition that brought about a further cascade of reciprocal pledges.[5]

Other internal compulsions may include thinking about a subject to the exclusion of almost everything else; going over and over past conversations; or being plagued by the need to imagine doing something terrible – killing someone, for example. People with OCD tend to be exceptionally good, in that they go to extreme lengths to avoid doing anything wrong. They are often obsessed with morality and are scrupulously honest. The need to be truthful may be taken to absurd lengths, as with this patient:

'If I was talking to you and I mentioned that I had seen someone in a red dress, as soon as I said it I would start to think: "Was it really red? Or could it have been another colour?" Once the idea had got into my head that I might have misled you – even if it didn't matter one bit what colour the dress was – I would start thinking: "Should I confess that I said the wrong colour or should I make myself live with the guilt?" So I try to avoid saying anything that could be wrong. I put "I think" or "I'm pretty sure" or "it could have been" in front of everything. It is a sort of ritual – a way of making sure I can never tell a lie.[6]

The behaviours associated with OCD tend to be the same the world over. The two most common are washing and checking. People who need to wash themselves may actually wear away the skin of their hands by constantly scrubbing them with soap and water. People who need to check things may find the activity takes up nearly all their time. A man who was compelled to check every car journey he made to make sure he had not knocked someone down had to get up at dawn to leave time to go over his journey to work two or three times to check for signs of an accident. The journey home was similarly repeated. And still he wondered, all day and all night, if in his minute scouring of the route he might somehow have overlooked a crushed corpse in the gutter.[7] Hypochondria (the need to check constantly for signs of bodily illness) and body dysmorphic disorder (the conviction that something is wrong with the way one looks) are variations on the theme. OCD is also thought to be the cause of about half the reported cases of compulsive hair-pulling.

These mental and behavioural tics are, like the physical jerks of the Touretter, fragments of preprogrammed skills. But in this case the memories are not personal ones picked up during the person's lifetime but those that are built into the species as instincts. The instinct to keep clean; to check the environment constantly for signs of anything untoward; the need to keep order and balance – all these things have a basis in survival. In OCD they have simply come adrift from the survival superstructure and appear as isolated, inappropriate and exaggerated habits.

As with Tourette's, what seems to be happening in OCD is that a particular neural pathway is overactive. This time it is one that runs between the frontal lobe (including the premotor area) to another part of the basal ganglia – the caudate nucleus. The caudate nucleus is joined to the putamen and they develop in the embryo as a single structure. The main difference between them is that, while the putamen is mainly connected to the premotor cortex, the caudate

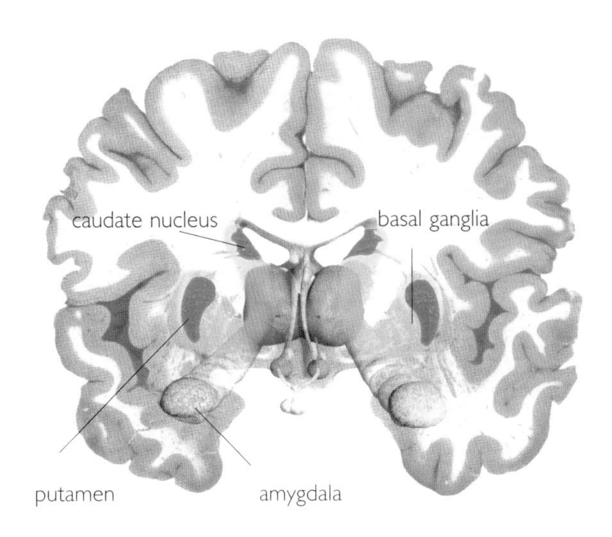

The caudate nucleus is closely connected to the amygdala, which gives rise to feelings of fear. The knock-on effect of caudate activity in OCD may partly explain why people with the disorder suffer such anxiety.

nucleus is in touch with the frontal lobes where thinking, assessing and planning – the highest forms of cognition – take place. In a normal brain the caudate nucleus looks after certain aspects of automatic thinking just as the putamen looks after automatic movements. The caudate nucleus is the part of the brain that automatically prompts you to wash when you are dirty; that reminds you to check the doors before you leave the house; and that alerts you to and focuses your attention on anything that is out of order.

It does all this by activating one particular area of the frontal lobe – a spot in the orbital cortex, the area of the frontal lobe just above the eyes. This is the area that lights up when something unexpected happens. It was first identified during monkey studies done by Professor E. T. Rolls at Oxford University. The animals were shown green and blue lights and trained to associate the blue light with a reward of fruit juice, and the green light with a salt drink. Once they had grasped the link between blue = juice and green = salt the drinks were switched. Suddenly the monkeys got a salt drink when they saw the blue light. When this happened an area of the brain that had been quiet until then leapt into life. The neurons in the orbital cortex that lit up were not simply responding to the saltiness of the drink – taste discrimination and the simple 'ugh!' reaction happen elsewhere in the brain. This particular area was clearly activated by the discovery that something in the world was not quite right – it was a 'hey! something is wrong here' reaction – a built-in error-detection device. Once the monkeys got used to getting salty drink occasionally, instead of the usual juice, the reaction disappeared.

Since then brain scans on humans have shown that this area is particularly lively in people with OCD. When a person with a hand-washing compulsion is told to imagine themselves in some filthy place their caudate nucleus and orbital frontal cortex fire away like mad. An area in the middle of the brain – the cingulate cortex – also responds strongly. This is the part of the brain that registers conscious emotion, and its involvement demonstrates the emotional discomfort generated by OCD.

A similar brain pattern can only be produced in a normal person by persuading them to think very hard about some major catastrophe like watching their home burn down with their family inside. After these imaginings have been engineered in the subjects' minds and when the researchers tell them they can relax and forget their terrible thoughts, the person with OCD continues to show a lit-up caudate nucleus and orbital frontal cortex. It doesn't matter to them that the laboratory and their hands are obviously shiny clean, the thought that they are contaminated by dirt just won't go away. Once they have left the scanner and slipped off to wash, the feeling may be reduced. A scan at that moment might show little caudate or orbital

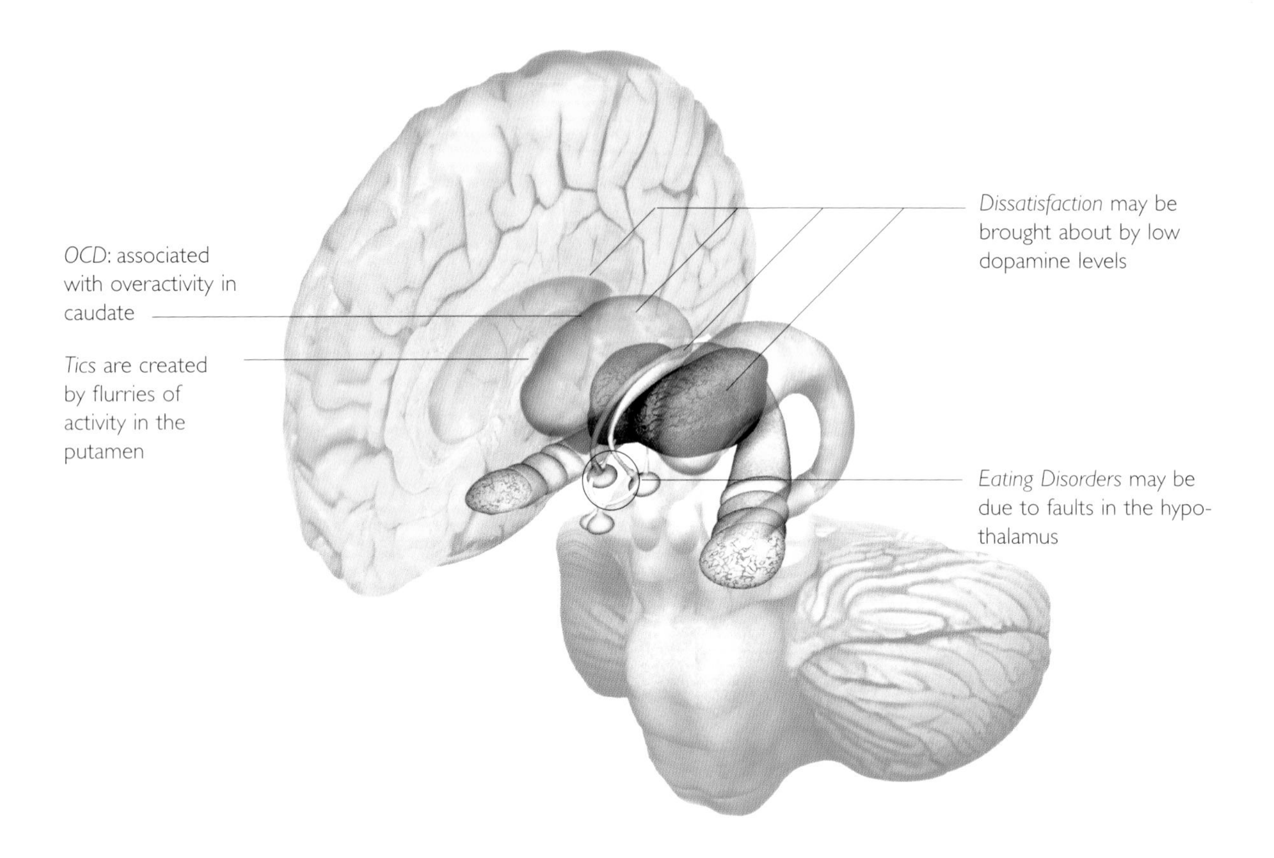

frontal activity. But soon the circuit will glow again, and the urge to wash will be rekindled. Their error-detection mechanism has somehow become stuck on alert, and no matter how often the appropriate turn-off action is carried out it continues to shriek its warning.

OCD is diagnosed when obsessions and compulsions are so severe that they disrupt normal life. About three people in every hundred come into this category, a proportion that appears to be more or less the same wherever statistics are available.[8] But you do not have to have OCD to be nagged by that feeling that everything in the world is not quite right. People who are forever cleaning and ordering their homes; people who can never relax unless they have double- and treble-checked that the doors are locked before they go to bed; people who insist on having twice the usual number of health checks and still worry they might have a terminal disease – these may all be cursed with an overactive error-detection system, a neural circuit that lights up too readily and stays lit up too long.

The same may be true of people who worry overly about the effect they are having on other people. Just as washing and checking doors stems from the need to ensure physical safety, so fretting about social behaviour may stem from a need to be secure within the group. This form of obsessiveness might involve, for example, going over and over each conversation, analysing what you said, or didn't say, and worrying about every unintended or imagined slight.

The drive to be 'good' and the need to follow

rituals may encourage many people with OCD to embrace formal religion or to enter sects where their need for security is satisfied by group adherence to a strict set of values that allows little room for doubt. Even Tourette's syndrome may have its mild echoes in normal people. The sniffers, the blinkers, the girls who repeatedly brush imaginery hair from their eyes – are these actions driven by the spasms of neural activity from the putamen?

Similarly, some people seem to live permanently in the grip of some seemingly unassuageable appetite. They can never get enough of whatever it is they are hooked on: sex, food, risk or drugs. US geneticists Kenneth Blum and David Comings have labelled this type of discontent reward deficiency syndrome, and they propose that an extraordinarily wide range of disorders can be clustered under its umbrella. Depending on which part of the reward system (and thus which part of the brain) is most affected, a person may display anything from mild anxiety, irritability or risk-taking to eating disorders, compulsive shopping and gambling, drug addiction and alcoholism.[9] As the name suggests, people with reward deficiency syndrome are unable to get satisfaction from life. Something in the way their brain works makes it impossible for them to turn off their appetites.

This type of discontent is very widespread: some surveys have concluded that one in four people suffer from one of the conditions included by Blum and Comings in reward deficiency disorder.[10] Many of these conditions are hard to treat because they engage the brain at all levels. Although all urges are ultimately connected to bodily needs, some of them bring about very complex behaviour that can become an end in itself. The urge to eat, for example, may set off an extraordinarily sophisticated train of behaviour: not just seeking food, but selecting it carefully and making it into a complicated dish. Once you have sampled the joys of double

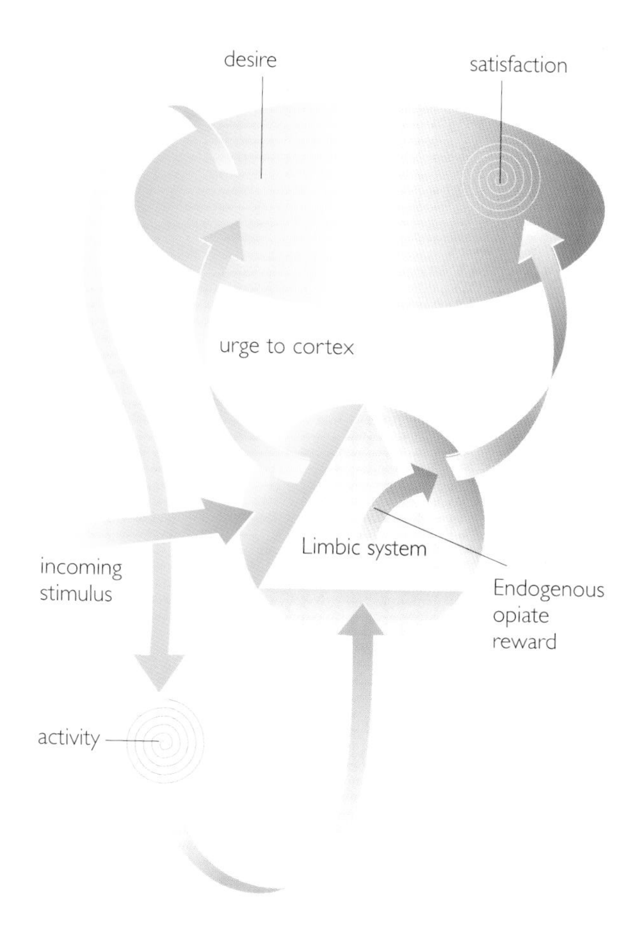

The brain uses a carrot-and-stick system to ensure that we pursue and achieve the things we need to survive. A stimulus from outside (the sight of food, say) or from the body (falling glucose levels) is registered by the limbic system which creates an urge which registers consciously as desire. The cortex then instructs the body to act in whatever way is necessary to achieve its desire. The activity sends messages back to the limbic system which releases opioid-like neurotransmitters which raise circulating dopamine levels and create a feeling of satisfaction.

chocolate chip cookie and yoghurt ice cream it requires a very faint urge, in this author's experience, to prompt a visit to the freezer. Therefore, urges involve both the most mechanical of brain functions – monitoring glucose levels, for example, in the case of hunger – and the most advanced. In physical terms this means everything from the brainstem to the frontal cortex.

This complicated system has evolved over millions of years and until recently it has worked for us pretty well. In a world of sparse resources the reward it delivers for the *act* of pleasure-seeking, as well as the goal, ensures that people will chase their supper or toil long and hard to gather sufficient food for survival. The trouble with evolution, however, is that it cannot keep pace with human ingenuity. Today we can get our dinner by pulling the plastic covering off an oven-ready pork chop rather than by chasing a wild boar to its death. No wonder our puny achievements give us so little satisfaction.

As the map of the brain takes shape, it is becoming possible to see where and how it might be possible to alter its architecture and function in a way that might cause it to serve us better than the design currently arrived at by evolution. Pharmaceutical companies are already investing vast sums of money in developing drugs to moderate urge-driven behaviour by affecting levels of the various neurotransmitters – 'behavioural pharmacology' is now a recognized vocational category.

After drugs comes genetic engineering. The system that once served us so well is encoded in our genes and by the dawn of the twenty-first century we will probably have the knowledge and techniques to tweak those genes to produce a brain that is better suited to its time.

Blum and Comings claim they have already identified a candidate for early genetic manipulation. It is called the D2R2 allele (an allele is a variation of a normal gene) and it is found in 50 to 80 per cent of alcoholics, drug addicts, compulsive eaters and pathological gamblers. The D2R2 gene prevents dopamine from binding to cells in the reward pathways, so the rush of pleasure the neurotransmitter usually gives when it is released is reduced. People who have the gene therefore feel driven to consume, or do, more and more of the things that stimulate dopamine in the hope that they will eventually get enough of it to feel satisfied. This hunger for dopamine, according to Blum and Comings, is what creates most of the neurotic and self-destructive behaviour that bedevils our society. If they are right, a bit of prudent engineering on a single bit of the human genome could save immeasurable suffering, disease and premature death.

Many people recoil instinctively from the idea of reinventing the species in this way – it reeks to them of cheating nature or playing God. Certainly, there are dangers – from the story of Frankenstein's monster to the development of antibiotic resistance, the consequences of scientific hubris are spelt out in neon.

Yet there are dangers, too, in ploughing on with what is in some ways outdated equipment. Evolution has produced marvellous mechanisms to aid our survival, but it cannot change us as fast as we can change our world. Perhaps the time has come to use the ingenuity with which aeons of natural selection have endowed us, to make ourselves more fit for the environment that we have created.

Hunger

A third of people in America are obese, one in four in Western Europe. If we go on getting fatter at the present rate, by 2020 everyone will be overweight.[11] Millions of us die early each year from clogged arteries and other complications of obesity. Our drive to pleasure is killing us.

As with all our urges and appetites, the mechanism by which simple, physical hunger is generated and satisfied centres on the hypothalamus. Information about the state of the body is constantly fed to the hypothalamus through a complex interplay of hormones, neuropeptides and neurotransmitters. If glucose, mineral or fat levels drop, the information is signalled from the blood,

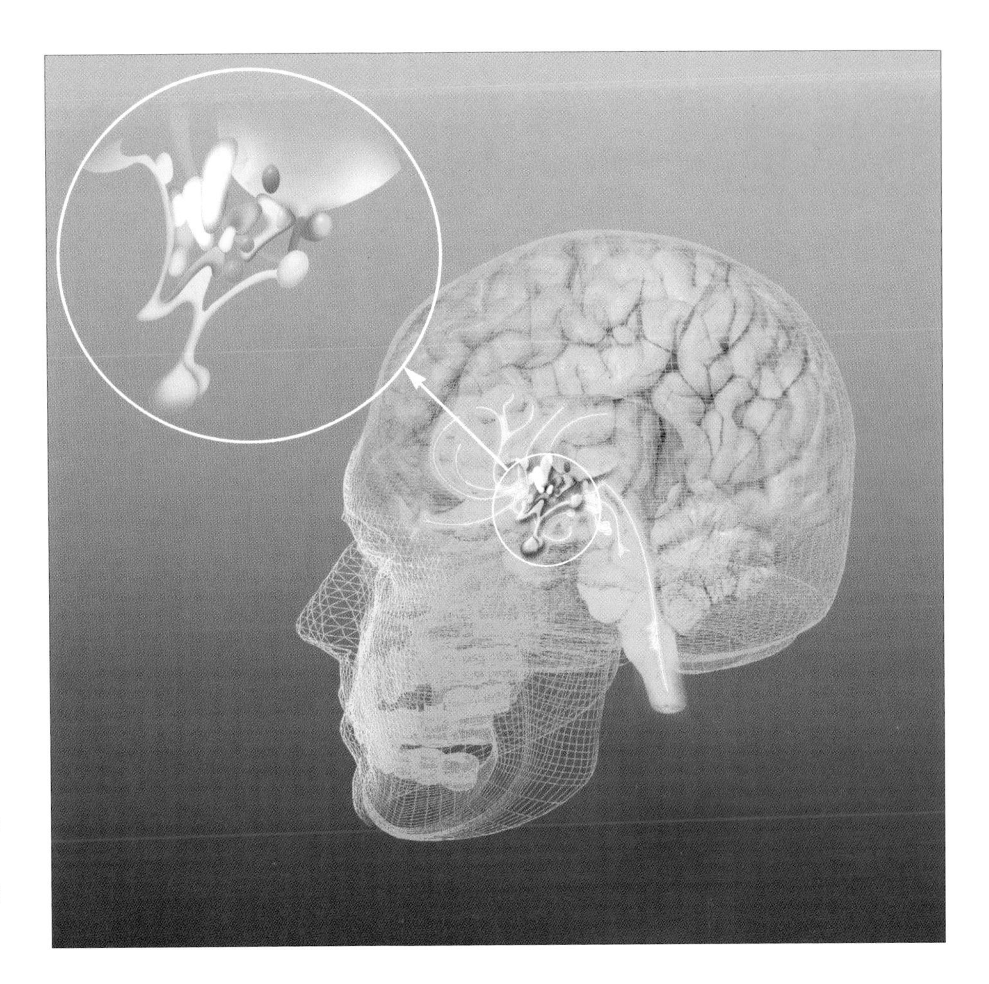

The hypothalamus is a cluster of nuclei, each of which helps to control our bodily urges and appetites. It is part of the diencephalon which means, literally 'in between brain', and acts like a bridge between the body and the brain. The hypothalamus is tiny – it weighs just one three hundredth of the total brain mass – but it has an enormous effect and even small dysfunctions in a single one of its nuclei can create serious physical and mental problems.

stomach, intestines and fat cells. The hypothalamus then feeds these signals up to the cortex where they excite the many areas that consciously register hunger and organize the finding, preparing and eating of food. As the meal is eaten the system goes into reverse: the body signals its satisfaction back to the hypothalamus, which passes the message on to the cortex, which then creates a conscious desire to stop eating.

This all sounds very straightforward and efficient. Yet clearly it is not fail-safe enough to prevent what amounts to a global epidemic of over- (and occasionally under-) eating.

One cause of eating disorder may lie in the hypothalamus itself. Two hypothalamic nuclei – the lateral and the ventromedial – play a central role in controlling appetite. The lateral nucleus senses falling blood glucose and signals hunger while the ventromedial responds to rising glucose levels and signals satiety. Animals with damage to the lateral nucleus thus eat very little while those with damage to the ventromedial nucleus gorge.

This dual-lobed mechanism is not a simple off/on appetite switch, however. Rats with ventromedial nucleus damage may overeat – but only if eating is made easy. If they have to press a lever to get food, they will do so less often than normal rats. An interesting experiment suggests that the same may be true of people. Two groups of obese subjects were each a given a dish of nuts and invited to nibble freely at them while they did some dull task. The nuts in one dish were preshelled while the others were in their shells and

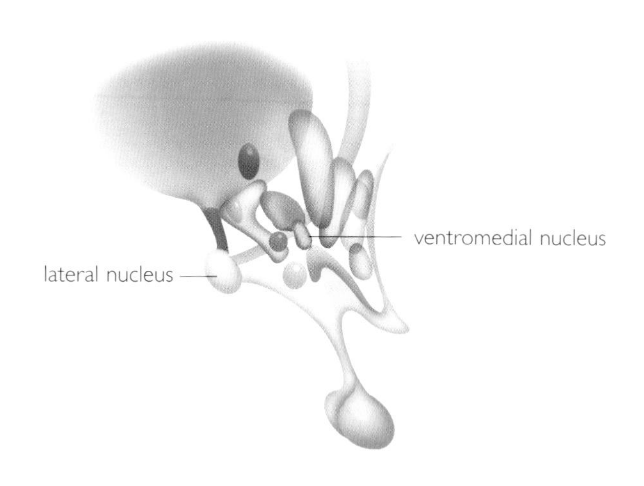

The lateral and ventromedial nuclei of the hypothalamus act like 'on' and 'off' switches for the appetite.

accompanied by a nutcracker. The subjects ate large quantities of the pre-shelled nuts but very few of those that had first to be shelled. This curious finding may, in part, explain why obesity is strongly correlated with consumption of processed, heat-and-eat meals and junk food.

Anorexics, like those who are obese, may also have impaired hypothalamic function. Brain imaging studies[12] show that activity in the limbic system does not seem to be transferred in the normal way to the cortex – for some reason the neural connections between the appetite sensors in the hypothalamus and the areas that register conscious hunger are not carrying signals effectively. This may explain why anorexics typically claim not to feel hunger, even when their bodies are starving.

Damage to deep limbic structures like the hypothalamus is relatively rare, so it seems likely that most hypothalamic dysfunction stems from a disturbance of the neurotransmitters that carry messages to and from it. Serotonin, for example, damps down activity in the lateral hypothalamus, so a high level of it is likely to reduce appetite,

The dopamine connection

Dopamine dysfunction is now recognized as playing a major role in a wide range of disorders.

Put crudely, too much dopamine seems to cause hallucinations and paranoia (the positive symptoms of schizophrenia), uncontrolled speech and movement (Tourette's), agitation and repetitive action (obsessive–compulsive disorder), overexcitement, euphoria and exaggerated convictions of 'meaningfulness' (mania).

Too little dopamine is known to cause tremor and the inability to start voluntary movement (Parkinson's disease and related disorders), and is implicated in feelings of meaninglessness, lethargy and misery (depression), catatonia and social withdrawal (negative symptoms of schizophrenia), lack of attention and concentration (adult attention deficit disorder), and cravings and withdrawal (addiction).

Other neurotransmitters – in particular dopamine's close chemical relatives serotonin and noradrenaline – are also implicated in these conditions. But dopamine is increasingly coming to be seen as the major player. Whether a person shows symptoms of mania or Tourette's syndrome, Parkinson's disease or catatonia, depends on which parts of the brain are affected by dopamine excess or depletion.

Dopaminergic cells are distributed through the brain in well-defined pathways. One pathway travels along from a nucleus in the brainstem called the substantia nigra (black area) to the basal ganglia. This is a cluster of nuclei that includes the putamen and the caudate nucleus (together these are sometimes called the striatum), which control automat-

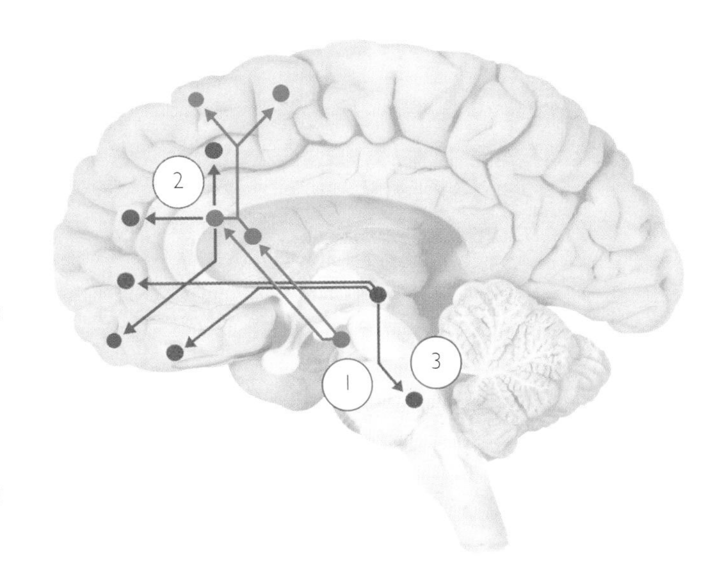

Path 1: Substantia nigra to the basal ganglia to the motor cortex – gives people 'get up and go' physically

Path 2: Caudate to the orbital prefrontal cortex and the premotor cortex – gives desire to act

Path 3: Ventral tegmental nucleus to basal ganglia to olfactory bulb/frontal lobes – gives feelings of pleasure, mental energy and drive

ic movement. These areas tell the body to carry out the sort of movements you make without thinking about them: putting one foot in front of another to walk, for example, or reaching out to pick something up. If dopamine is prevented from activating this area, the result is tremor and reduced ability to get moving. This is what happens in Parkinson's disease (in which the dopamine-producing cells in the substantia nigra die off) and related movement disorders.

The basal ganglia are in turn connected by dopamine pathways to various cortical areas. The putamen is mainly connected to the premotor and motor cortex and overactivity in this pathway is thought to account for the physical tics in Tourette's syndrome. The caudate nucleus has more connections to the orbital cortex – an area concerned with higher order planning of action. Overactivity in this pathway is thought to result in obsessive–compulsive disorder.

The basal ganglia–frontal lobe pathway is important for attention and consciousness. Lack of dopamine here is thought to be a factor in attention deficit disorder (particularly in adults), negative symptom schizophrenia and the lethargy and mind-numbing effects of depression.

A third set of dopamine pathways starts in a midbrain area called the ventral tegmental nucleus, which is densely packed with dopamine-producing cells. One path trips off to an area in the brainstem called the locus coeruleus (blue area), which is a major producer of the excitatory neurotransmitter noradrenaline. Another goes the other way, running deep into and through the limbic system, then on through the old 'smell brain' to the olfactory bulb. A branch of it travels up to the frontal lobes, the cortical area concerned with emotion. Too little dopamine here is thought to be one cause of depression and the negative symptoms of schizophrenia. Too much is thought to be a factor in mania and the positive symptoms of schizophrenia.

The pathways meet, intersect and merge with each other along the way, so a bout of overactivity in one is likely to have a knock-on effect on another.

while a low level is likely to increase it. This is borne out by studies that show that anorexics do indeed tend to show abnormally raised serotonin, while people with bulimia have unusually low levels.[13] This may eventually translate into actual damage because, if a brain area is not fully stimulated, it may wither and shrink.

Although limbic area dysfunction is increasingly seen to play a part in eating disorders, it is clearly only part of their underlying cause. Cultural influences are at work, too. An anorexic's ambition to be thinner-than-thin, or the bulimic's determination to prevent their body absorbing what they have eaten, is conceived in the conscious part of the brain and clearly has a lot to do with the way that a person thinks. Anorexics, for example, tend to be disciplined, persevering and introvert, while bulimics are more inclined to be extravert, distractable and impulsive.[14] This makes these disorders open to treatment by psychological methods, and – although there are now several fairly effective drugs that work directly on the brain – most therapies are currently based on altering consciousness rather than altering physiology.

One day brain mappers may deliver a detailed enough blueprint of the food appetite system to allow direct modification, rather as is being tried now for people with OCD. For the moment our food-seeking genes continue to dispatch our bodies, clogged with fat, to their early graves.

Sex

'The male subject was strapped to a chair-like device with his head painlessly immobilized. In this position it was possible to insert a fine microelectrode into the hypothalamus. The female subject was then strapped into another chair that was positioned several feet away.

'The male was provided with a button that he could push to bring the chair bearing the female adjacent to his own.

ADDICTION

Drug addiction is caused by a similar train of events to hunger.

However, unlike most types of food, addictive drugs cause changes in the receptors to which they bind, making them less sensitive. This creates tolerance (the need to take sequentially larger amounts of a substance to produce the same effect) and addiction (feelings of unease when the substance is removed). Most addictive drugs work by altering levels of neurotransmitters, mainly serotonin, dopamine, endorphins and noradrenaline in the brain's reward circuitry centred on the limbic areas known as the ventral tegmental/nucleus accumbens. However, many other brain areas are also involved and each type of drug works in a slightly different way to produce its characteristic effects.

Ecstasy, for example, stimulates serotonin-producing cells, which turn on the areas of the prefrontal cortex that give feelings of euphoria, meaning and affections.

This is similar in effect to antidepressants, but Ecstasy causes the cells to produce a greater rush of neurotransmitter than antidepressants and its mechanism puts regular users in danger of 'burning out' the cells, creating temporary withdrawal symptoms and a long-term risk of chronic depression.

Hallucinogenic drugs like LSD and magic mushrooms also stimulate serotonin production or contain chemicals that mimic its effects. As well as stimulating the brain's pleasure centres, these drugs also activate areas in the temporal lobes that give rise to hallucinations. Bad trips may be the result of stimulation of the amygdala, which produces feelings of fear.

Cocaine increases the amount of dopamine available to cells by blocking the mechanism that usually gets rid of excess dopamine. It also blocks re-uptake of noradrenaline and serotonin. The rise in these three neurotransmitters causes the feeling of euphoria (dopamine), confidence (serotonin) and energy (noradrenaline) associated with the drug.
Amphetamines release dopamine and noradrenaline. This creates energy but may also produce feelings of anxiety and agitation.
Nicotine activates dopamine neurons by mimicking the effect of dopamine in binding to receptors in the cells' surface. Its initial effect is therefore similar to a dopamine rush. Nicotine quickly desensitizes the cells it works on, however, so the initial effects are no longer felt. Nicotine also affects neurons that produce a neurotransmitter called acetylcholine. This is one of the chemicals involved in alertness and is known to boost memory.
Opioids like morphine and heroin fit into receptors that normally take endorphins and enkaphelins. This triggers the reward circuit that creates the dopamine rush. The disengagement from pain produced by these drugs is thought to be caused by the deactivation of an area in the cortex called the anterior cingulate gyrus, which concentrates attention on adverse internal stimuli. The withdrawal effects of opioids are associated with a steep rise in stress hormones that activate the urge-making areas of the brain.
Alcohol and tranquillizers like benzodiazepines decrease neural activity through action on GABA (gamma amino butyric acid) neurons. Drugs that block the receptors in these cells reduce the pleasure of drinking and are effective in weaning addicts off alcohol. Alcoholism – and probably all drug addiction – is closely associated with genetic factors. Children of alcoholics, for example, are four times more likely than other people to become addicted to drink – even if they are brought up away from their natural parents.

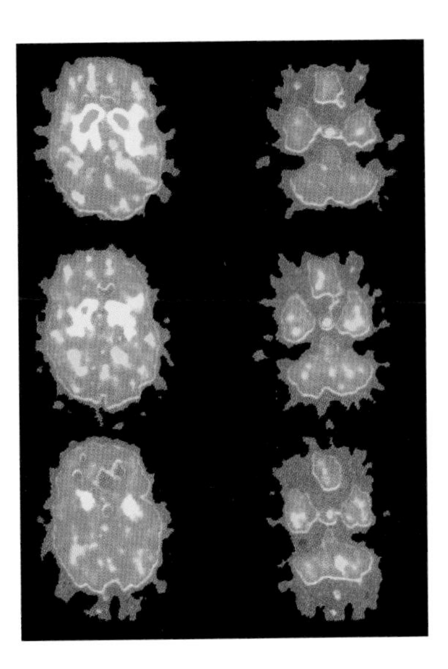

Cocaine creates euphoria by blocking receptors in brain cells which normally mop up excess dopamine. This leaves more dopamine free to excite the areas of the brain which affect mood. The scans above show levels of dopamine absorption in a normal brain (red = high absorption, yellow = medium; green = low)
Top: *after a placebo. Middle: after a low dose (0.1mg per kg body weight) of cocaine.* Bottom: *after a high dose (0.6 mg per kg). As you can see, after a high dose hardly any dopamine is locked up – it is all free to work on other neurons.*

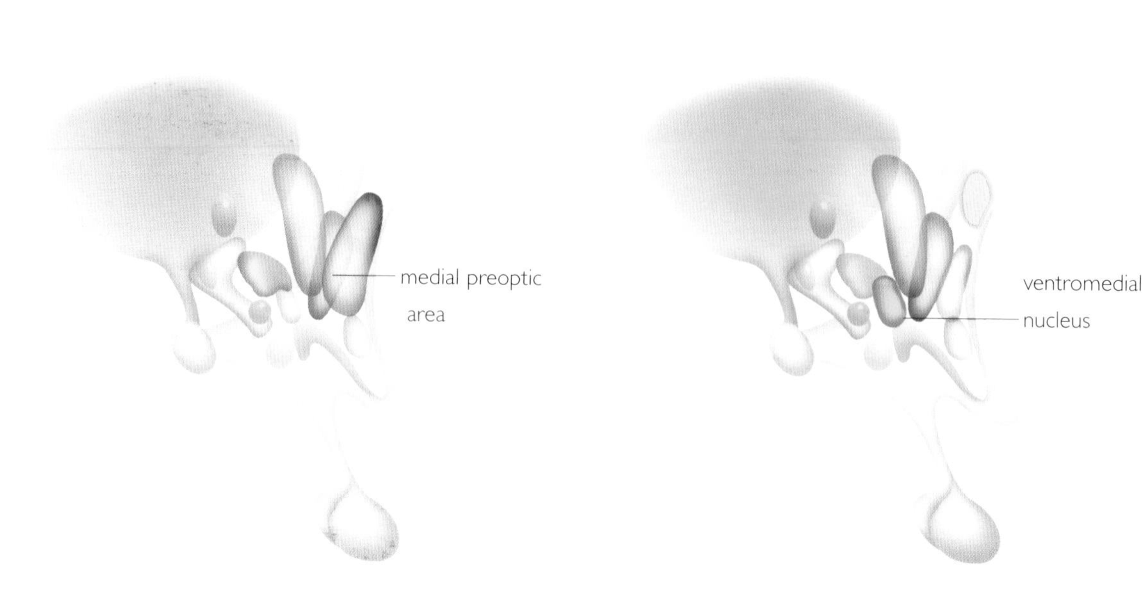

Typical male/female sexual responses are brought about by separate parts of the hypothalamus.
Left: *The medial preoptic area orients the sex drive towards female. Signals from here are sent to the cortex, producing conscious excitement, and down to the penis, to produce an erection.* Right: *Female typical sexual behaviour is prompted by the ventromedial nucleus. This is the same cluster of cells that are involved in turning on the appetite for food. When it is stimulated in a sexual context it encourages lordosis – genital display. This gesture is also made by some animals as a sign of submission.*

In this position the two were able to copulate without the male having to move his head. Recordings of neuronal activity were therefore obtained from the moment the male saw the female to completion of copulation.

'The highest neuronal activity (50 impulses per second) was recorded from a neuron in the medial preoptic area of the hypothalamus as the subject pressed the button to bring the female towards him. During copulation the discharge rate dropped and after ejaculation it ceased almost entirely. The specifically sexual nature of this activity was confirmed by a control experiment in which the female was replaced by a banana.' – *Report of experiment*[15]

Brain researchers are forced to go to extraordinary lengths to study sex, particularly in humans.

There is, for a start, the delicate problem of finding suitable subjects. If you can overcome that there is a practical difficulty: imaging equipment requires the subjects' heads to be rock-still in order not to dislodge the electrodes (in EEG experiments) or fuzz up the picture (in PET and MRI scanning). This is usually achieved by strapping the head into a brace, or getting the subject to clamp their teeth around a rubber 'bite bar'. Then there is the design of the scanning machine – definitely for one person only. Even if you can persuade someone to become sexually aroused while their teeth are clamped around a rubber bar in a clanging metal cigar tube, the confined space seriously limits the nature of the subsequent activity.

Studies of human brains in advanced states of sexual arousal are therefore few and far between. Most human brain-sex studies have involved correlating reported feelings or behaviour with brain structure rather than actually

The sexual brain

The main structural differences so far observed between men's and women's brains are:
* The hypothalamic nucleus INAH3, in the medial preoptic area, is on average 2.5 times larger in men than in women. This nucleus is responsible for male-typical sexual behaviour. It contains more cells that are sensitive to androgens (male hormones) than any other part of the brain. Some studies[16] have found a correlation in women between assertive (male-typical) and excessive heterosexual behaviour and small breasts, low voice, acne and hirsutism. The physical characteristics generally signify unusually high androgen levels, and it may be that the women's behaviour is caused by their hormones stimulating INAH3.
* The corpus callosum – the band of tissue through which the two hemispheres communicate – is relatively larger in women than in men.[17] So is the anterior commissure – a more primitive connection between the hemispheres that links the unconscious areas of the hemispheres only.[18]
This may explain why women seem to be more aware of their own and others' emotions than men – the emotionally sensitive right hemisphere is able to pass more information to the analytical, linguistically talented left side. It may also allow emotion to be incorporated more easily into speech and thought processes. Women also have more tissue in the massa intermedia, which connects the two halves of the thalamus.
* Men lose their brain tissue earlier in the ageing process than women, and overall they lose more of it. Men are particularly prone to tissue loss in the frontal and temporal lobes.[19] These areas are concerned with thinking and feeling,

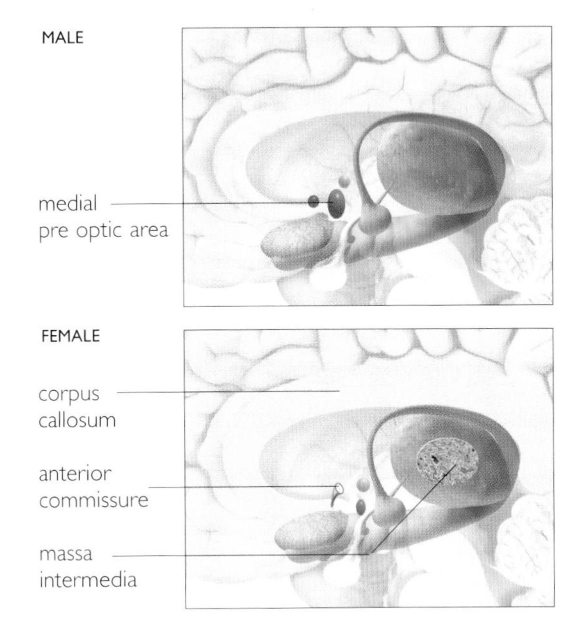

and loss of tissue in them is likely to cause irritability and other personality changes. Women tend to lose tissue in the hippocampus and parietal areas. These are more concerned with memory and visuo-spatial abilities, so women are more likely than men to have difficulty remembering things and finding their way about as they age.
Imaging studies show that men and women use their brains differently, too. When they do complex mental tasks there is a tendency for women to bring both sides of their brain to bear on the problem, while men often use only the side most obviously suited to it. This pattern of activity suggests that in some ways women take a broader view of life, bringing more aspects of the situation into play when making decisions, for example. Men, on the other hand, are more focused.

observing the brain during sex. Animals' brains, on the other hand, have been studied going the whole way from the first sight of an attractive partner to post-copulatory snooze.

The experiment above was carried out in Kyushu University and the subjects were – you may be relieved to know – macaque monkeys. These primates have very similar brain structure to humans and what happens in the limbic areas of their brains during sex is thought to be similar to what happens in ours.

Sexual drive centres on the hypothalamus but, like other urges, it radiates out to encompass a wide range of other brain areas in both the limbic area and the cortex. Again, like other urges, it can be split into various elements, each of which is localized. The clumps of tissue that produce each aspect of sexual feeling and behaviour are activated by various neurotransmitters in conjunction with sex hormones. The basic urge-reward-relief circuit operates in this area too: sexual drive is created by excitatory neurotransmitters; the intense 'reward' of orgasm is caused by a massive rush of dopamine and the feeling of relaxation that follows is due to a hormone called oxytocin.

The brain areas connected with sex differ between males and females. The differences are created by hormones and can be modified by behaviour and environmental factors. However, the underlying layout is determined – much of it prenatally – by genes. These physical differences are mirrored by observed differences in behaviour. Within any group of significant size – whether it be people, monkeys or rats – these behavioural distinctions are on average great enough for it to be accurate to talk of male-typical and female-typical sexual behaviour.

Male-typical sexual behaviour is more assertive than that of females; it is more closely linked to aggression and it involves taking the penetrative role. Female-typical is more submissive; it typically involves displaying the genitals (lordosis) and taking the receptive role in intercourse. The physical layout of male and female brains explains, in part, how these behavioural differences come about.

The medial preoptic area of the hypothalamus, which was so active in the copulating male monkey, seems to be the centre of male-specific sex behaviour. It is the area that has the largest number of androgen (male hormone)-sensitive neurons and it is bigger in men than in women. When this area is stimulated male monkeys become fanatically interested in any female that is around – providing she is in heat. However, if a female is not in heat the male is not interested. Conversely, if the medial preoptic area is removed the monkey loses interest in all females – but not in sex *per se*. Monkeys will continue to masturbate after this area of brain is removed and may show increased female-typical behaviour like lordosis.

This suggests the main function of the medial preoptic area is to respond to hormonal signals given off by receptive females. These signals come in from various areas. The olfactory system is a main source of input in monkeys, underlying the importance of smell to their sexuality, but the role of smell in humans is less certain.

The medial preoptic area also receives signals from two nuclei in the amygdala (corticomedial and basolateral), both of which are concerned in one way or another with producing aggressive or assertive behaviour. This connection may explain why sex, in the male, is sometimes coupled with aggression – the excitement of the medial preoptic nucleus in the hypothalamus may have a knock-on effect on the areas in the amygdala that generate aggression and vice versa.

Once it is aroused by a suitable sexual stimulus, the medial preoptic hypothalamus passes the signal up to the cerebral cortex, which does (or tries to do) whatever is required to get the body it is attached to into an appropriate position for intercourse. At the same time signals are sent down to the brainstem, which produces

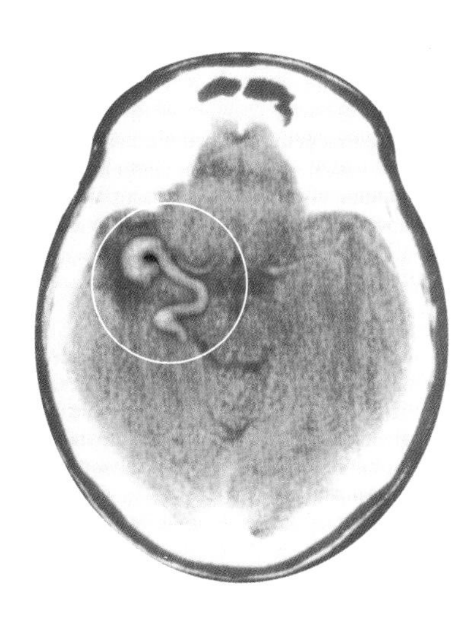

Direct stimulation of the temporal lobe can produce strong erotic feelings. The swollen blood vessel seen in this brain scan produced flurries of activity which caused the patient to experience orgasms at entirely inappropriate times. The unwanted orgasms happened about once a fortnight for three years before the cause was uncovered.
[Ref: Reading J.P. and R.G. Will 'Unwelcome Orgasms', The Lancet 350 page 1746, 13 December 1997)

penile erection. Once intercourse begins, the motor cortex comes into play, producing appropriate thrusting movements. Finally, another hypothalamic nuclei, the dorsomedial nucleus, triggers ejaculation.

Normally, all of this takes place in an ordered sequence, but sometimes – through brain damage or dysfunction – one or another element may get out of synch, fail, or arise without the rest. If the dorsomedial nucleus is stimulated during an epileptic fit, for example, ejaculation may take place without any other sign of sexual arousal, and, in men, stimulation of the septum – an area of the limbic system that lies adjacent to the hypothalamus – has been known to produce orgasm without pleasure. This has been held to demonstrate that orgasm is, at its root, a form of 'reflex epilepsy'. Septum damage may also cause

Is there a 'gay brain'?

In 1991 the prestigious journal *Science* published a study showing that the brains of a group of homosexual men who had died from Aids were structurally different from the brains of heterosexual men. The nucleus in the hypothalamus that triggers male-typical sexual behaviour was much smaller in the gay men and looked more like that in the brains of women. The author, Simon LeVay, then Associate Professor at the Salk Institute for Biological Studies and Adjunct Professor of Biology at the University of California, was immediately attacked by gay activists who feared that the recognition of homosexuality as a physical-based condition might lead to it being re-stigmatized. LeVay, who is himself gay, then went on to discover that the corpus callosum differs between gay and straight men, too – in gays it was found to be bigger. Three years later a study led by molecular biologist Dean Hamer of the National Institute of Health in Washington, DC, found evidence to suggest that a specific gene – carried on the maternal line – influenced sexual orientation in men. Put together, these studies provide strong evidence that homosexuality is rooted in biology – and hostility to the idea has largely disappeared.

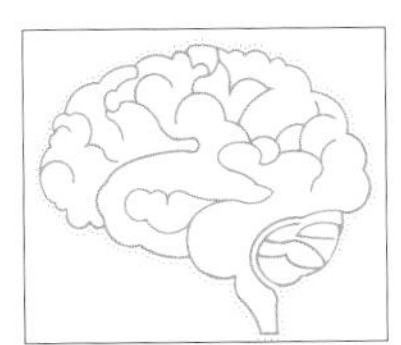

priapism – a condition in which a man has a permanent erection. Conversely, lack of stimulation to one or another of the sex nuclei may result in impotence: if sexual signals from the hypothalamus fail to reach the brainstem, for example, erection is impossible.

Female-typical sex behaviour is located in the ventromedial nucleus of the hypothalamus – the same one that plays a crucial part in hunger. This area is rich in oestrogen-sensitive neurons, and it is this hormone that seems to excite the ventromedial nucleus to produce lordosis – the genital display that is typical of female sexual invitation in many species. In rats lordosis is a reflex action – it occurs whenever the skin on a female animal's rump is grasped, as it would be by a mounting male. Unlike simple knee-jerk reactions, however, lordosis seems to be partially under conscious control – it is exhibited much more strongly when the stimulus comes from a male rat rather than the fingers of a laboratory researcher. In humans lordosis is generally kept firmly and completely under conscious control, though a glance through any selection of hardcore porn shows that it has by no means died out.

Although the female sex hormones seem to dictate the *type* of sexual behaviour exhibited they do not have much effect on the *strength* of the sex drive. This is controlled, in both sexes, by the action of adrenaline and testosterone. These chemicals work very widely throughout the brain, and it seems likely that sex drive is brought about by the interaction of many different brain areas rather than just one.

Sex permeates the entire brain, especially the human brain. Connections from the sex-sniffing, sex-seeking and sex-reactive areas of the limbic system radiate to almost every corner of every cortical lobe, feeding the urge to our conscious minds. The cortical area most closely associated with sexual feelings is the right frontal cortex. A rare imaging study of orgasms in humans found that cerebral blood flow to this area was increased during sexual activity,[20] and a curious report of a woman who experienced spontaneous (and unwelcome) orgasms because of a vascular bulge in her frontal cortex suggests that, in this at least, human sexuality is similar in both sexes.[21]

The sexual traffic in the brain goes in both directions. As lascivious promptings push up from below, so the conscious brain pushes sexually stimulating information gathered in from the environment down to the limbic system. Between them the two interactive brain levels keep humans warmed up and ready for sex at any time. Other species show this degree of interest only when the female is physically ready to conceive. Humans alone have elevated sex from a periodic spree to a full-time occupation and in doing so they have turned it into something very elaborate. Sex involves practically every type of brain activity – from the high level cognition involved in romantic love, through visual and physical recognition, down through emotion and straightforward bodily function. Because of this any fault, anywhere in the brain, may bring about some type of sexual dysfunction along with other problems. The infiltration of sex into the frontal lobes – the place where humans construct their most sophisticated, abstract ideas – has entangled it with our deepest notions of morality. Hence sexual obscenity and disinhibition are characteristic of people with the sort of frontal lobe damage that causes the destruction of a person's 'higher' cerebral functions.

Frontal lobe damage may also bring about erotomania – an obsessional delusion with a strong sexual element. Erotomaniacs believe that another person – usually someone famous – is in love with them and passing them secret messages of adoration. Sometimes they report making love to the object of their obsession, even though they have never even met them.[22] Stalkers are often erotomaniacs. They are usually harmless, but their attentions can drive their victims to exasperation. One recent case involved a man who had broken into the home of a female neighbour

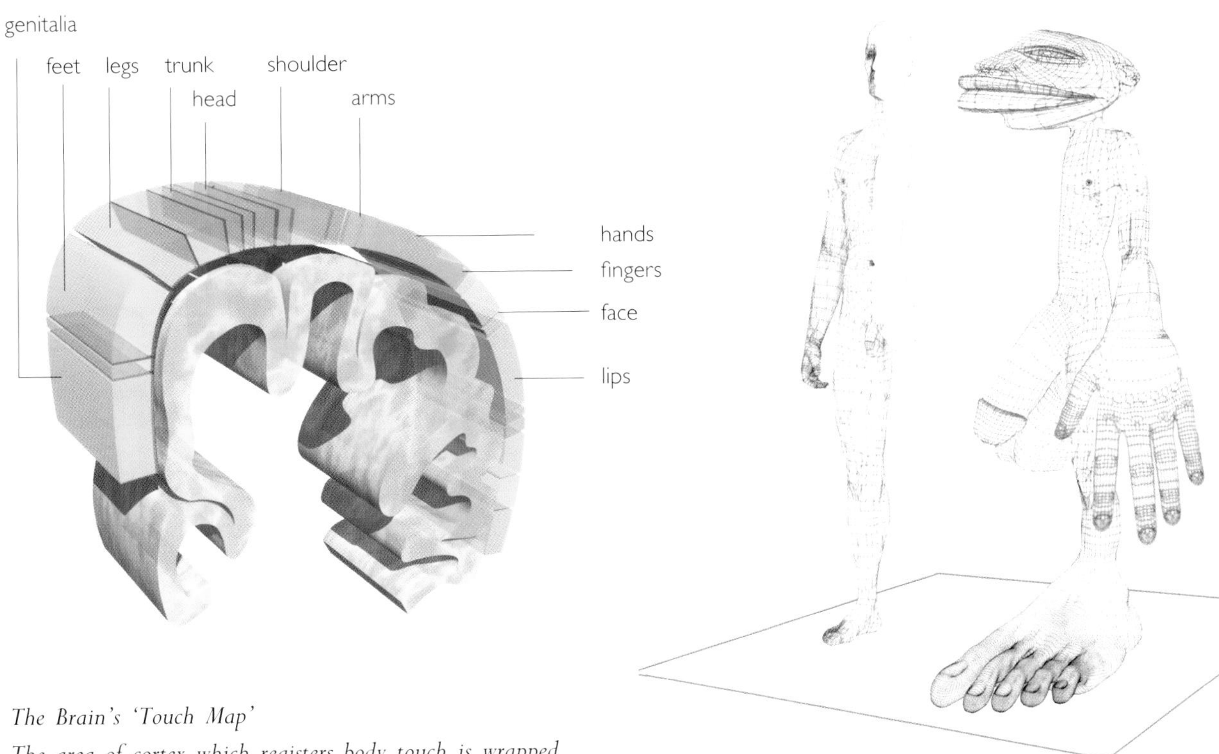

The Brain's 'Touch Map'
The area of cortex which registers body touch is wrapped around the cerebrum like an Alice band. The area given over to the genitals is about as large as the rest of the chest, abdomen and back put together.

If the area of each body part was proportionate to its sensitivity people would look very different ...

– whom he had followed constantly but never spoken to – and began to move in his belongings. When the police came to arrest him he informed them that the woman and he were engaged and were about to start living together. An MRI scan showed a massive (benign) growth in his left frontal lobe. When it was removed his behaviour returned to normal.[23]

The temporal lobe is also implicated in sexual function. Damage to the cortex towards the front of that lobe, and the amygdala, which lies below it, results in a condition known as Kluver-Bucy syndrome, in which the patient tries to stuff anything in the vicinity into their mouth and/or make love to it. One unfortunate man with this condition was arrested while trying to make love to a pavement.[24]

The reason for this bizarre behaviour may be that the cortical area of the temporal lobe that is damaged in Kluver-Bucy syndrome normally sends inhibitory signals to the ventromedial nucleus in the hypothalamus. Part of the ventromedial nucleus, as we have seen, switches on the compulsion to put easily available food in the mouth, while another part triggers female-typical sexual behaviour. So cortical damage that allows the ventromedial nucleus to fire away without inhibition may create a continuous urge to eat and copulate. The strange lack of discrimination about what is eaten and copulated with may be due to disruption of the ability to recognize categories of object – another function that is localized in the temporal lobe. Put these two malfunctions together and it is quite feasible that a pavement may appear quite appetizing, as well as sexually alluring.

At the junction of the frontal and parietal lobes lie the sensory and motor cortices – the 'body map' in the brain where each individual area of the body has a corresponding area of cortex devoted to it. Here lies another demonstration of how important sex is to humans: the area of sensorimotor cortex given over to the genitals is rather more than is provided for the entire surface of the chest, abdomen and back put together. Stimulation of the area that represents the genitals produces sensations and/or action in the parts themselves, and epileptic fits that centre on this area of the brain produce strong sexual sensations. During such seizures people sometimes make thrusting movements, as in sexual intercourse.

But sex – human sex particularly – is not all about thrusting and ejaculating. In humans it gives rise to a complex package of feelings and thoughts that we label love. Romantic love is born out of the evolutionary success of pair bonding as a reproductive strategy. Our brains have evolved to feel pleasure in sexual bonding and discomfort at separation. It is brought about by an even more elaborate than usual interplay of hormones and neurotransmitters. So far, only the crudest moves in this chemical concerto have been charted. We have a good idea of the substances associated with the various stages of falling in love, but it is not yet known precisely which brain areas each of them activates.

The feelings of euphoria associated with early stages of love seem to be brought about by a combination of dopamine and a chemical called phenylethylamine (PEA). These probably work on the reward pathways leading from the limbic system to the cerebral cortex. The drive to make love comes from testosterone (in men and women) and oestrogen (in women) working on the hypothalamus. Bonding – both sexual and between parent and child – seems to be brought about largely by the action in the brain of a hormone called oxytocin.

Oxytocin is thought to be a relatively recent (in evolutionary terms) mutation of a much older hormone called vasopressin, to which it is chemically very similar. Vasopressin is an antidiuretic and its main function is to control blood volume and pressure. However, the drug is also known to help lay down new memories, and is widely used (or abused, depending on your view) as a 'smart drug' or cognitive enhancer. Oxytocin is made in the hypothalamus and released as a result of stimulation to the sex and reproductive organs. It floods the brain during orgasm and in the final stages of childbirth, and in doing so it produces a warm, floaty, loving feeling that encourages pair bonding. In the short term oxytocin seems to blunt the memory but it is possible that it has 'inherited' from vasopressin the ability to enhance the forging of new memories, so the impression left by a person who causes oxytocin to flow may be particularly strong. The mechanism may be similar to addiction: oxytocin is closely linked to endorphins – opiate-like brain chemicals – and the agitation typically felt by lovers when they are separated from the ones they adore may in part be due to their desire to push up their oxytocin level.

Countless psychological studies have shown that people in the throes of this hormonal storm are more than usually divorced from reality, particularly when it comes to making assessments about the person they love. They are – famously – blind to the other's faults and often wildly over-optimistic about the future of the relationship. Looked at coldly, romantic love is a chemically induced form of madness and a terrible basis for social organization, as the divorce rate in the Western world demonstrates.

From the brain's point of view, however, it is just about the biggest thrill there is. So long as the limbic system is still in the driver's seat sexual love will continue to disrupt, delight and occasionally ambush us when we are least expecting it. It may not keep the world going round but it certainly makes it a more interesting place to live.

IS AUTISM AN EXTREME FORM OF MALE BRAIN?

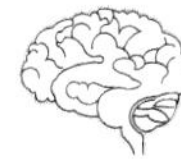

DR SIMON BARON-COHEN
Department of Experimental Psychology, University of Cambridge

After decades of research into whether and how the sexes differ psychologically, some differences are repeatedly found that, though not true of every individual, certainly emerge when groups are compared.

Women
* score better than men on some language tasks
* show a faster rate of language development
* have a lower risk of developmental dysphasia
* score better than men on some tests of social judgement, empathy and co-operation
* are better at matching items
* are better at tests that involve generating ideas.

Men
* perform better than women on mathematical reasoning tasks (especially geometry and mathematical word problems)
* score higher at tests that involve distinguishing between figure and background
* find it easier to rotate objects in their mind's eye
* are better at hitting targets

I am not arguing that one sex is better than another, simply that there seem to be different cognitive styles associated with being male or female – that is, not every male will have a spatial advantage but the likelihood of having this advantage is increased if you are male. Such sex differences could, of course, be the result of differential socialization, different

[1]

[2]

[3]

Priests, Politicians, Pundit, Physicists, Psychological, Parent, Pedagogue, Police, Philosopher, Paedology, Polymath, Palaeontologist, Palimpsest, Pagan.

[4]

[5]

biological predisposition or both. I suggest that such psychological differences are in part the result of biological differences in brain development, which is itself the product of genetic and endocrinal differences. My main reason for pursuing this line of reasoning is the evidence from autism.

First, consider a model. A brain could be more developed in terms of 'folk psychology' (understanding people in terms of mental

[6]

states) than in terms of 'folk physics' (understanding objects in terms of physical causality and spatial relations) or vice versa. In this box I will operationally define the male brain type as an individual whose folk physics skills are in advance of his or her folk psychology skills, and the reverse will form the operational definition of the female brain type. Those with roughly equal folk physics and psychology skills will be called the cognitively balanced brain type.

Autism is a strongly heritable psychiatric condition characterized by abnormal social and communicative development, narrow interests, repetitive activity and limited imagination. Four times more males than females suffer from autism and nine times more from Asperger's syndrome ('pure' autism without other handicap). I will now argue that autism and Asperger's syndrome are extreme forms of male brain type.

Children with autism perform better than mental age-matched controls on the Embedded Figures Test – a test in which normal males perform better than normal females. They fare worse than controls on selective aspects of social cognition, especially on tests involving ascribing mental states to other people – an area in which normal females outstrip normal males. Indeed, even parents of children with autism show the male brain type more strongly than sex-matched controls. I suggest that this is no

On average....

1) Men are better than women at rotating images mentally

2) While women are quicker at spotting when two images are alike

3) and better at generating words

4) Men are more accurate in targeting tasks

5) Women can recognise missing objects

6/8) Spotting when a particular shape is embedded in a complex pattern is done more easily by men

7) but women can do fine manual tasks, like placing pegs in a board, more easily

9) mathematical calculation favours women

10) yet mathematical reasoning comes more readily to men.

[7]

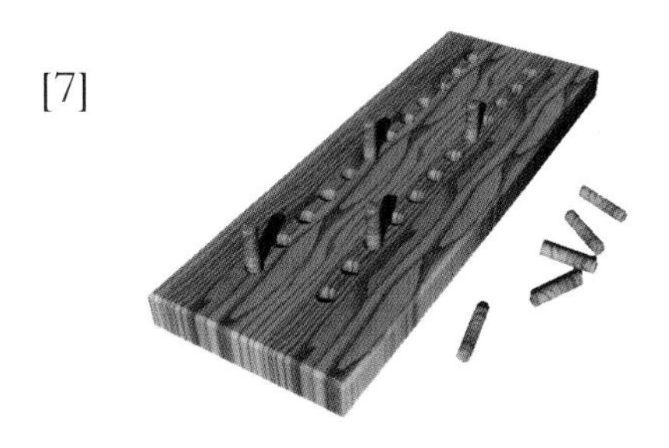

coincidence, rather that it might reflect the existence of sex-linked neurodevelopmental processes in the population.

What might these neurodevelopmental processes be? Here we consider the much-discussed foetal testosterone model:

In a male embryo the XY genotype controls the growth of testes, and at around eight weeks, the testes are formed and release bursts of testosterone. Testosterone has frequently been proposed to have a causal effect on foetal brain development such that by birth clear sex differences are evident. Some psychologists claim that at birth human female babies attend for longer to social stimuli such as faces and voices while male babies appear to be more interested in spatial stimuli such as mobiles.

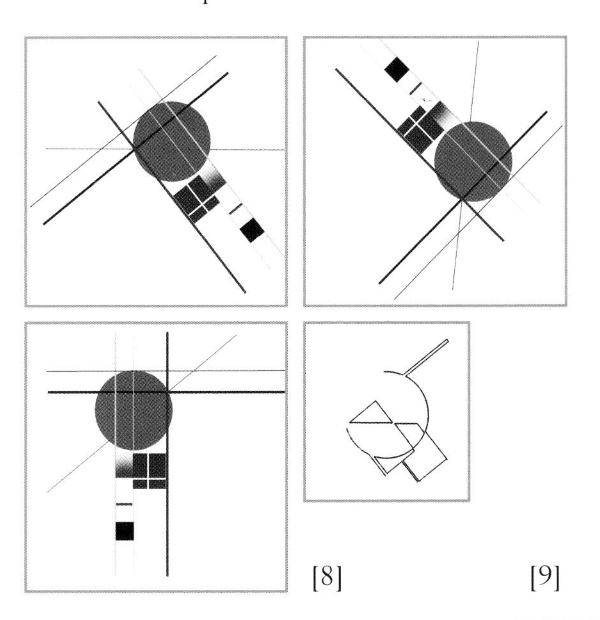

[8] [9]

Levels of prenatal testosterone have been found to predict spatial ability at age seven.

Regarding the two brain types mentioned earlier, precisely which structures distinguish these two brain types is still controversial. Some people have found that the right hemisphere cortex is thicker in male babies than in female babies. Others show that the corpus callosum is larger in females (which could explain their superiority in verbal fluency) than in males and even smaller in individuals with autism. There is also evidence that exposure to androgens increases spatial performance in human females (and, in rats, castration decreases it), which is consistent with the notion of a male or female brain type being a function of the levels of circulating hormones during critical periods of neural development.

So one theory is that people fall into a continuum as regards male and female brain type and that autism and Asperger's syndrome are extreme forms of male brain type. Many questions have yet to be answered. What happens to cause someone at the extreme end of this postulated continuum to develop autism? Is it early hormonal events? Do these result from genetics? What does it mean in neurobiological terms for someone to be an extreme form of male brain? And if some sex differences arise for neurodevelopmental reasons, what sort of evolutionary factors have shaped such sexual dimorphism?

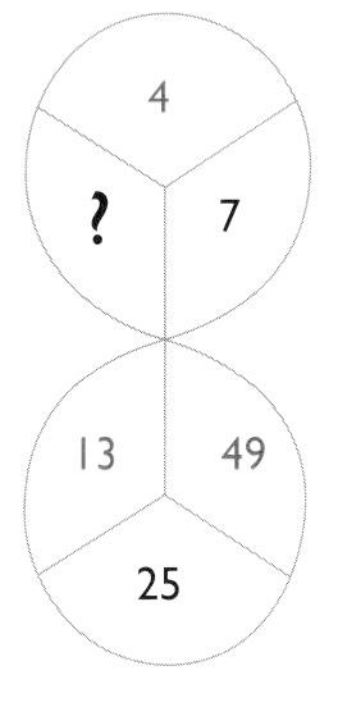

$$18-246 \div 8 =$$

$$4(18+36) + 10-\frac{15}{5} =$$

[10]

CHAPTER FOUR
A Changeable Climate

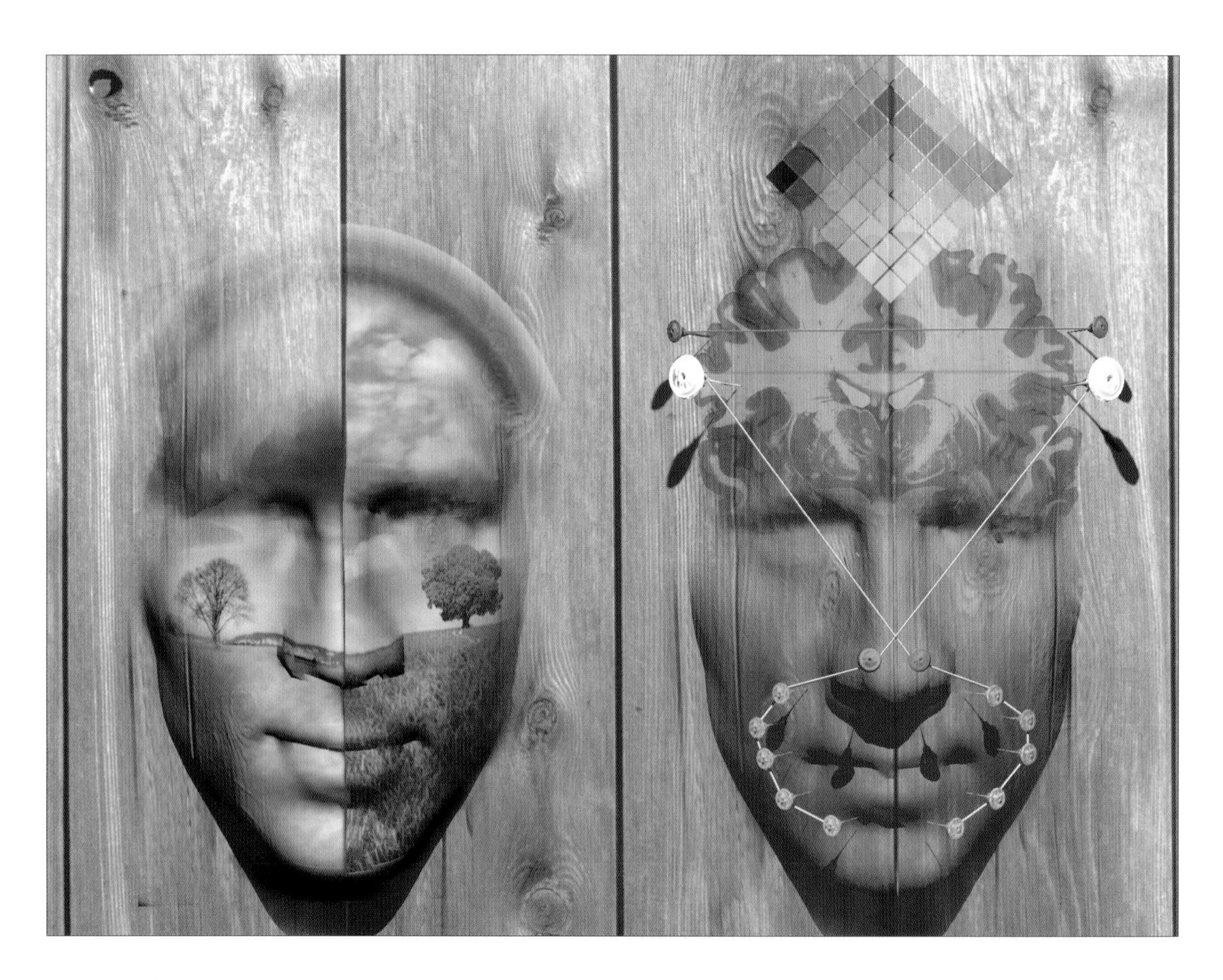

We feel as well as act … The sunshine and shadows that play over the landscape of the mind are generated by chemicals that turn the modules of our brains on and off, creating the neural patterns that spell out our moods. The limbic structures stab out urgent messages of fear and anger, and the cortex responds by flooding our consciousness with emotion.

THERE IS A SET OF PHOTOGRAPHS THAT IS SOMETIMES USED by psychologists to provoke emotional responses. The photos show people writhing with pain in the aftermath of gory accidents, flailing helplessly in flood water, or screaming from the windows of burning buildings.

In one particular experiment the participants' bodily reactions were monitored while these images were presented to them, one by one, on a big screen and in full colour. The subjects were uniformly disturbed by what they saw. Their pulses went up, their blood pressure increased and stress hormones started to course in their blood. Asked afterwards how they felt most said they were mournful or nauseated or anxious. The odd one or two admitted to a perverse thrill of excitement.

One subject, however, felt none of these things. Elliott had an operation some years ago to remove a fast-growing tumour from an area near the front of his brain. A large chunk of surrounding tissue also had to be removed, and with it went his capacity to feel emotion. Antonio Damasio, the neurologist called in to assess Elliott, describes him thus:

'He was always controlled, always describing scenes as a dispassionate, uninvolved spectator. Nowhere was there a sense of his own suffering ... he was not inhibiting the expression of internal emotional resonance or hushing inner turmoil. He simply did not have any turmoil to hush.'[1] A life without joy or love, sadness or anger, sounds dull beyond belief. But there is something, you might think, to be said for it. At the very least it would seem likely to equip a person with a major requirement for success – the ability to make rational decisions, even in the midst of a crisis.

In fact, quite the opposite is true. The reason that Elliott was referred to Professor Damasio was because after his operation he seemed unable to function efficiently in almost any capacity. His IQ was the same as before his operation, his

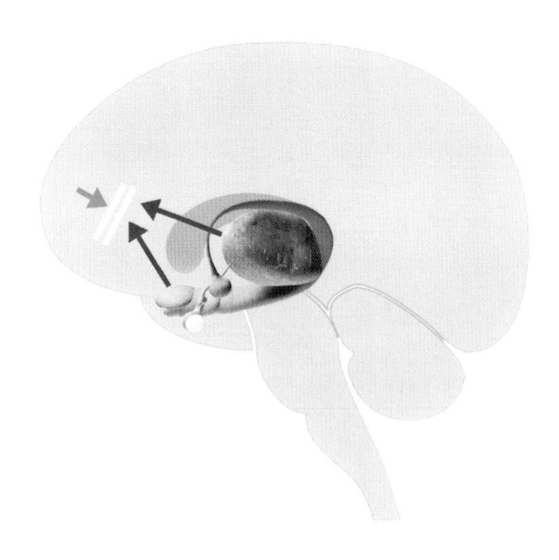

If the neural pathways from the limbic system to the cortex are blocked or severed emotions cannot register.

memory was fine and his powers of calculation and deduction were unaffected. Yet he found it hard to make the simplest decision or to pursue any single plan to a fruitful conclusion. He needed prompting to get up in the morning, and once at work he might fritter away a whole day either trying to decide what to do first or attending diligently to some unimportant detail while urgent tasks went unheeded. When he lost his job he threw himself into one wild new enterprise after another and finally went bankrupt.

A battery of behavioural and neuropsychological tests, including the ghastly picture-show, finally revealed the root of Elliott's problem: emotions no longer registered and without them he was unable to weigh up or evaluate one thing over another. Faced with a situation that called for decisive action he could generate a full range of appropriate responses – but none of them felt any more 'right' than any other. The result was that he could not choose between them. He had no 'gut feelings' to warn him away from dodgy enterprises and no instinctive sense of who to trust. One reason he went bankrupt was because he became involved with a character most people

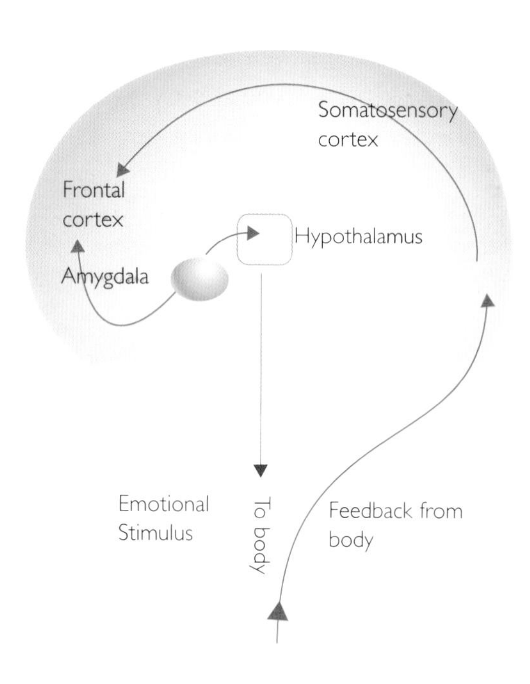

Emotional stimuli are registered by the amygdala. Conscious emotion is created both by direct signals from the amygdala to the frontal cortex, and indirectly. The indirect path involves the hypothalamus, which sends hormonal messages to the body to create physical changes like muscle contraction, heightened blood pressure and increased heart rate. These changes are then fed back to the somatosensory cortex, which feeds the information forward to the frontal cortex where it is interpreted as emotion.

would have recognized immediately as an unsuitable business partner. It was not that he did not know about normal emotional reactions and he acknowledged that something serious was missing in his responses. After seeing the pictures of people in distress he said: 'I know this is horrible – I just don't feel the horror.'

The curious dissociation between knowing and feeling was caused, in Elliott, by a severance of some of the neuronal connections that run between the frontal cortex, where emotions are consciously registered, and the limbic system, the brain's deeply buried unconscious core, where they are generated. His case and others like it demonstrate how our perceptions and behaviour are informed by brain processes of which we are not even conscious. These in turn are informed by purely visceral reactions.

Without feedback from our bodies emotions are indistinguishable from thoughts. People with high spinal injuries who cannot feel anything below the neck typically report this damping down of emotion. One such patient explained: 'Sometimes I act angry when I see some injustice, I yell and curse and raise hell because if you don't do it sometimes I have learned people take advantage of you. But it doesn't have the heat it used to [before the injury].'[2] Our emotional vocabulary – the aching heart, the lump in the throat and the pain in the neck – reflects the direct connection between body states and felt emotion.

The limbic system does not have sovereignty over our emotional reactions, however. The emotional traffic between it and the cortex is two-way. Just as impulses from below mould our conscious thoughts and behaviour, so the way we think and behave can affect the reactions of the unconscious brain. But there are more connections running up, from the limbic system to the cortex, than there are going in the other direction – at the moment emotion is in the driving seat.

What exactly is this thing that informs us so completely? We think of emotion as feeling, but the word is misleading because it describes only half of the beast – the half that we do, indeed, feel. Essentially, emotions are not feelings at all but a set of body-rooted survival mechanisms that have evolved to turn us away from danger and propel us forward to things that may be of benefit. The mental component – the feeling – is just a sophistication of the basic mechanism. Emotion researcher Joseph LeDoux, of New York University, calls it 'a frill – the icing on the cake'.

Human emotions are rather like colours: there seems to be a handful of primary ones and a wider range of more complex concoctions created by mixing the primary ones together. Various

Just as three primary colours can produce an almost infinite range of hues, so a handful of basic emotions are mixed to produce complex feelings.

researchers claim to have identified the primary emotions, usually as disgust, fear, anger and parental love. These are the responses that seem to be displayed by nearly all living things of any complexity. Primary emotions do not require consciousness – they can trigger a person to turn away or lunge forward in the complete absence of their conscious volition. Sometimes, as we shall see, the result of this can be catastrophic.

Complex emotions, on the other hand, are sophisticated cognitive constructs that are arrived at only after considerable processing by the conscious mind and an elaborate exchange of information between the conscious cortical areas of the brain and the limbic system beneath. Take, for example, the pleasure-tinged-with-guilt-streaked-with-affection-and-irritation you might feel when you receive a birthday card from a friend whose own birthday you neglected to acknowledge. There is nothing in the sight of the card itself to produce an emotional reaction. It comes about only when your recognition of the handwriting on the card triggers a confluence of related thoughts and memories that together provide the raw ingredients for the emotion. You recall, for example, your friend's birthday and your decision to ignore it. This will produce guilt – which is itself a mixture of fear (of retribution) and disgust (of self). Then there is the background to your friendship as a whole – enter affection, a moderated form of love. You may also recall why you did not send your friend a card – you were too busy. Hence the irritation – a mild form of anger. Put all of these together, stir well, and you have that complicated emotional mix, the like of which (we presume) only humans can experience.

The separate perceptions that bring about emotions are registered consciously in the cortex where they are assembled into a single, many-faceted concept. But this does not of itself guarantee emotion. So long as it is a purely cognitive assembly it is mere knowledge. A person whose emotional processing went only this far could look at the birthday card and say: 'This should produce guilt-affection-irritation.' But, like Elliot, they would not actually *feel* the emotion. For that another step is required.

Once the conscious mind has perceived that the situation calls for some emotional reaction it sends signals down the line to the limbic system – effectively calling for appropriate action to be taken. The limbic system obliges in its usual way – it sends messages to the body (via the hypothalamus) to make certain changes. Neurotransmitters are released and inhibited; hormones are pumped out; vital processes like heartbeat and blood pressure are altered. These changes are then monitored by the hypothalamus and the message flies back up to the cortex: 'We have emotion.'

The bodily changes themselves are not specific to the precise mix of guilt-affection-irritation that has been assembled in the cortical association area. A shot of adrenaline, for example, will make people feel angry or elated according to their situation. But when the shakes and tingles, butterflies, breathlessness and muscle tension come to the attention of the conscious brain it interprets them in accordance with its preconceived notion. The sequence is: 'I think I should be feeling angry,' followed by 'Yes! I seem to be feeling something,' followed by 'It must be anger.'

Even at this stage an emotion has not come to full flower. To do its work as a survival mechanism it has to be expressed, and that requires yet another round of cognitive processing. Emotional expression requires some sort of bodily action. It may be a sob, a blow, a dash for safety or just the insertion of a slightly jarring tone in an otherwise neutral vocal statement.

It is quite common for people to be able to feel emotion but to be quite unable to express it. These people, whose condition is known as alexithymia, are in a different situation to those like Elliot, who cannot express emotion because they do not feel it to start with. The inability to express felt emotion probably arises when there is some disruption in the neural connections between the cortical (conscious) emotional processing areas and the brain regions that control facial expression, speech and the other physical means by which emotions are displayed. If, for example, the break lies between the emotional brain and the speech areas in the left hemisphere, the result may be a curious flatness of voice. Such people may say, in quite a neutral tone: 'I am very angry.' Then, aware that the statement has some shortcoming, they might add: 'and I mean it.'

Alexithymia strips those who have it of an important social tool – the ability to convey, swiftly and economically, how they feel. It is an unfortunate affliction for anyone, and can be disastrous in those who seek to influence people on a large scale. Think of the mildly alexithymic voice of the former British prime minister John Major. 'Oh yes,' Mr Major would say after some trenchant statement, 'oh yes.' But still, very often, the words failed to hit home.

And hitting home, ultimately, is what emotions are for. At their crudest they force us to hit, flee, or scream. But at the level on which, most of the time, we experience them, their purpose is largely to bring about a corresponding emotional change in other people, which causes others to behave in a way that is beneficial to us. People who describe themselves as 'too emotional' may feel they suffer from an overload of feeling, but it is more likely to be those around them who get battered on the rollercoaster of their passions. Emotions exist to help us manipulate, influence or bully those around us – whether we consciously desire it or not.

Expressions

In order to do their job of influencing other people, emotions need to be expressed. Imagine for one moment that you work for a fast food company that insists you dispense a bright smile

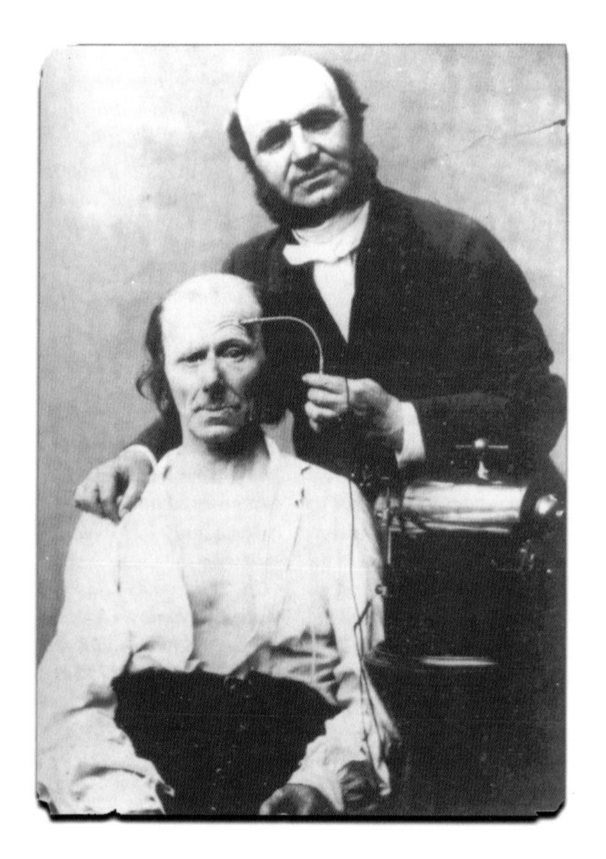

French neurologist Guillaume Duchenne made this patient 'smile' by putting an electric current through his facial nerves as part of his investigation into the anatomy of emotional expression.

with every order. You have a large family to support. And you are about to serve the two hundredth customer of the day.

Feel those muscles contract? Well done. You have composed your face into the social smile, one of the 7,000-odd facial expressions that this species has in its repertoire.[3] Between them these expressions provide us with a formidable battery of tools for easing social intercourse. Some of them – in particular the social smile – have a very specific role: they allow us to lie about our inner feelings. Other animals do not have this dubious ability because their facial expressions are out of their control.

The social smile is quite distinct from the genuine smile of pleasure. For a start, the one you conjured up just now has almost certainly by now disappeared without trace. A spontaneous smile lingers, and fades more evenly and slowly.[4] But the difference is more fundamental than that: the two types of smile are brought about by a different set of facial muscles, which are in turn controlled by entirely separate brain circuits. The spontaneous smile, called the Duchenne smile after the French anatomist who first identified it, emerges from the unconscious brain and is automatic, while the 'Have a Nice Day' version comes from the conscious cortex and can be summoned at will.

The conscious brain can produce a wide range of expressions to order, but they are never quite the same as those produced automatically because some facial muscles are outside cortical control. The Duchenne smile, for example, contracts a number of tiny muscles that run around the orbit of the eye socket, something the social smile rarely achieves. A smile directed at someone who is loved or sexually attractive also involves dilation of the pupils – one reason why low lighting (which also causes the pupils to dilate) is so conducive to romance.

Facial expressions of emotion are very similar the world over, suggesting that the neural circuits that create and respond to them are hard-wired into the brain rather than moulded by culture. The basic expressions are sadness, happiness, disgust, anger and fear. The thousands of other faces we make are blends of these.

Children respond appropriately to facial expressions almost from the time they are born, but they get progressively better at it as they get older.[5] The improvement matches the pattern of maturation of the frontal lobes – the cortical area concerned with emotion.

Expressions of fear[6] are picked up and identified by the amygdala – a tiny piece of tissue in the unconscious limbic area of the brain. One

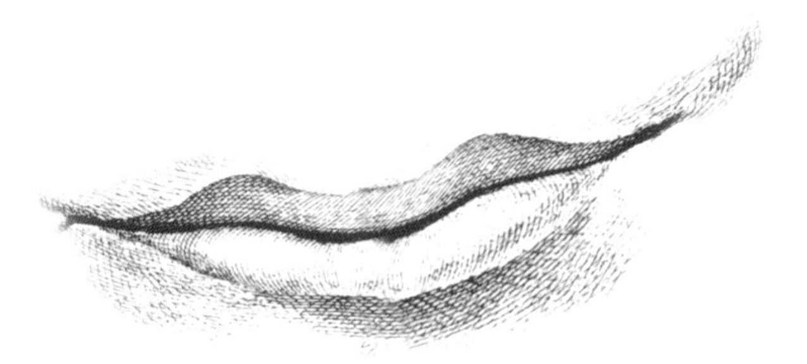

part of the amygdala responds to facial expression and another is sensitive to tonal qualities in the voice – the give-away rasp of anger or quiver of fear. The left amygdala seems to respond more to vocal expression while the right amygdala is more sensitive to facial movement. It follows from this that people whose amygdala is oversensitive may be quick to take offence or be easily hurt, while those whose amygdala is slow to react may seem rather dull and detached.

Disgust – which means, literally, 'bad taste' – is expressed in the face by a distinctive wrinkling of the nose, narrowing of the eyes and pursing of the lips. Brain scans show that watching a person displaying this expression activates the anterior insular cortex – an area of brain that is also stimulated by offensive tastes. The sight of an expression of intense disgust also lights up a circuit in

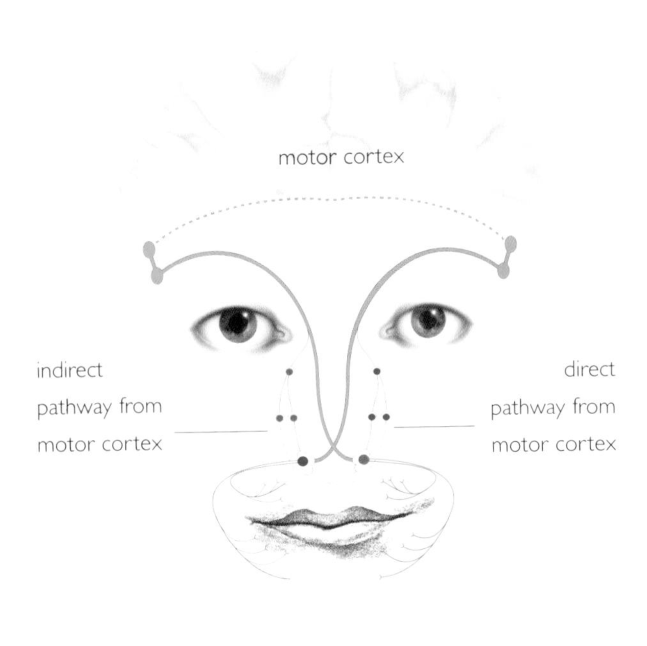

VOLUNTARY SMILE CIRCUIT

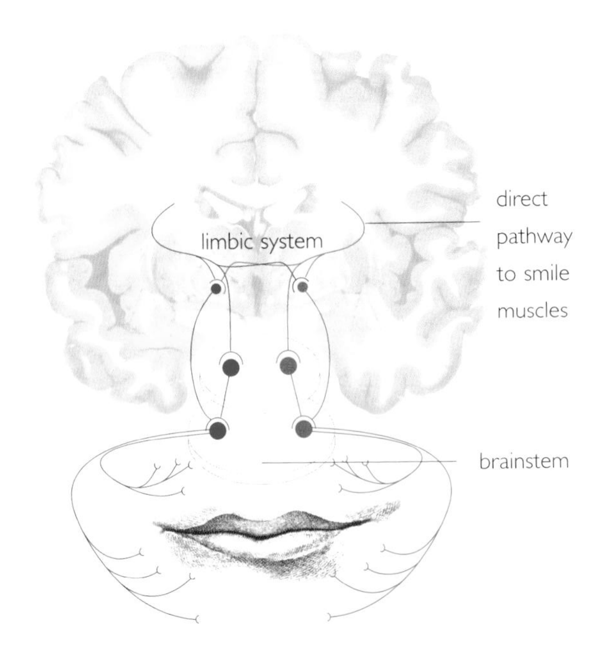

SPONTANEOUS SMILE CIRCUIT

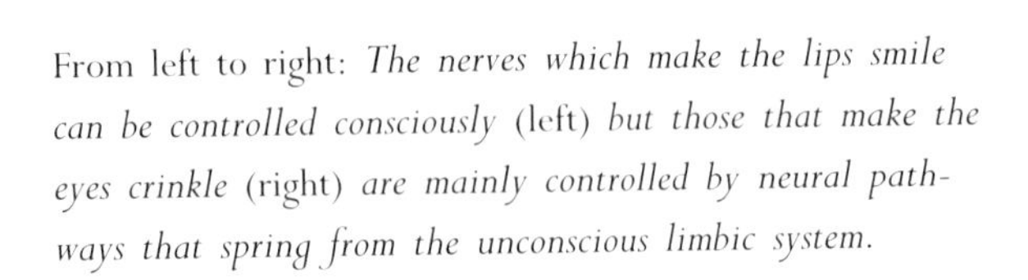

From left to right: *The nerves which make the lips smile can be controlled consciously* (left) *but those that make the eyes crinkle* (right) *are mainly controlled by neural pathways that spring from the unconscious limbic system.*

the observer's brain that connects the cortex with the limbic system. This suggests that when you look at a person who is showing mild disgust, you register it with your conscious brain only. But if the person's expression changes to intense disgust, your emotional brain kicks in – effectively causing you to *experience* disgust, as well as recognizing it in the other person.[7]

Certain physical gestures – the Gallic shrug of contempt, the thrust-out pelvis signifying aggression, the drooping shoulders of resignation – seem to be processed by the brain in a similar way to facial and tonal expressions. Clowns, mime artists and cartoonists exaggerate these gestures and when they do it well the messages they send via body posture and movement speak volumes. In everyday life, though, people tend to damp down their body language. Some people are exquisitely tuned in to it, however. In the words of one such person:

'I sometimes find that the shift of a leg or the swing of a hip hits me like a frown or a giggle. When that happens the emotional response is very fast and subconscious. I will unexpectedly find myself echoing an emotion I know in reality the other person never meant to send. Smiles are the most common emotions for me to receive in that way, but occasionally I get others. About once a month I'll get a hearty laugh and of course, in a knee-jerk fashion, I return it. If someone asks about it (which seldom happens), I just say: "Oh, sorry. For a moment there my mind was somewhere else."'[8]

Some people are so sensitive to emotional expression that they are effectively equipped with a built-in lie detector. Facial expression systems based on objective measurements of muscle tone are at best 85 per cent accurate at detecting when a person is lying,[9] but a few people can spot when someone is fibbing every time.

Most of us, though, are surprisingly poor at detecting dishonesty from facial expression. In one experiment a group of nurses was shown a film in which people were shown suffering horrendous injuries. Another group was shown a pleasant film. Both groups were then interviewed about what they had seen and how they felt about it. The group that had seen the pleasant film was instructed to answer all questions truthfully. The other nurses were instructed to smile brightly and pretend that they had seen the pleasant film. They were told that this was a test to see if they could keep up a nothing-to-worry-about façade in the event of a real disaster occurring.

The nurses' interrogators included psychologists, judges, detectives, customs officers, secret service agents, non-specialist adults and college students. The only group that showed a significant ability to tell those who were lying from those who were not were the secret service agents, and they mainly used techniques – cleverly constructed questions and so on – that did not depend on judging facial expression.[10]

Given the sophistication of our emotional expression system, it seems odd at first sight that we should be so bad at interpreting it. One possibility is that our ability to overlook or misread emotions is actually the most sophisticated part of it, in that it allows us to rub along together and enjoy life – laughing at jokes we have heard before; swallowing convenient white lies; and finding TV soap operas credible despite the poor acting.

Another possibility is that we use expressions to generate feelings as much as to display them, and that a little bit of deception is necessary for this to work.[11] If, for example, you contract your forehead into a frown, the nerves enervating the muscles feed back a message to the brain that effectively says: 'Something wrong here – we are worried.' This sows a seed of real worry – a feeling that in turn feeds back to the frown muscles and intensifies the expression. The feedback loop then sends a stronger message back to the brain: 'Things getting worse!' At this point the seed may sprout into a full flood of anxiety.

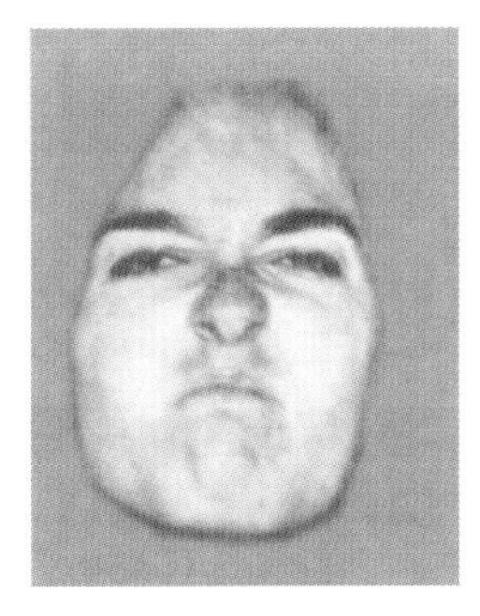
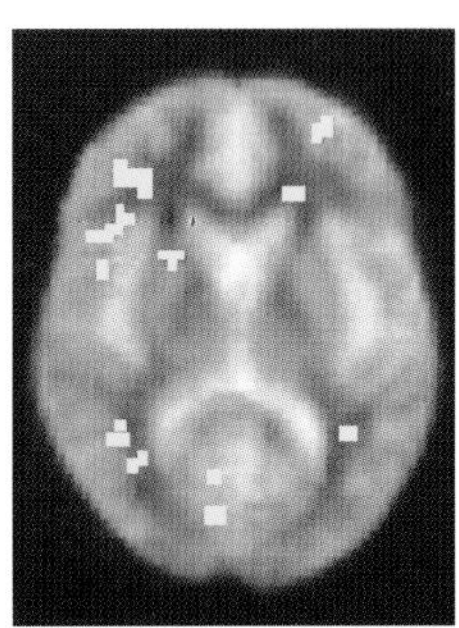

Looking at this picture (left, above) of someone expressing extreme disgust triggered brain activity (right) in the viewer in the same area that lights up when disgust is experienced directly. The more intense the expression, the more the brain reacts.

Once the feeling is there, the brain then casts around for a reason for it. There is always a reason for worry if you look hard enough, and once you have found one it then fuels the feeling of concern even more, creating an ever-escalating emotional spiral. Behaviour therapy – one of the more effective forms of psychotherapy – teaches people to exploit this feedback mechanism in order to turn negative feelings, like worry, into positive ones simply by replacing the frown with a smile.

Expressions can also transmit emotions to others – the sight of a person showing intense disgust, remember, turns on the observer's brain areas that are associated with feeling disgust. Similarly, if you smile, the world does indeed smile with you (up to a point). Experiments in which tiny sensors were attached to the 'smile' muscles of people looking at faces show that the sight of another person smiling triggers automatic mimicry – albeit so slight that it may not be visible. This tiny muscular twitch may be enough to trigger the feedback mechanism, so the brain concludes that something good is happening out there and creates a feeling of pleasure.

This, presumably, is why people serving hamburgers are instructed to smile. Whether

Just talking to someone can be an ordeal for people who have difficulty spotting facial and tonal expression. This is how one such man describes it:

'I have learnt to watch the mouth of the person I am talking to and to note when they show their teeth. That is how I know they are smiling. When I see their teeth I try to remember to smile back. I also look at their eyes. When people smile their eyes crinkle around the edges.

'The trouble is, by the time I have worked it out there is a time-lag. The conversation has moved on and I always return the smile that bit too late. People find this upsetting – I think it must seem to them as though I am not really connecting.

'Normal people must exchange an enormous amount of information through their expressions. I have worked this out because I know I miss out on a lot when I listen in to other people's conversations – things get transmitted without words. It makes you feel left out.

'But the worst thing is that a lot of the time people don't take me seriously. If you can't see other people's expressions, you never learn to do them yourself, so – unless I make a really big effort – I don't put out expressions with my own face. Because of this people don't think I mean what I say. Sometimes they even think I am lying.

'I used to find this very hurtful. Now I make sure to emphasise what I say with extra words. If someone gives me something to eat and I want to tell them I'm enjoying it, I make sure to say something particular about it, like "The herbs in this are really interesting," rather than just "This is good." If I want to show someone I am angry, I might swear at them. I don't like doing this but sometimes it is the only way to get the message across and now it has become a habit.

'Because of all this I find dealing with people is hard work. Sometimes I feel too tired to try and I shut myself away for a while. It can be lonely.[23]

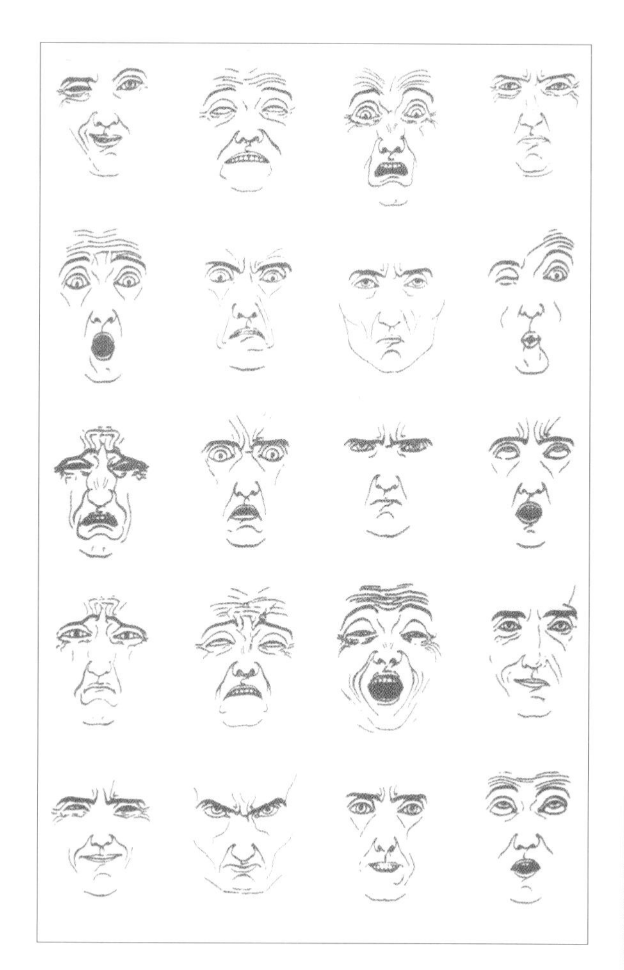

Human faces can express a vast range of emotions. Expression from the Art of Pantomime *by Charles Aubrey, 1927*

Mona Lisa's smile (top left) is a gentle Duchenne variety. It loses some of its mystery when it is morphed (top right) to show an exaggerated Duchenne smile or (bottom) into a typical 'social' smile.

activating their major zygomatic muscles all day long keeps them in the state of perpetual cheeriness that might be expected from the foregoing studies has yet to be determined.

Anger

Emotional expression is not always as benign as a smile in a fast food joint. Rage and fear bring about a wide range of reactions – road rage, bullying, phobias – that have no survival value in today's world and serve only to make it a bleaker place in which to live.

Just like smiling, these reactions may either be consciously connected with some appropriate outside cue, or they may be brought about by a spasm of activity in the unconscious mind over which the person may have little or no control. It may even be possible to murder someone as a result of a such a reflex.

One in three killers claims to remember nothing about the moment when they committed the crime. Patrick – a case reported by US neurologist Richard Restak – is typical. This forty-two-year-old man shot his wife after sixteen years of reasonably happy marriage in what appears from the outside to be a fit of jealous rage. Yet Patrick claims to remember nothing of the episode: just a 'rushing-numb-out-of-control' feeling, followed by a blank space and then the sight of a dead body and a smoking gun.

There are several ways of viewing the selective forgetfulness of killers. The psychoanalytic approach holds that knowledge of one's appalling deeds is too much for the ego to bear and is thus repressed. The cynical view is that their amnesia is a transparent attempt to secure a soft sentence. And the latest (and most controversial) idea is that these people do indeed not remember their crimes because, effectively, they were absent when they were committed.

Can it really be possible for an unconscious person to pull a gun, do whatever needs to be done to prime it, take aim and shoot – all while the victim presumably shouts and screams and flails? If you are to believe the perpetrators the answer is yes. Some even claim to have carried out long and seemingly calculated assaults in this state, including rape. Recent research into the neurobiology of what is probably our most powerful emotion – anger – suggests that some of them, at least, are telling the truth.

The amygdala, as we have seen, is the brain's alarm system – the central generator of states of mind that evolved to aid survival under threat. Stimulate one part of the amygdala and you get the typical fear reaction – a feeling of panic combined with flight. Stimulate another and you produce what people have described as a 'warm, floaty feeling' and excessively friendly behaviour – appeasement. Activity in a third region of the amygdala results in outbursts of rage.

Packing the trigger mechanisms for all three basic survival strategies – flight, fight and appeasement – into a single small nugget of tissue is advantageous in that it allows swift transition from one to another. If a bully is not seen off by a smile (or bottom-baring, if you are a monkey) it takes only a tiny amplification of activity in the amygdala to trigger flight. And if flight is impossible the resulting uplift in activity provides the thrust, coupled with the subjective feeling of anger, to attack.

The drawback is that we live in a world where physical flight or fight is likely to be more catastrophic than the original threat. If you are facing an aggressive boss in a boardroom, for example, appeasement is the only non-disastrous option and even that may have its downside (there is, as we all know, a particularly derogative word for those who are excessively conciliatory towards their superiors). It is therefore essential that the emotional responses generated by the amygdala are mediated by the 'thinking' part of the brain, the cortex.

Controlling emotions is effectively the reverse of the process required to feel them. The amygdala receives emotional stimuli first via what Joseph LeDoux has termed 'the quick and dirty route': a fast track that produces an almost instantaneous automatic response – smile, jump back or lunge forward – is begun. A quarter of a second later, however, the information reaches the frontal cortex where it is placed in context and a rational plan of action is conceived to cope with it. If good sense dictates that one of the three basic survival strategies is in fact appropriate, the bodily reaction already begun will be continued. But if the rational decision is to respond verbally rather than physically, the cortex sends a 'damp things down' message to the hypothalamus, which in turn signals the body to halt or reverse the changes it has started to make. This lowering of bodily arousal is in turn sensed by the hypothalamus via a loop-back system, and the hypothalamus then sends inhibitory messages to the amygdala, calming activity there, too.

In this way, emotions are held in check by the 'higher' functions of the brain and in most people the mechanism works pretty well. So why is it that a minority is given to apparently uncontrollable outbursts of rage?

There are two fairly obvious ways in which emotional control might break down. One is if the signals sent from the cortex to the limbic system are too weak or undirected to override the activity arising from the amygdala. The other is if the amygdala is activated in the absence of any outside stimulus that would simultaneously arouse the cortex.

Of these the first is commonplace. It is the relative weakness and diffusion of cortical signals that cause children to have far more emotional outbursts than adults. Infants cannot control their emotions because the axons that carry signals from the cortex to the limbic system have yet to grow. And the cells in the prefrontal lobe, where rational processing of emotion takes place, do not mature fully until adulthood. The amygdala, by contrast, is more or less mature at birth and thus capable of full activity. The young brain is therefore essentially unbalanced – the immature cortex no match for the powerful amygdala.

Cortical maturity can be accelerated by use – children who are encouraged to exhibit self-control are likely to become more emotionally continent than those whose tantrums are allowed full rein. This is because, in general, constantly stimulating a particular group of brain cells (like those required to inhibit the amygdala) makes them more sensitive and thus more easy to activate in future. It is rather like keeping a TV set on standby. By the same token, children who rarely activate the emotional control centre in their brains may grow up to be poorly controlled adults because the necessary brain equipment was not nourished during the most critical stage of development. One of the saddest demonstrations of this concerns the children from Romanian orphanages who were adopted by Western families in the late 1980s. Until they were adopted the children had been given little or no stimulation, no individual attention from adults and nothing that could be construed as normal love. Despite being doted on in their adoptive homes, many of them have grown up with profound social and emotional problems. One mother, speaking of her ten-year-old adopted daughter, says:

'Nicola has no idea what love is. We have treated her exactly like our other children, who are normal and affectionate, but she has just never got the idea. She doesn't seem to connect to us any more than to anyone else – she will go and sit on a stranger's knee when she wants attention just as soon as come to one of us. She's quite intelligent, but she can't learn to show concern for other people. She never flushes the lavatory when she has used it, for example. We tell and tell her but she just can't be bothered. It is not that she wants to upset us – it is more that she doesn't seem to take on board that we exist.'[12]

Harry Chugani, of the Children's Hospital in Michigan, has scanned the brains of some of these children and found that nearly all of them show distinct functional eccentricities in various areas connected with emotion. 'There is just a short window of time during development when children have to be emotionally stimulated if they are to feel those emotions in later life,' says Chugani. 'These children missed out on that, and their brains carry the evidence.'[13]

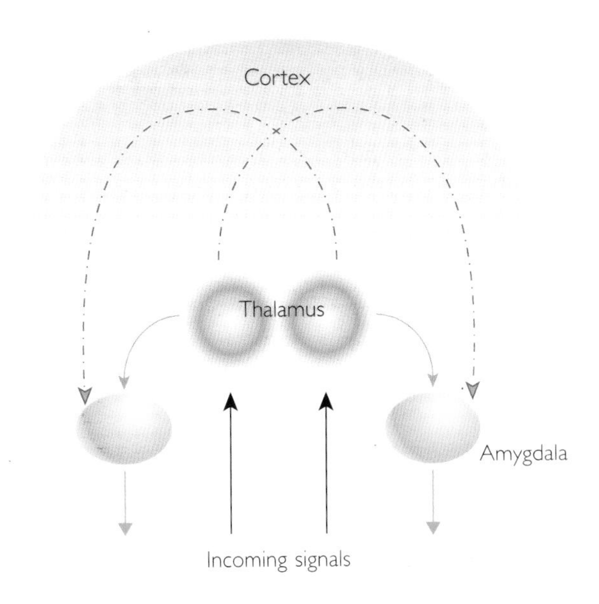

Emotional information is sent to the conscious brain and the amygdala via two routes. The path to the amygdala is shorter, so emotional reactions are faster than conscious ones.

Damage to the emotional cortex may also reduce its ability to inhibit activity in the amygdala. As we have seen, activity in this area may produce three distinct types of reaction: appeasement (that nervous, over-anxious friendliness that most of us recognize as insincere), fear or rage. Small irritative lesions in the area have been found to produce anger and aggressive behaviour and a number of studies of murderers have found evidence of brain damage and dysfunction. A brain scanning study of forty-one convicted murderers (thirty-nine men and two women) found that the majority showed reduced frontal lobe activity, which, as we shall see in Chapter Eight, may severely compromise a person's ability to control their impulses.[14] The value of studies like these is questionable because there is still so much to be learnt about the practical implications of brain dysfunction. But they certainly suggest that a brain with weak cortical activity in the frontal lobes is one that is more than usually subject to rage.

Frontal lobe dysfunction does not, however, explain what happened to Patrick. What mechanism could cause a person to carry out a sustained act of violence and then have no memory of it?

One possibility is that such a person's actions are the result of epileptic seizures. The amygdala is a particularly responsive part of the brain and very little electrical stimulus is required to make its cells fire. This gives it a low threshold for epilepsy – fits often begin there and radiate outwards. The stirring of activity in the amygdala prior to such a fit is probably what causes the feelings of fear and foreboding so often reported by epileptics just before they black out.

In his book *The Mind Machine*, Colin Blakemore, Waynflete Professor of Physiology at Oxford, reports the case of Julie, a twenty-year-old woman who developed panic attacks and curious dream-like interludes in which she did not know what she was doing. 'This strange feeling would come over me,' she said. 'Stranger and stranger than hell. A frightening feeling...you had no control over how your body reacted.' One day, in such a state, she knifed another woman through the heart. Julie was subsequently studied by Boston neurosurgeon Vincent Mark. He inserted electrodes through Julie's scalp and deep into her brain, so that they terminated in and around the amygdala. Then he sent a small electrical current into each one. When the stimulation was directed at one spot, and one spot only, Julie suddenly started to thrash around and beat the wall in a semblance of wild anger. As soon as the stimulation ended she was back to normal, with no recollection of what had happened. Mark identified the spot as the basolateral nucleus of the amygdala. He burnt it out, and Julie's rages disappeared. The inference is that they had been caused by small, brief seizures focused on just one part of the amygdala.

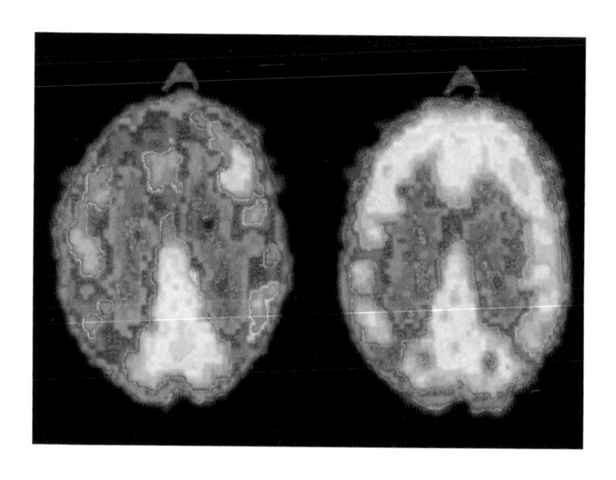

The brain of this murderer (above, left) shows significant lack of activity compared to a normal brain. Studies show this is typical of violent criminals. (From Raine et al "Selective reductions in prefrontal glucose metabolism in murderers "Biological Psychiatry Vol 36, September 1st, 1944)

If acts of 'mindless' violence really are that – no more conscious than the kick of a leg when the knee is tapped – it seems pointless, as well as unfair, to punish the perpetrators. In the absence of any better way to deal with them it might be justified, but if brain mapping fulfils its promise of revealing exactly what is going on in rage-driven brains, it should also be able to suggest more appropriate means of dealing with them.

Fear

Phobias are among the most life-restricting of conditions. It is not so bad if your particular fear is something you can be sure to avoid: flying, for example, may limit your choice of job and holidays but you can live a perfectly full life without having to confront it. Others are more problematic.

Josephine, for example, is scared of chicken legs. In order safely to attend any social function involving food she has first to convey this curious item of information to her hosts. On the one occasion when the message failed to get through Josephine was confronted unexpectedly with a plate of *coq au vin*, which produced a reaction of such dramatic intensity that both she and her host ended up in the emergency department of the local hospital. Josephine rarely dines out nowadays.

How do phobias like this arise? And why are they so very hard to control?

The potential for certain fears seems to be hard-wired into the brain, like a faint memory trace of those things that have proved harmful to our species in the distant evolutionary past. Animal studies and careful observation of human babies show an instinctive shrinking from certain stimuli. This reaction does not necessarily occur the first time the object is encountered, but if during that first experience there is the slightest hint that the object might be something to worry about, a profound and permanent fear of it will be forged.

Baby monkeys born in captivity will not show automatic fear of a snake, for example. But if a snake is presented to them and at the same time they are shown a film of another monkey looking fearful, the snake will become an object of terror. This does not happen if the object that is presented is, say, a flower. Fear of snakes, it seems, is pre-programmed into primates' brains but may lie dormant until an appropriate signal triggers it into life.[15] The most common phobic objects are those that once posed the greatest danger to the species. Among humans these are snakes, spiders, big swooping birds, dogs, heights and reptiles. Their roots in our evolutionary past are obvious – phobias about things that are dangerous today, like cars and guns, are far less common.

This is not to say that the newborn brain contains a rogues' gallery of potentially fearsome objects. Rather it is equipped to respond to certain fairly crude stimuli: large objects above; a certain slithering movement on the ground, base notes like those in the growl of a dog and so on. Certain human postures and gestures also seem to be inherently recognizable as fearful: the characteristic configuration of a human body bent in the pain of a heart attack, for example,

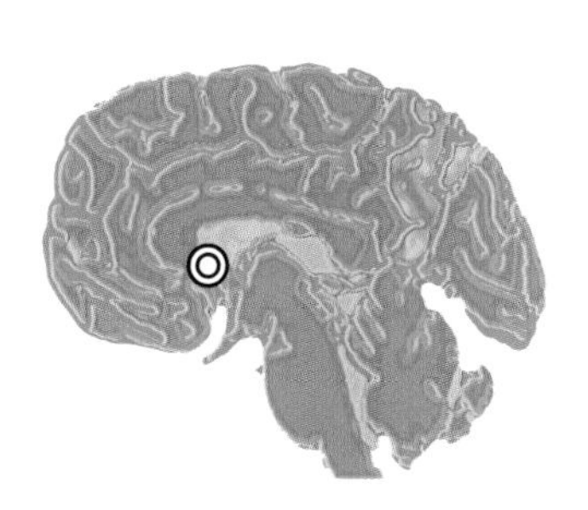

The amygdala is underactive in people with psychopathic tendencies

'Are you Married to a Psychopath?' asked a recent headline. The answer from the nation (mainly wives) was a resounding 'yes'. Thousands wrote to say their partners exhibited classic signs: emotional coldness, bullying, deceit, lack of remorse and love of risk. A more controlled study (published by Oxford Psychologists Press in 1996), found that almost one in six UK managers fulfilled the diagnostic criteria for psychopathy or, officially, Anti-social Personality Disorder.

Brain scans of psychopaths (Professor Robert Hare, University of British Columbia) suggest their behaviour may be partly due to an amygdala malfunction, particularly in the right hemisphere. A normal amygdala is activated by emotional stimuli. Psychopaths' amygdala show little response at the sight of another's distress. Some studies show they do not react to threat stimuli either. Scans show that psychopaths process emotional information unusually: in most people the right hemisphere lights up most in an emotional situation, but psychopathic brains are equally active in both hemispheres.

The inability to sense emotion in others and to generate it themselves leaves psychopaths immune to remorse and punishment. Some think this is due to brain damage; others that lack of maternal bonding may be responsible: close interaction between infants and mothers is necessary to stimulate and maintain normal function in the amygdala.

sends a chill to a person's heart even if they have never witnessed it before and have no conscious idea of what is going on.

Such 'natural' fears are not phobias, though. Once we recognize that the snake and the spider are harmless most of us can control our fear of them. Phobics cannot. Their fear is almost beyond conscious control. It bears no relation to the actual threat and may even create danger by preventing a person from acting sensibly. Someone with a phobia about heights, for example, might be literally paralysed with fear if required to climb down a ladder from the window of a burning building.

Phobias have no survival value. So what turns a fear into a phobia? Freud maintained that irrational fears arose because the feared object had come to symbolize something that really was frightening but that was for some reason too embarrassing or awful to acknowledge. One of his most famous case studies concerned little Hans, a boy who had been frightened of horses ever since he saw one fall down in the street. Freud concluded that Hans's fear came from an unconscious Oedipal complex – he secretly desired his mother but was terrified his father would castrate him for it, and this fear was displaced on to the horse.

Explanations such as this have now, at last, been shown to be ludicrously elaborate. Phobias can be produced by manipulating quite basic brain mechanisms – there is no need to involve sophisticated cognitive machinations like symbolism, guilt and covert desire.

The root of phobias lies in conditioning, the process demonstrated, famously, by the Russian physiologist Ivan Pavlov nearly a hundred years ago. Pavlov showed that dogs will salivate at the ringing of a bell once the sound of the bell has been associated in their minds with food. Countless Pavlovian-style experiments have since shown that fear can be induced by an unrelated but associated stimuli just as simply as salivation. The latest research, mainly from Joseph LeDoux, is uncovering the neural mechanisms that underlie conditioned fear. In doing so it is revealing the causes and pointing the way to new treatments for phobias, anxiety, panic and post-traumatic stress disorders.

Conditioned fear (as opposed to 'ordinary' fear, which has a rational basis) is a special type of memory. Unlike most memories, it does not need to be consciously recalled in order to have an effect. It does not even have to be registered consciously at the time it is laid down. To understand how this happens it is necessary to look at what happens to potentially fearful information once it enters the brain.

Everything coming in through the senses goes first to the thalamus where it is sorted and shunted onward to appropriate processing areas. In the case of emotional stimuli – the sight of a snake in the grass, for example – the information is split in two and sent on its way via two separate pathways. Both paths end up at the amygdala, the brain's alarm system and generator of emotional responses. The routes they take, however, are quite separate.

Pathway number one goes to the visual cortex at the back of the brain which analyses it and then sends on what it finds. At this stage it is just information – a long, thin, wriggly thing with patterns on its back, here, now. Next the recognition areas of the brain get to work on it, deciding what this long wriggly thing is. The information, now tagged as a snake, triggers the release of stored knowledge about snakes – animal/different types/dangerous? – from long-term memory. These elements are added together to create a message: 'Snake! Here, now, aargggh!' (or something to that effect.) This is then sent to the amygdala where it stirs the body to action.

As you can see, pathway number one is a long and winding one with several stops along the way.

Given the urgency of the situation, on its own it would be dangerously slow – a quick-response system is needed. This is provided by the second pathway to emerge from the thalamus. The thalamus is close to the amygdala and is linked by a thick band of neuronal tissue. The amygdala in turn is closely connected to the hypothalamus, which controls the body's fight or flight response. These connections form LeDoux's 'quick and dirty' route along which information can zap from eyes to body in milliseconds.[16]

Conditioned fear seems to be formed out of information that takes this short cut. Most memories are initially encoded by the tiny but extremely important nucleus in the limbic system called the hippocampus. This is where all recent conscious memories are stored, and where those of them that are destined to become permanent brain furniture are dispatched to long-term memory. This takes a long time to do – it may be three years or so before a memory is firmly lodged in the cortical long-term store. People who have suffered serious damage to the hippocampus (a rare event) cannot recall anything in their immediate past or lay down anything new – a terrible condition to be in, as we shall see later when we meet some people to whom this has happened.

However, the hippocampus does not seem to be responsible for all memory acquisition. There is a famous case study involving a woman who had such severe hippocampal damage that she was unable to remember anything or anyone for more than a few seconds. Thus every time she saw her doctor he had to introduce himself afresh. He usually did this by shaking her hand. One day he hid a pin in his hand and when the woman grasped it she received a nasty prick. She appeared to forget all about it within a few seconds, and at the next meeting she seemed to be as unaware of the doctor's identity as ever. However, this time when he held out his hand she refused to shake it. She couldn't explain why, she was just scared of doing it. Clearly, at some level the pinprick had made a lasting impression.

Recent research suggests that unconscious memories like this are stored in the amygdala – an area of the brain never before thought of as a memory store. LeDoux for one thinks the amygdala lays down unconscious memories in much the same way that the hippocampus lays down conscious ones. Similarly, when an event is recalled the hippocampal system will come up with conscious recollections while the amygdala-based system will produce a sort of physical reminiscence, reconstituting the body state – pumping heart, sweaty palms and so on – that arose with the original experience.

If a memory is burnt into the amygdala with enough force, it may be almost uncontainable, and trigger such dramatic bodily reactions that a person may re-experience the precipitating trauma, complete with full sensory replay. This condition, post-traumatic stress disorder, is quite clearly linked to a particular experience and so are most frightening memories. Sometimes, though, the amygdala-based unconscious memories flood in without the corresponding conscious recollections that could pin them to a specific event. The irrational fear felt then may be vague – a thin cloud of anxiety – or it may be sudden and intense – a panic attack. If the feeling is provoked by a conscious stimulus, it may show itself as a phobia.

Unconscious memories are particularly likely to be formed during stressful events because the hormones and neurotransmitters released at such times make the amygdala more excitable. They also affect the processing of conscious memories.

During a trauma attention is very narrowly focused and whatever happens to be the centre of attention – whether it is relevant or incidental – will be laid down as a particularly sharp 'flashbulb' memory. However, if the trauma is

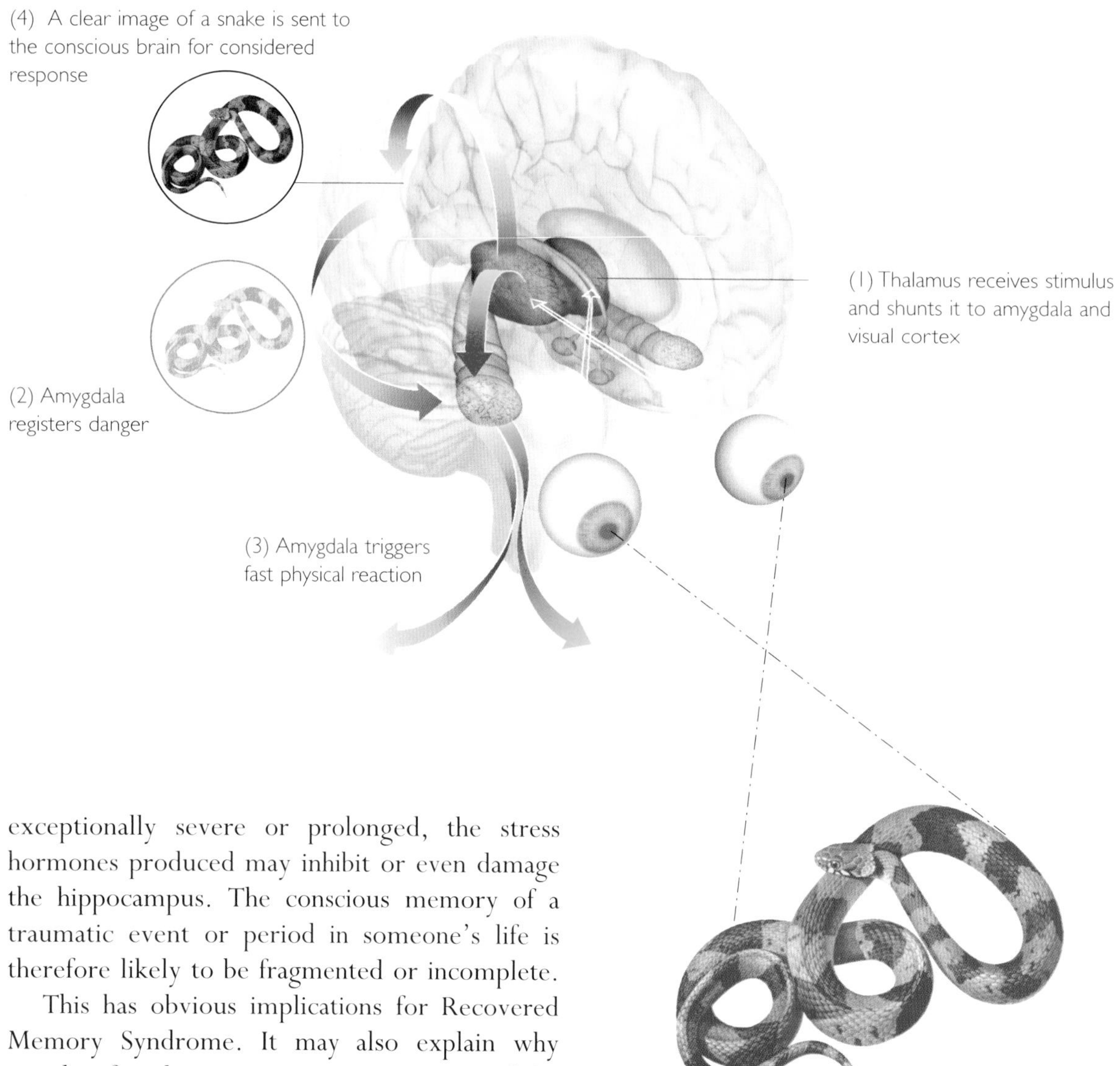

exceptionally severe or prolonged, the stress hormones produced may inhibit or even damage the hippocampus. The conscious memory of a traumatic event or period in someone's life is therefore likely to be fragmented or incomplete.

This has obvious implications for Recovered Memory Syndrome. It may also explain why people often have no conscious memory of the crucial part of some terrifying experience. Someone who is held up at gunpoint, for example, may recall precisely what the gun looked like but be quite unable to recall the robber's face. Later, however, they might find they have developed an aversion to beards, crooked noses or bright blue eyes – some characteristic that, although they do not know it, was exhibited by their tormentor.

LeDoux has shown that it is not necessary – in rats, at least – for a conditioned (fear-inducing) stimulus to be consciously registered at all. In one experiment he first repeatedly played a particular musical tone at the same time as giving a rat a mild shock. After a while the animal, true to Pavlovian conditioning, showed fear when the tone was played even if there was no accompanying shock. At that point LeDoux removed the animal's auditory cortex – the part of the brain that hears – while leaving the rest of its auditory mechanism, ears and so on, intact. An equivalent operation on a human would leave them without any conscious hearing whatsoever, and, to the extent that animals are conscious at all (itself a contentious issue), LeDoux's rat was left in a similar condition.

When it had recovered from its operation he played the tone to it again. The rat could not, presumably, hear anything. Yet it still exhibited fear. The noise appeared to register, unheard, in the thalamus and amygdala, creating an emotional reaction even though the animal probably did not know to what on earth it was reacting.[17]

It is easy to see from all this how apparently irrational fears and phobias may be created. It follows, too, that they will become more troublesome in times of stress, when the amygdala is excited by circulating stress hormones. This same overexcitability may explain why people with a phobia often develop other irrational fears when they are anxious or chronically stressed. LeDoux called the short-cut route to the amygdala dirty (as well as quick) because only crude information can travel along it. His deafened rat, for example, could not easily distinguish between the fear-inducing tone and a similar sound. In the same way, memories laid down in and recalled from the amygdala are likely to be less precise than those processed by the hippocampus, and one fear may easily bleed into another when stress hormones excite the amygdala to a frenzy.

Having seen how irrational fears may be created, how about eradicating them? Normal fears can be thought away if they turn out to be non-protective, but phobias are resistant to simple common sense because they do not involve the thinking part of the brain at all.

Recent research has, however, pointed towards the possibility of preventing fearful memories from taking root. Experiments with mice show that – in them at least – a particular brain protein is required in order for fearful memories to be retained. The protein, known unmemorably, as Ras-GRF, is made by a single gene. Researchers Riccardo Brambilla and Rudigar Klein, of the European Molecular Biology Laboratory in Heidelberg, bred a line of mice that did not have the Ras-GRF encoding gene. The mutant mice appeared to have quite normal brains, yet their behaviour proved to be very strange. Brambilla and Klein put the mice, along with normal animals, into a cage where they received a frightening shock. They were then removed. Half an hour later all the mice were invited to re-enter the cage. None of them – mutant or normal – would go in. The next day the researchers offered all the mice another opportunity to enter the torture chamber. The normal mice declined – they were clearly still very frightened. The mutant mice, however, marched right on in. The memory of the day before seemed to have vanished.[18]

Ras-GRF may not work the same way in humans, and there is certainly no likelihood of human gene manipulation aimed at preventing people from developing long-term fears. However, the fact that a single chemical has been found to play such a clear and important part in memory creation suggests that one day we may have drugs that could moderate or even eradicate memories that are too painful to bear.

Meanwhile, conditioned fears are notoriously difficult to extinguish. The time-honoured way to treat them is to force the person (or animal) to confront the feared object time and again until eventually a new association is made: feared object = safety, rather than feared object = danger. This, however, is a conscious association, and as such it is formed in the cortex (the middle of the prefrontal cortex, to be precise, an area directly behind the middle of the forehead). Cortically based notions can override amygdala-based tenets but they cannot eradicate them. For this reason, when stress hormones start circulating and the amygdala sparkles with activity, a person with a usually well-controlled phobia may succumb again to dreadful fear.

EMOTION — ICEBERG OF THE BRAIN

JOSEPH LEDOUX
Henry and Lucy Moses Professor of Science, New York University

In his book* The Emotional Brain *neuroscientist Joseph LeDoux describes how he has shown experimentally that emotions are survival mechanisms.

There is no such thing as an 'emotion' facility in the brain and no single system dedicated to this phantom function. If we want to understand the various phenomena we call emotion, we have to focus on specific types. Each system evolved to solve different problems that animals face and each has a separate neural basis. The system we use to defend against danger is different from the one used in procreation; the feelings we get when these two are activated – fear and sexual pleasure – do not have a common origin.

Brain systems that generate emotional behaviours are rooted deep in our evolutionary past. All animals, including people, have to do certain things to survive as a species. At the least they have to eat, defend themselves from danger, and reproduce. This is as true of insects as it is of fish and humans – and the neural systems that achieve these ends are fairly similar in all species with brains. This suggests that if we want to know what it is to be human, we should find out how we resemble animals as well as how we differ.

No one knows if animals are conscious and so no one knows whether they 'feel'. Certainly, it is not necessary for animals to have conscious feelings for emotional systems to do their job, and it is not necessary for the functioning of basic emotion systems in humans either.

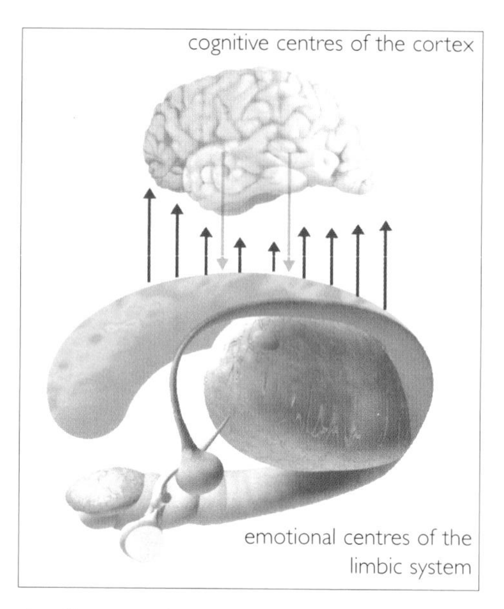

More neural traffic rises up from the limbic system than down from the cortex. This means the emotional part of our brain has more power to influence behaviour than the rational part.

Emotional responses are for the most part generated unconsciously. Freud was right on the mark when he described consciousness as the tip of the mental iceberg.

Conscious emotion is in a way a red herring: the feeling and behaviour it prompts are surface responses the initial mechanism orchestrates ... Emotions are things that happen to us rather than things we make happen. We try to manipulate our emotions all the time but all we are doing is arranging the outside world so it triggers certain emotions – we cannot control our reactions directly. Anyone who has tried to fake an emotion knows how futile it is. Our conscious control over emotions is weak, and feelings often push out thinking, whereas thinking fights a mainly losing battle to banish emotions. This is because the wiring of the brain favours emotion – the connections from the emotional systems to the cognitive systems are stronger than the connections that run the other way.

The darkest place

When Jennifer was at her most depressed she thought she was dead. Her doctors pointed out that her heart was still beating, her lungs were still pumping and her skin was still warm, but she was not much impressed. With the frustratingly circuitous logic of the truly deluded, Jennifer insisted that these things did not necessarily signify life because they manifested themselves in her, and she was dead.[19]

The earliest report of a person believing themselves to be dead dates from 1788 when the French physician Charles Bonnet told of an elderly lady who insisted on being dressed in her shroud and placed in a coffin. At first her daughter, who cared for her, resisted, but the woman was so insistent that eventually she got her way. Once in her coffin the woman fussed about with her shroud, complaining that it was not quite the right colour. Then she fell asleep. Her daughter and servants removed her from the coffin and put her to bed, but when she woke up the woman was furious at the move and demanded to be put back in the coffin and buried. Burial was refused, but the woman was returned to her coffin where she stayed, until, several weeks later, the feeling (though not the lady) passed away.[20]

The conviction that one has died is common enough to have a name: Cotard's delusion, after a French psychiatrist who collected a dozen or so cases. Modern analysis of those studies, and of more recent cases like Jennifer, suggests that they all have some brain damage (usually in the right temporal lobe) that disrupts their perception of the world so as to give it an air of unreality. This alone does not explain their delusion, however. Many people sustain this sort of brain damage but very few think they are corpses. A second factor seems to be required to bring about this particular delusion. One thing that nearly all the dead–alive have in common is a background of severe depression.

Clinical depression is singularly life-diminishing. Its symptoms of despair, guilt, exhaustion, anxiety, pain and cognitive retardation often make sufferers wish they were dead, and one in seven of those who are severely afflicted fulfil that desire through suicide. In extremis, perhaps, depression may extinguish the life-force to such an extent that it is easier to believe in one's own death, despite the self-evident absurdity of the notion, than it is to believe that life can consist of such misery.

Does our zest for life, and ultimately our belief in our own 'aliveness', have some physical basis in the chemistry or anatomy of our brains? The question would have seemed silly twenty years ago, but since then millions of people have been transported from death-inviting despondency to joyous rebirth via antidepressant drugs. The causal connection between brain activity and moods is no longer in serious doubt.

The feelings are different, too. Depression is more than just a mood – it also brings on physical symptoms like fatigue, pain, sleep and appetite disturbance. Memory is affected and thinking is slowed. Anxiety, irrational fear and agitation may be present and a depressed person

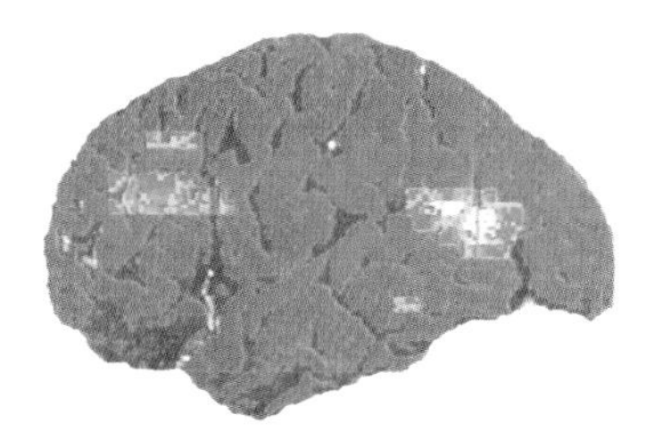

The brain of a depressed person (below) shows less activity than that of someone who is healthy (above).

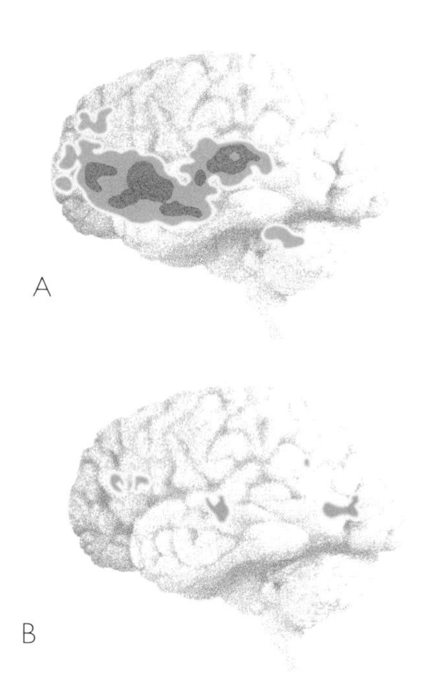

When asked to think of something sad women (A) generate more activity in their emotional brains than men (B). This suggests that women may have stronger emotional reactions to self-generated thoughts and memories.

typically feels guilty, worthless, unloved and unloving. Life seems pointless and the 'meaning' is missing from things. A favourite piece of music or a painting, for example, is no longer perceived of as beautiful or significant. It is just a familiar sequence of notes or a pattern of marks. At its very worst – like Jennifer – a person may feel as though they are dead.

Major depression is not a single disorder but a symptom of several different conditions, each of which probably has a slightly different brain abnormality at its root. The picture is still incomplete, but brain imaging studies are starting to reveal what they are.

The brains of people who are depressed are generally much less active than normal – there is simply less going on than there should be. This probably accounts for the general feeling of slowness, lethargy and lack of excitement that such people feel. Scans carried out by researchers at the Wellcome Department of Cognitive Neurology in London have revealed that people with depression show a similar pattern of brain activity as negative symptom schizophrenics (those who are withdrawn and apathetic rather than flamboyantly deranged). In both conditions parts of the frontal lobes were found to be drastically underactive. The area most notably affected was that which generates self-willed actions and which is thought to be responsible for the feeling of 'agency' that goes along with self-willed behaviour. This spot seems to create the essential feeling of 'aliveness' that we usually take for granted and it seems likely that it is this spot that is turned off in people who develop Cotard's delusion.

Other 'dead' areas in depressed people include parts of the parietal and upper temporal lobes that are associated with attention, and particularly with attending to what is going on in the outside world. This suggests that the depressed brain is turned inward, tuned in to its own thoughts rather than to what is going on around it. This may explain why depressed people are less reactive to outside stimuli and more 'bound up in their own problems'.[21]

Another notable area of underactivity, found by Wayne Drevets and colleagues at Washington University School of Medicine, lies right at the front of the brain, along the bottom inside edge of the central chasm that runs from the back to the front of the brain. This area is older, in evolutionary terms, than the outer surface of the brain, and is welded to the limbic structures beneath by thick neural connections. It is thus on the receiving end of dozens of pathways that run up from the unconscious brain as well as down from the cortical areas that process thoughts. A huge amount of traffic passes along these routes: urges, desires and wordless memories from beneath, and news of plans, ideas and fantasies from above.

Its location makes the cingulate cortex an excellent candidate for the brain's emotional control centre, which is what it seems to be. High activity in this area is associated with mania – a condition of overexcitement, euphoria and buoyant confidence – the very opposite of depression. One of the hallmarks of mania is an increased sense of meaningfulness. People in an advanced manic state see significance in every little thing and often think they have insight into some Grand Scheme in which each incident and thing is bound together in a mystical wholeness. The feeling of interconnectedness and enhanced significance is also experienced in paranoia, into which mania sometimes tips. Paranoia is one of the symptoms of certain types of schizophrenia, which in turn is associated with fluctuations in dopamine, the neurotransmitter that activates the prefrontal cortex. The finding of prefrontal abnormality in depression therefore fits well with other findings.[22]

Although the brains of depressed people are generally underactive, certain areas show overactivity. One is the outside edge of the prefrontal lobe. Michael Posner and Marcus Raichle also at Washington University School of Medicine, found that this is activated in normal people when they work on a memory test that involves pulling things out of long-term memory. It also lights up in normal people when they are asked to think about sad things that have happened to them. This suggests that its role is to hold long-term memories in consciousness. A second area is the amygdala, which, as we have seen, is responsible for negative feelings. A third is the upper middle part of the thalamus, which is known to stimulate the amygdala. And a fourth is the anterior cingulate cortex – another spot on the inside edge of the central chasm, but higher up and further back from the other. The anterior cingulate lights up when we concentrate on things, and it is particularly active when we register things that are generated inside our heads – like pain – rather than those that are outside.

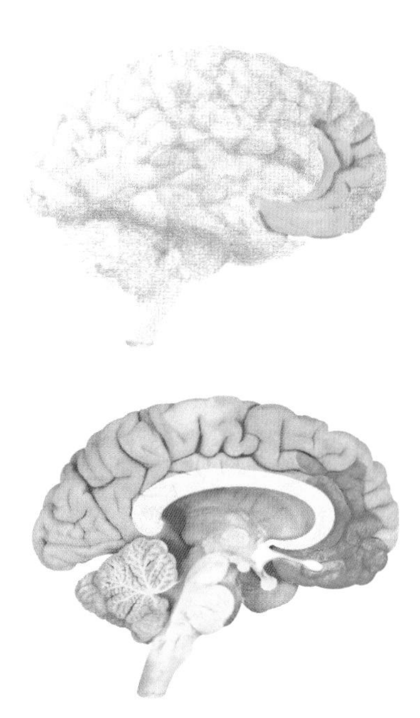

People who are depressed show abnormal patterns of activity in the areas of the frontal cortex coloured above. The side section of the frontal cortex – which is concerned with generating actions – is underactive while the middle part – which registers conscious emotion – is overactive. As a result the depressed person is without drive or desire to do anything, yet abnormally fixated on their intense emotional state.

These areas are all joined by neural pathways in such a way that when one is activated the others are stimulated, too. Posner therefore speculates that depression is caused by the firing of a circuit in which the amygdala feeds negative feelings to consciousness, the prefrontal lobe pulls out long-term memories that match the feeling, the anterior cingulate cortex fastens on to them and prevents attention from shifting to anything more uplifting, and the thalamus keeps the whole circuit alive and firing.

This interpretation of the brain imaging studies, if correct, explains many otherwise puzzling features about depression. The action of the anterior cingulate, in locking attention on miserable memories, explains why the usual remedies for unhappi-

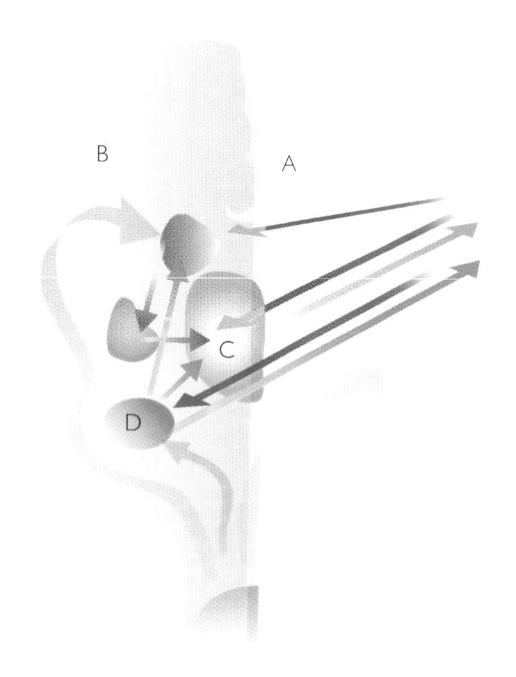

Some brain areas are overactive in depression – they seem to form a vicious circle of negative feeling.
(A) *Anterior cingulate cortex – locks attention on to sad feelings.* (B) *Lateral prefrontal lobe – holds sad memories in mind.* (C) *Middle Thalamus – stimulates Amygdala.* (D) *Amygdala – creates negative emotions.*

ness – that holiday, say – do not work. The lack of activation in the subgenual medial prefrontal cortex (what I have called the 'emotional control centre') explains the deadening lack of meaning and low mood brought on by depression. The involvement of the amygdala explains why sadness is so often experienced without any apparent foundation – the amygdala, as we have seen, does not convey concepts, it simply creates emotional feelings. It also explains why drugs that raise neurotransmitter levels work: they turn on the areas that should be on and turn off those that should not be on.

The brain imaging studies done by Posner and Raichle, combined with other studies, also help to explain why depression may follow an adverse life event (so-called reactive depression) or just come out of the blue. Drevets's study of people with familial depression – a sort that tends to occur spontaneously – shows that the subgenual medial prefrontal cortex is smaller in people in whom depression runs in the family – an architectural abnormality that might well prevent them from feeling the same degree of positive emotion as most people.

One thing it does not explain is why our brains should be constructed in such a way that the vicious neural circuit that creates depression should be so easily triggered. Its consequences – which include withdrawal from normal social intercourse and a high incidence of suicide – would, you might think, have long ago knocked out the genes responsible for making people susceptible to it.

It is conceivable that depression once had some survival value. Something that looks very much like depression occurs in animals trapped in nasty situations over which they have no control. And in the wild normally dominant animals may become subservient – and again appear to be depressed – if they are successfully challenged by a stronger adversary. In these cases depression may help the animals survive, either by conserving energy, in the first example, or, in the second, by forcing withdrawal from situations in which the animal is over-faced and therefore at risk. This might explain why today depression is so commonly triggered by events that undermine a person's self-belief – the human equivalent of losing control over one's circumstances or being beaten on one's own ground by a stronger adversary.

In today's world, though, depression is very rarely helpful when a person is stuck in a bad situation or confronted by an overwhelming challenge; it usually makes things worse. If it was ever a survival mechanism, it is now one that should be sloughed off along with full body hair and prehensile toes. Natural selection is a slow worker, but human ingenuity in inventing effective psychological and pharmaceutical therapies for depression is mercifully starting to overtake it.

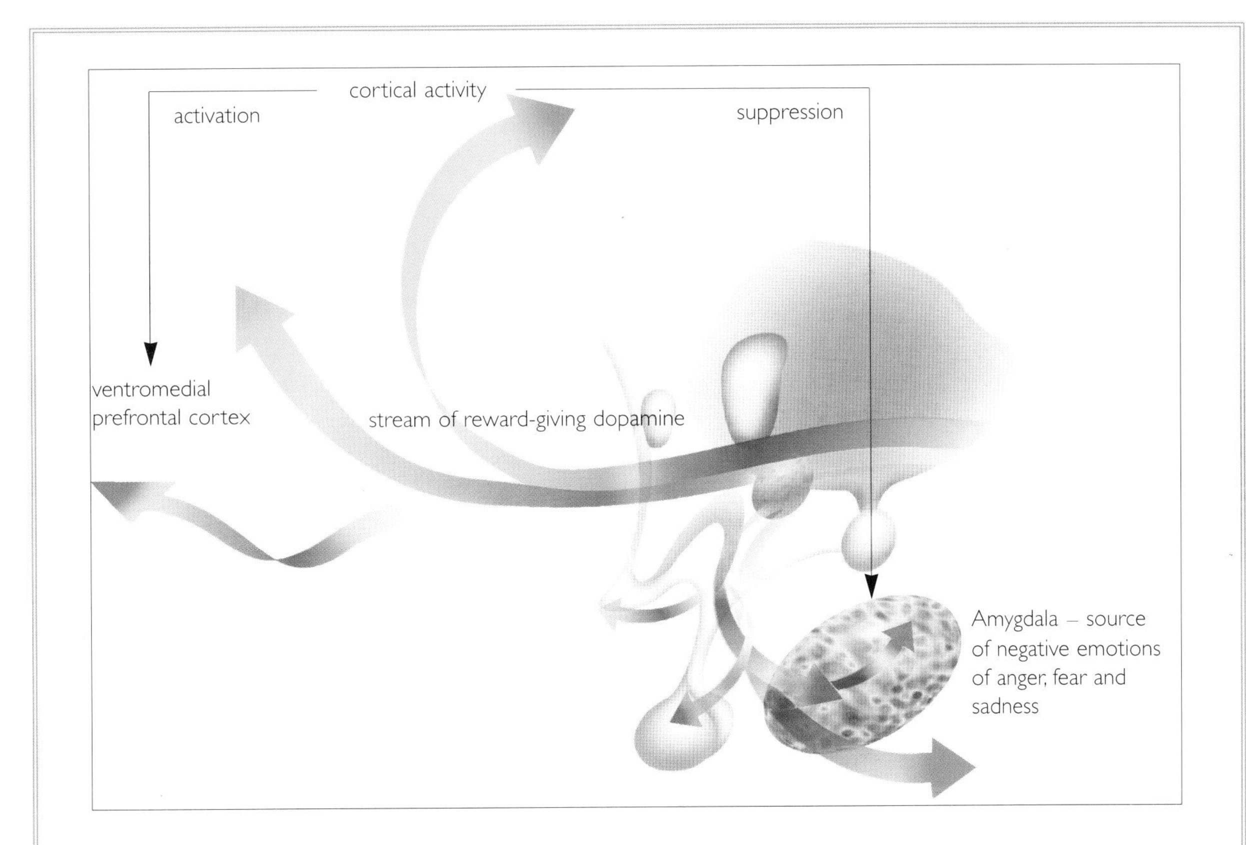

Anatomy of joy

Happiness is not a single, or a simple, state of mind. Its main components are:
* physical pleasure
* absence of negative emotion
* meaning.

Pleasure is the result of a rush of dopamine in the reward system. It can be brought on by a simple sensory or sexual thrill, or by a more complex route – the sight of someone you love, perhaps. It lasts, however, only as long as the neurotransmitters continue to flow.

Absence of negative emotion is essential for happiness because as soon as strong fear, anger or sadness enters, pleasure is reduced. The amygdala is responsible for generating negative emotions, so to prevent them flooding the brain this part of the limbic system must be quiet. Working hard on non-emotional mental tasks inhibits the amygdala, which is why keeping busy is often said to be the source of happiness.

Absence of sorrow and pleasure are still insufficient to create an all-pervading sense of well-being. For this, activity is required in the ventromedial area of the prefrontal cortex – one of the areas that is deadened in depression. The ventromedial cortex creates a feeling of cohesiveness – without it the world seems pointless and fragmentary. Overactivity in this area is associated with mania.[23]

The right hemisphere seems to be more sensitive to negative emotion, while high activity in the left hemisphere is associated with happiness.

KAY REDFIELD JAMISON
Professor of Psychiatry,
Johns Hopkins University School of Medicine

Author *Touched by Fire* (Free Press, 1996)

For years, scientists have documented a connection between mania, depression and creativity. In the late nineteenth and early twentieth centuries, researchers turned to accounts of mood disorders written by prominent artists, their physicians and friends. Their work strongly suggested that renowned writers, artists and composers and their first-degree relatives were more likely to experience mood disorders and to commit suicide than was the general population.
Alfred, Lord Tennyson, who experienced recurrent depressions and probably hypomanic spells, often expressed fears that he might inherit the 'taint of blood' in his family. His father, grandfather and two great-grandfathers, as well as five of his brothers, suffered from insanity, melancholia, uncontrollable rage or what is today known as manic-depressive illness. One brother was confined to an asylum for nearly sixty years. Lionel, one of Alfred's sons, displayed a mercurial temperament, as did one of his grandsons.
People with manic-depressive illness and

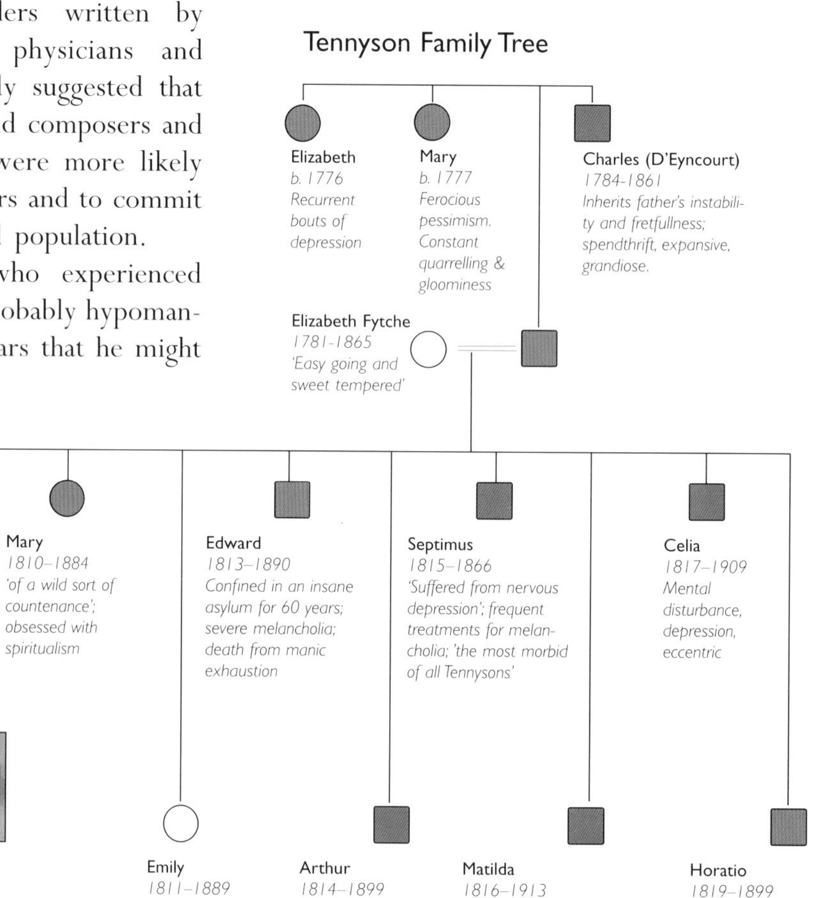

those who are creative share certain features: the ability to function well on a few hours' sleep, the focus to work intensively and an ability to experience depth and variety of emotions. Where depression questions, ruminates and hesitates, mania answers with vigour and certainty.

Robert Schumann's musical works, charted by year and opus number, show a striking relation between his moods and his productivity. He composed most when hypomanic and least when depressed. Both his parents were clinically depressed, and two other relatives committed suicide. Schumann himself attempted suicide twice and died in an insane asylum. One of his sons spent more than thirty years in a mental institution.

In the late 1980s, while on sabbatical in England, I began a study of forty-seven writers and artists: painters and sculptors were Associates of the Royal Academy; playwrights had won the New York Drama Critics Award, or the *Evening Standard* (London) Drama Award. Half the poets were in *The Oxford Book of 20th-Century Verse*. As against the 5 per cent of the general population who met the diagnostic criteria for a mood disorder, I found that 30 per cent of these artists and writers needed treatment. And 50 per cent of the poets – the largest fraction from any one group – had needed much extensive care.

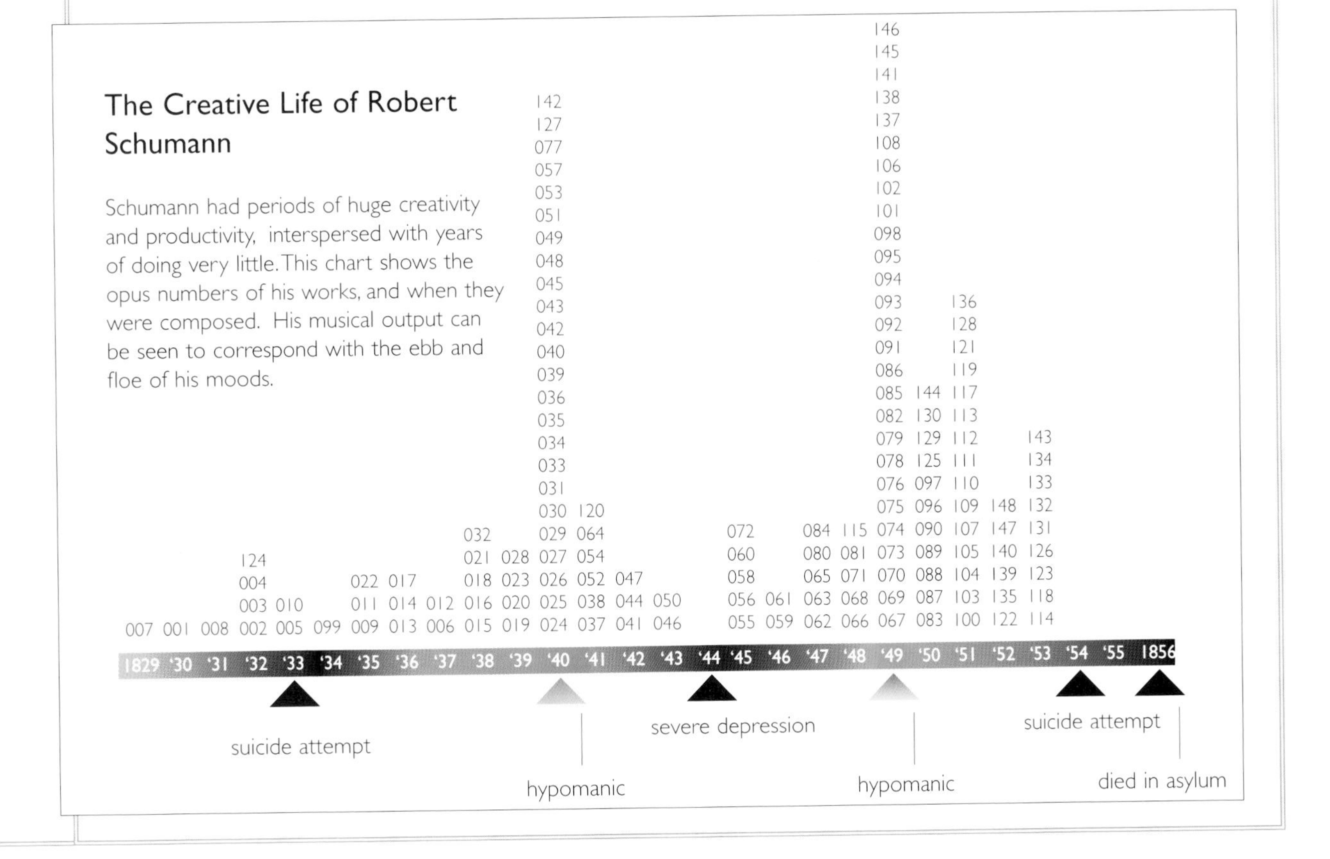

The Creative Life of Robert Schumann

Schumann had periods of huge creativity and productivity, interspersed with years of doing very little. This chart shows the opus numbers of his works, and when they were composed. His musical output can be seen to correspond with the ebb and floe of his moods.

CHAPTER FIVE
A World of One's Own

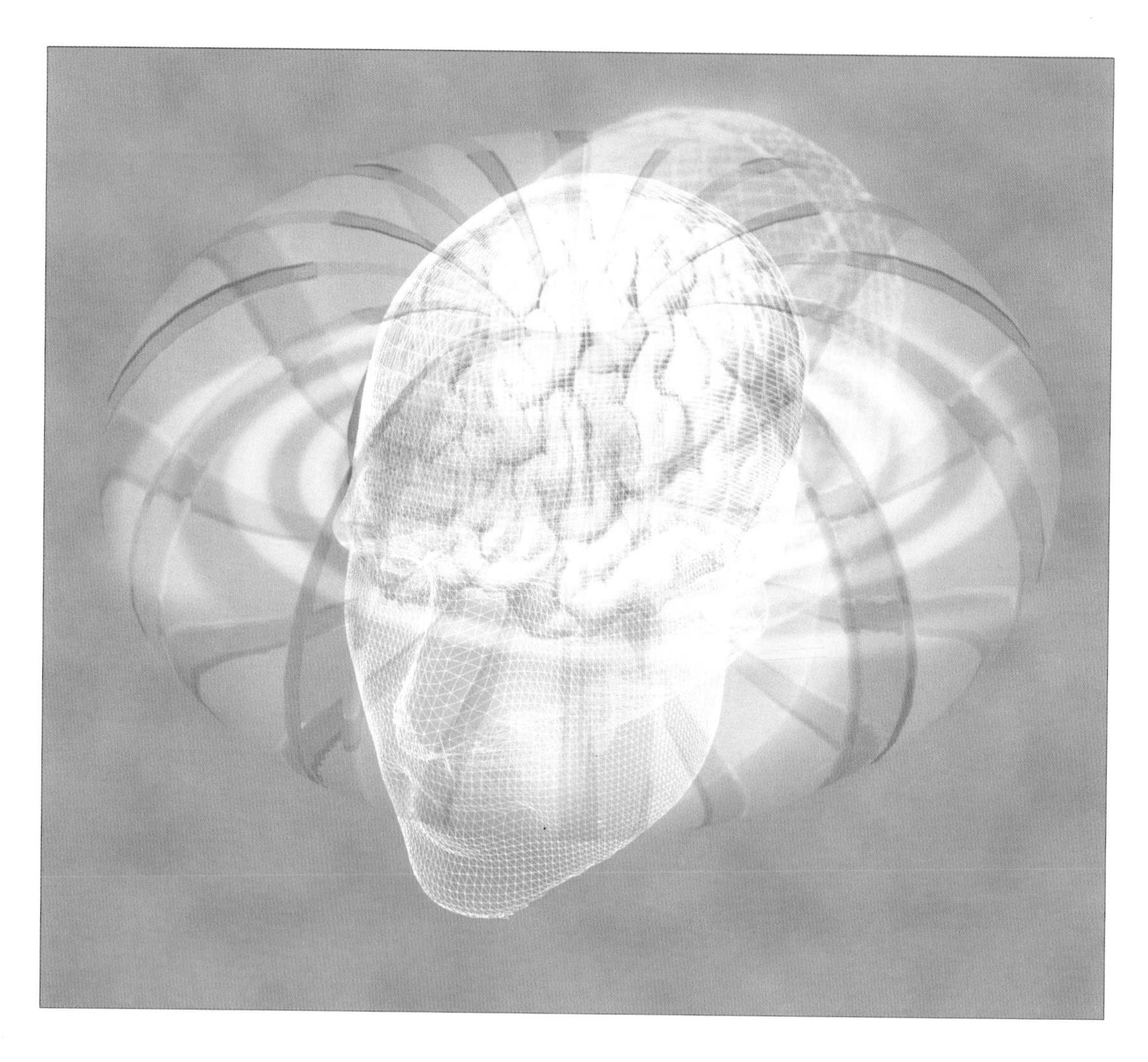

The brain is a factory with many products. Its raw material is information: the length of light waves hitting the retina; the duration of sound waves pulsing in the ear; the effect of a molecule in the olfactory canal. From this the sensory areas of the brain create an idea of what lies outside. But that basic perception is not the brain's finished product. The final construct is a perception that is invested with meaning. The meanings we attach to our perceptions are usually useful: they transform mere patterns of light into objects we can use, people we can love; places we can go. But sometimes they are misleading: the pool of water in the desert turns out to be a mirage; the axeman in the dark corner a mere shadow...

'YOUR NAME, RICHARD, tastes like a chocolate bar – warm and melting on my tongue.' It is the sort of thing one might say to a lover as an extravagant declaration of affection, but in fact the woman who said it was speaking literally. At the time, she was the subject of a research study, and the chocolate sensation produced by the name of Richard owed nothing to her feelings for neurologist Richard Cytowic, to whom the remark was addressed. Any Richard would have the same effect, because it is the *word* 'Richard' that tastes to her like chocolate – not figuratively but in the way that a cup of cocoa tastes to anyone else.

The woman has a condition – or a gift, as you prefer – called synaesthesia. It causes sensory perceptions – sound, sight, touch, taste and smell – to blend. Some synaesthetes see sounds; others smell sights. Practically every combination has been reported. One boy found words had distinct 'postures' that he was able to demonstrate by contorting his body into various poses. Another saw tastes: 'We'll have to wait a bit before we eat – the chicken hasn't got enough spikes on it yet,' he announced once to his bemused dinner guests.

Most people know what it is like to get a fleeting suggestion of colour or pattern when they hear a piece of music, and everyone can imagine how synaesthesia feels; our sensory similes – the 'soupy' sound of an oboe, the 'sharp' taste of lemon – all depend on drawing parallels between various senses. But only about one in 25,000 people consistently experiences two or more senses together.[1]

Synaesthesia is of more than idle interest because it undermines some of our most basic assumptions, both about sensory perception and about the nature of the external world.

What makes sound sound, vision vision and smell smell? Wave amplitude and molecular structure? Think again. If one person experiences the effect of light waves as music and another tastes chocolate in response to the sound waves made by a spoken name, who is to say that light waves create vision rather than taste or that molecules, rather than sound waves, create smells? All we seem to have by way of authority on the subject is a majority vote.

Nonetheless, in a standard issue, one-sense-at-a-time brain, a particular type of stimulus consistently registers as sound while another is experienced as vision. How come?

The obvious place to look for clues is at our sense organs – eyes, ears, nose, tongue and somatosensory receptors in the skin. Each one is intricately adapted to deal with its own type of stimulus: molecules, waves or vibrations. But the answer does not lie here, because despite their wonderful variety, each organ does essentially the same job: it translates its particular type of stimulus into electrical pulses. A pulse is a pulse is a pulse. It is not the colour red, or the first notes of Beethoven's Fifth – it is a bit of electrical energy. Indeed, rather than discriminating one type of sensory input from another, the sense organs actually make them more alike.

All sensory stimuli, then, enter the brain in more or less undifferentiated form as a stream of electrical pulses created by neurons firing, domino-fashion, along a certain route. This is all that happens. There is no reverse transformer that at some stage turns this electrical activity back into light waves or molecules. What makes one stream into vision and another into smell depends, rather, on *which* neurons are stimulated.

In normal brains incoming sensory stimuli follow well-worn neural paths from the sensory organ to specific brain destinations. As the stimulus passes through the brain it is split into several different streams which are processed in parallel by different brain modules. Some of these modules are in the cerebral cortex – the wrinkled outer grey skin where sights and sounds are put together and then made conscious. Others are in the limbic system where the stimuli generate the bodily reactions that give them an emotional quality – the thing that turns noise into music and a pattern of lines and contrasts into a thing of beauty.

The cortical area for each sense is made up of a patchwork of smaller regions, each of which deals with a specific facet of sensory perception. The visual cortex, for example, has separate areas for colour, movement, shape and so on. Once the incoming information has been assembled in these areas it is shunted forward to the large cortical regions known as association areas. Here the sensory perceptions are married with appropriate cognitive associations – the perception of a knife, for example, is joined with the concepts of stabbing, eating, slicing and so on. It is only at this stage that the incoming information becomes a fully fledged, meaningful perception. What we now have in mind was triggered by stimuli from the outside world, but it is not a faithful reflection of that world – rather it is a unique construction.

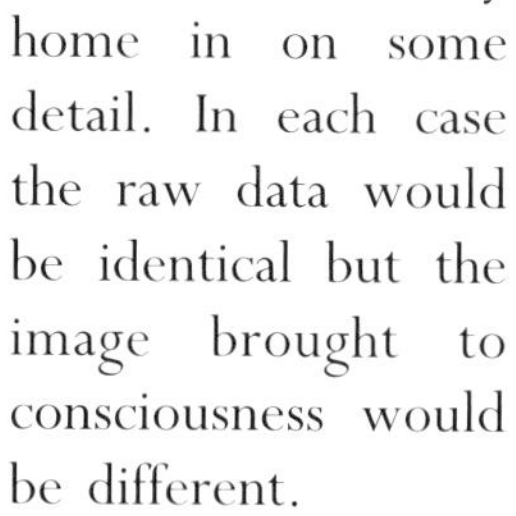

Every brain constructs the world in a slightly different way from any other because every brain is different. The sight of an external object will vary from person to person because no two people have precisely the same number of motion cells, magenta-sensitive cells, or straight line cells. For example, one person – someone with a particularly well-developed colour area (V4), say – may look at a bowl of fruit and be struck by the gleaming colours and the way they relate to each other. Another – with a more active depth discriminatory area (V2) – may be caught instead by the three-dimensional form of the display. A third may notice the outline. A fourth may home in on some detail. In each case the raw data would be identical but the image brought to consciousness would be different.

Sometimes a person with a particularly striking way of looking at things manages to convey it to others by representing their view in a work of art. That person's view may be seen as more beautiful than our own and by absorbing it – and perhaps stimulating our own visual pathways in such a way that they start to function more like the artist's – we may start to see things that way for ourselves. One reason that new works of art are often so shocking is that they present the world in a way that clashes with our own view. In time, when we, too, have learnt to look at things as if through the artist's eyes, the works become less startling.

An individual's view is formed both by their genes and by how their brain has been moulded by experience. Musicians, for example, have been found to have, on average, 25 per cent

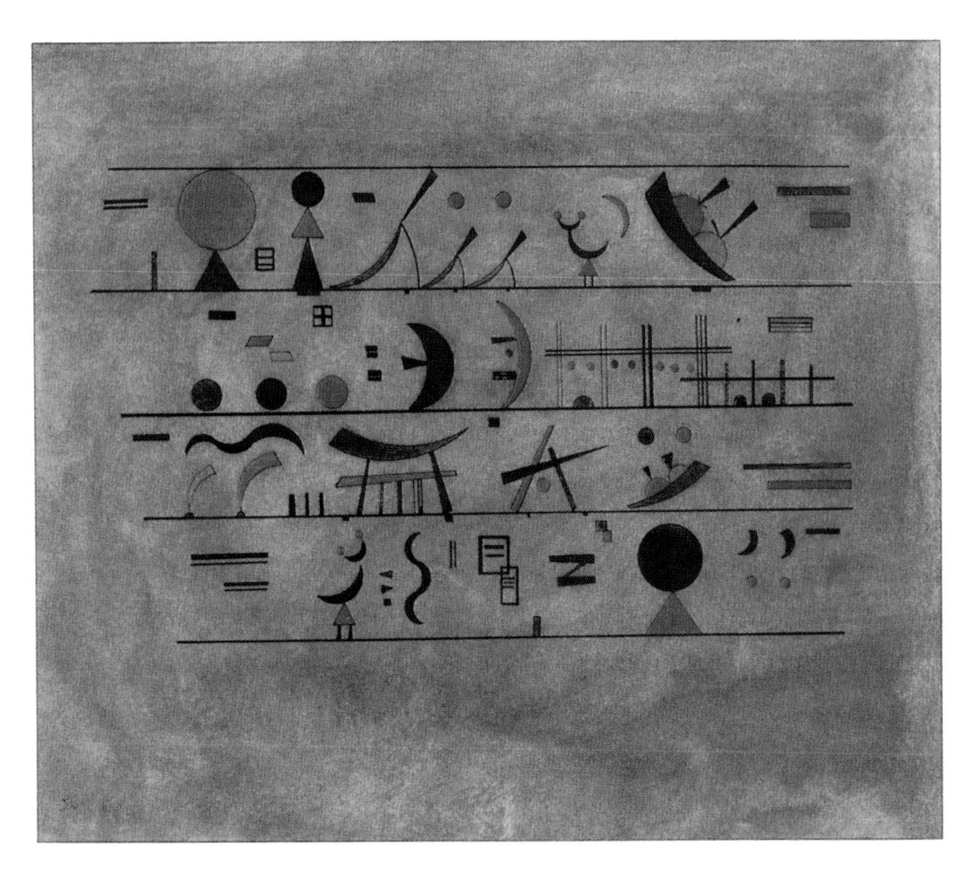

Right: *Rows of Signs by Wassily Kandinsky, 1931 (Photo: Giraudon). Artists have often tried to represent synaesthesia: Rimbaud assigned colours to the five vowels, translating impressions into visual ones and Whistler and Mondrian were among those who tried to paint sound. Wassily Kandinsky here places images on musical staves.*

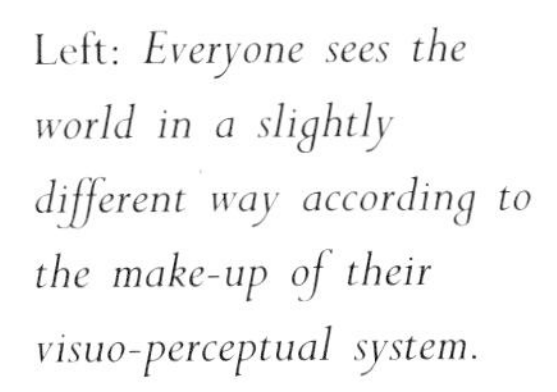

Left: *Everyone sees the world in a slightly different way according to the make-up of their visuo-perceptual system.*

more of the auditory cortex given over to musical processing than other people. The greatest amount of extra 'music' area is found in those who started to play earliest, suggesting that the difference is at least partly acquired by experience.[2] Animals that are deprived in infancy of the sight of one particular visual element – horizontal lines, say – are bad at or even incapable of discerning them in adulthood because the cells that would normally detect that visual component fail to develop if they are not stimulated at a particular time in infancy.[3]

Something similar happens with all our faculties. There is a story concerning Captain Cook that claims he came across a group of islanders who seemed to be unable to see the great hulk of his ship moored just a few yards off their shore. An object that vast had simply never entered their lives before, and thus they did not have the conceptual equipment to take it in. The tale is apocryphal but it reflects a real truth: there is no definitive picture of 'out there', only a construction in our heads triggered by the external elements we are best equipped to register.

Most differences between brains are too subtle to be clearly seen by imaging machines but the extraordinarily different sensory-processing styles of people with synaesthesia show up quite clearly. Brain imaging studies of a group of synaesthetes showed that when they listened to words their visual cortex lit up in addition to the brain area associated with hearing. Normal people only activated the auditory area.[4] The nearness of the active visual area to the auditory cortex suggests that in these people the experience of vision may be due to a knock-on effect – the neurons in the auditory cortex excite their neighbours, and this 'spillover' creates the visual effects.

However, another theory holds that synaesthesia reflects a more fundamental difference in sensory processing. Richard Cytowic has found that when people experience synaesthesia the overall level of activity in the cortex drops quite dramatically while that in the limbic system increases. Other scans show a similarity between this pattern of sensory processing and that seen in infants. This suggests that synaesthesia may be a type of prototype sensory perception that is done at subcortical level. According to this view, any stimulus – be it light waves, molecules or sound waves – has the potential to create a multi-sensual experience, and in the limbic system this is what they do. As babies we experience everything this way, but as our cortex develops it effectively hijacks the incoming information and subjects it to a ruthless categorizing, shunting each stimulus into one or another sensory modality only. It is the result of this selective process that we are conscious of rather than the holistic perception beneath.[6]

If this theory is correct, our strict categorization of information into its different sensory modalities probably developed in order to speed up our identification of incoming stimuli. If a wasp is experienced as a taste and a smell as well as a buzzing yellow thing, it must take that little bit longer to work out that you should swat it, so limiting perception of it to sight and sound may once have had survival value. Like many of our cortex-derived sophistications, however, it also impoverishes us slightly. Synaesthetes undoubtedly experience a richer sensory world than those of us who take our pleasures in one modality at a time. If, like them, we could switch off our cortical categorization process from time to time we might discover a richer sensual world by far. At the moment, unfortunately, the only reliable way to do it is by taking drugs, most of which are illegal. Perhaps when the mechanisms of synaesthesia are better understood, someone will invent a safer way of opening this particular door of perception.

A FINE CASE OF COLOURED HEARING

The writer Vladimir Nabokov describes himself as a 'fine case of coloured hearing' in his autobiography *Speak, Memory*:[5]
'The long "aaa" of the English alphabet has for me the tint of weathered wood, but a French "a" evokes polished ebony. This black group [of sound] includes hard "g" (vulcanized rubber); and "r" (a sooty rag being ripped). Oatmeal "n", noodle-limp "l", and the ivory-backed handmirror of "o" take care of the whites. I am puzzled by my French "on" which I see as the brimming tension-surface of alcohol in a small glass. Passing on to the blue group there is steely "x", thundercloud "z", and buckle-berry "k". Since a subtle interaction exists between sound and shape, I see "q" as browner than "k", while "s" is not the light blue of "c", but a curious mixture of azure and mother-of-pearl.'

Nabokov 'saw' the sound of each letter as a different colour or texture...

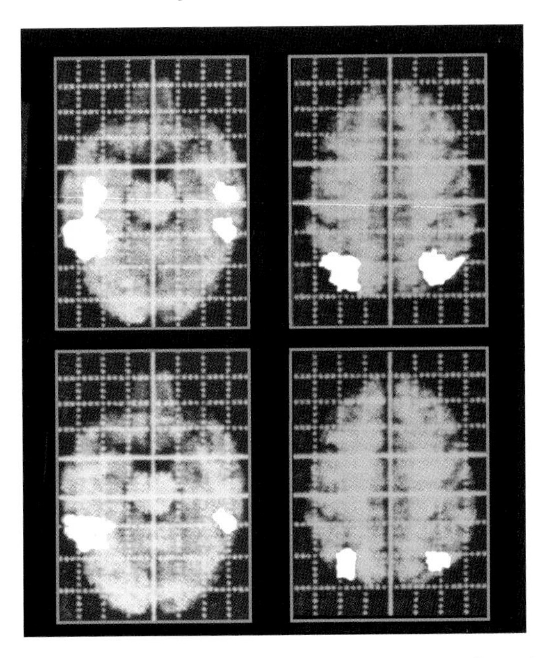

These are scans of neural activity in the brains of people with synaesthesia as they listen to words. Top left: *a large area of auditory cortex lights up, not just the language areas and* (top right): *a significant amount of visual cortex is active, too. The scans below show the extra extent of activity in these areas compared to that seen in normal people listening to the same words.*

Recognition

However well constructed a sensory perception may be it is meaningless until the brain recognizes it.

There are two distinct types of recognition: one is the inner 'ah-ha!', a cerebral snap of the fingers that happens when you hear a familiar piece of music or see someone you know. Getting a joke is a form of this type of recognition – a good punchline delivers a sudden jolt of recognition. So is coming up with a solution to some problem that you just 'know' is right. The penny drops. Got it. Eureka!

This type of recognition is quite different from the other – the conscious acknowledgement of a correct answer that you arrive at if, say, you add a string of figures together. You 'know' the sum of the figures because your conscious brain created the knowledge by a process of deduction. The deduction involves a sequence of rule-bound cognitive moves in which other knowledge – the decimal system, the rules of arithmetic, the way to use a pocket calculator – is brought to bear. It is a process, and those who are unpractised at it may fancy they can almost hear cogs and wheels in their brains creaking to turn out the answer. Others may do this type of task very easily. But (with the possible exception of *idiot savants*) everyone has to arrive at this sort of knowing by a conscious effort. Automatic recognition, by contrast, is instant, effortless and unavoidable.

The automatic type of recognition happens when one of the many parallel streams of incoming information passes through the limbic system. Modules here register the emotional content of the information, including familiarity. It happens so fast that the unconscious brain recognizes something is known to it before the conscious brain has even decided what the thing is.

This form of recognition does not extend to consciousness. It may make itself felt, at its strongest, as a vague feeling. But in order to *know* that you recognize something, and to say what it is, the conscious brain has to be brought into the process.

Conscious recognition takes place along the pathway that runs from the appropriate cortical sensory area to the association area that abuts it. Here the stimulus starts to take on identity. If you are looking at an object, say, the association areas in the lower temporal lobes begin the process of classification, starting with rough divisions like living/non-living, and perhaps human/non-human. The left temporal lobe then assigns a name to it. Meanwhile, higher up in the brain, in the parietal lobes, the object is fixed in space. If you are listening to something, a similar process – language/animal cry?; near/far? – takes place in the auditory association

SENSING THE WORLD

Vision

Light from a visual stimulus is inverted as it passes through the lens. It then hits the retina at the back of the eye, where light-sensitive cells turn it into a message of electrical pulses. These are carried along the optic nerve from each eye and cross over at the optic chiasma – a major anatomical landmark. The optic track then carries the information to the lateral geniculate body, part of the thalamus. This shunts it on to V1 at the back of the brain. The visual cortex is split into many areas, each processing an aspect of sight, such as colour, shape, size and so on.

Layout of visual cortex:
V1 – general scanning
V2 – stereo vision
V3 – depth and distance
V4 – colour
V5 – motion
V6 – determines objective (rather than relative) position of object
'Where?' path: V1-V2-V3-V5-V6
'What?' path: V1-V2-V4

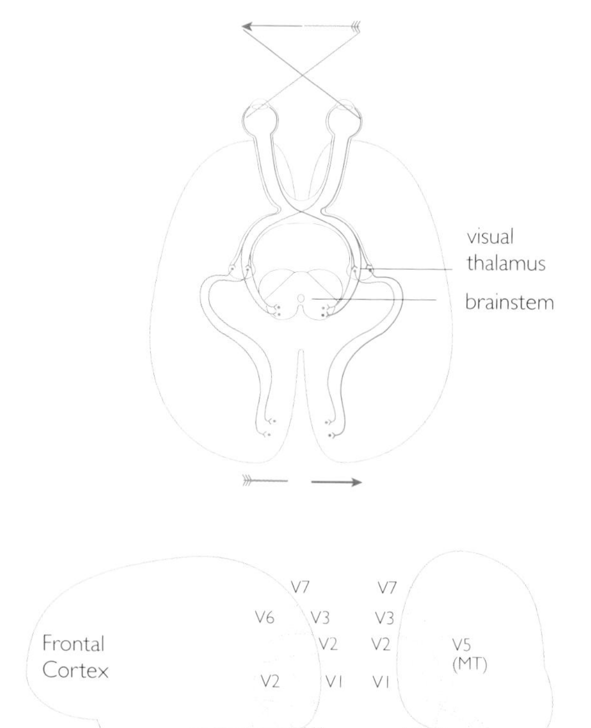

Above: *Awareness of sighted objects is conveyed to the limbic system but is not consciously visual.*
Below: *Each visual element is processed by a separate brain area.*

V1 mirrors the world outside in which each point in the external visual field matches a corresponding point on the V1 cortex. When a person stares at a simple pattern like a grating the image is reflected by a matching pattern of neuronal activity on the surface of the brain.

The 'map' is distorted, as the neurons responding to the central area of the visual field take up a much greater cortical area – so the 'picture' painted on V1 is a little like that seen through a fish-eye camera lens.

The centre of the retina, the fovea, is much more densely packed with neurons and sees far more detail. The eyes therefore dart around, in a series of leaps called saccades, in order to scan the visual field in detail. Saccades are triggered by the attention system of the brain and are not generally under conscious control.

You do not need eyes to see. Blind patients have been fitted with a device that turned

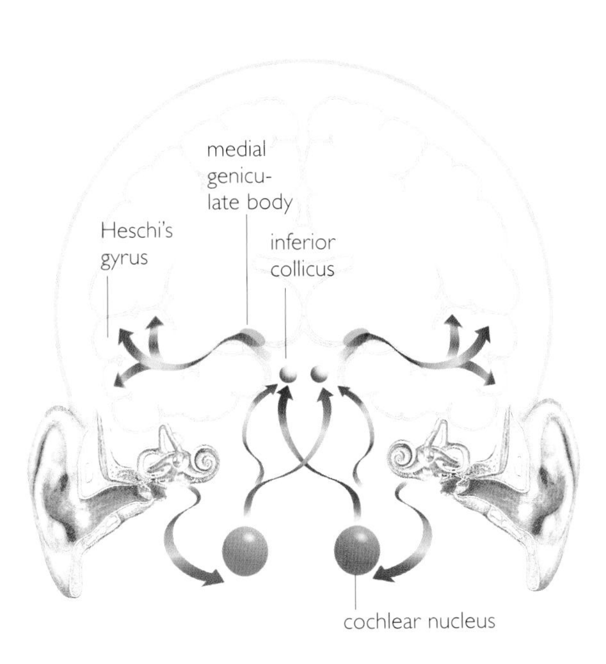

The neural pathways that convey sound information to different parts of the brain.

low level video pictures into vibrating pulses that could be 'read' tactilely, rather like Braille. A camera mounted next to the subjects' eyes, spread the pulses – which felt like a grid of tingles – over their backs, so they had continuous sensory input from the visual world. The patients soon started to behave as though they were 'really' seeing. They ceased to be aware of the tingles and their 'point of view' shifted to the camera. One of the devices had a zoom lens, and when an experimenter – without warning – operated the zoom, causing the image on the subject's back to expand suddenly as though the world was looming in, the subject ducked and raised his arms to protect his head.

However, there seemed to be a limit to the impact of visual information presented in this way. After the (male) subjects became practiced 'viewers', an erotic picture was projected – the subjects were able to describe it accurately but were unmoved by it.

Hearing

The neural pathways carrying sound information from each ear divide into unequal parts once they leave the ear.

On each side the broader path goes off to the brain hemisphere opposite the ear from which it came, so sound from each ear reaches both hemispheres – but most of the left ear's signals go to the right hemisphere and vice versa.

Both hemispheres have a distinct role in sound processing, and this means that sounds are dealt with (and therefore experienced) slightly differently according to which ear they enter. For example a person deaf in the right ear will receive most sound signals in the left auditory cortex (the side of the brain opposite the 'good' ear). This is the side that deals mainly with the identification and naming of sounds rather than their musical quality, so rhythm and melody perception may be blunted.

Conversely, a person deaf in the left ear may find that words are more difficult to distinguish than music, irrespective of loudness.

Smell

Flavour perception seems to be processed separately from either smell or taste. In one study students who learnt new words while sniffing an unusual smell and then sniffed the smell again when they had to recall the words showed a 20 per cent boost in memory power.

Whether we find a smell nice or nasty depends crucially on what memories we associated with it. For one person the smell of a bonfire may bring back happy memories of fireworks and winter barbecues, for another it may bring melancholy recollections of summer's end. The first person is likely to find the smell pleasant and the second to dislike it. Scanning studies suggest that pleasant odours (a) mainly light up the frontal lobes' smell area, particularly on the right-hand side. Unpleasant odours (b) activate the amygdala and the cortex in the temporal lobe (insula). Unlike other senses smell passes directly to the limbic system. This fast route to the brain's emotional centre gives smell its power to elicit strong emotional memories.

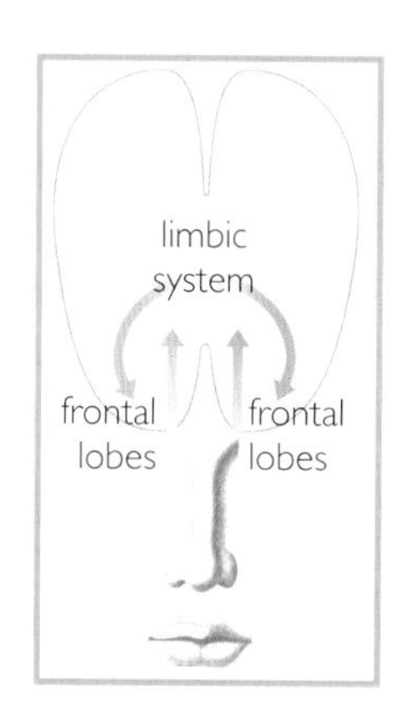

Smell is different from other senses because it goes straight to the limbic system – a fast route to the brain's emotional centre. Unlike other senses it does not cross from nostril to opposite hemisphere.

Taste

Damage to the frontal lobe of the right hemisphere may turn ordinarily hungry people into fanatic seekers of fine food. Gourmand syndrome has been identified by Swiss researchers who first suspected it when two of their patients developed foodie obsessions after sustaining brain injuries. The researchers subsequently scanned thirty-six gourmands: thirty-four of them had lesions in the right frontal lobes. The mechanism causing the new interest in food has yet to be revealed – serotonin levels in the frontal lobe may play a part.

Sensation

Sensation travels along several different types of nerves to the brain. Pain is carried by two types of nerve – fast, which carries sharp pain; and slow, which carries deep, burning pain. Stimulation of one type blocks messages from the other by closing a 'gate' in the spine. That is why 'rubbing it better' is effective.

The anterior cingulate cortex – an area primarily associated with emotion and attention – is essential for conscious pain. Opioid-type analgesics (including morphine and codeine) are the most effective type of painkillers. They block the receptors in brain neurons normally be filled by enkephalins – the brain's own painkilling chemicals, which are released by acute pain stimuli. Opioids also damp down activity in the anterior cingulate cortex.

The importance of the anterior cingulate cortex in pain perception is demonstrated by brain scans showing that people with cardiovascular disease appear to get angina – the chest pains associated with lack of oxygen to the heart – only when the anterior cingulate cortex is active. In some people, it seems, the anterior cingulate cortex lights up as soon as the heart is short of oxygen. This creates conscious pain, warning them to stop doing whatever is straining the heart. In others the heart can be severely short of oxygen before the anterior cingulate cortex is activated. These people can develop potentially dangerous heart disease without angina, making them vulnerable to surprise heart attacks.

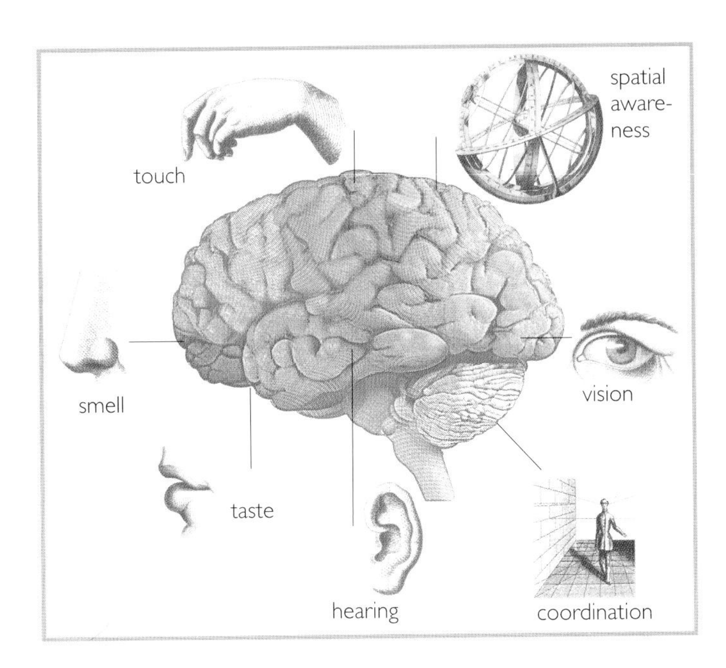

Left: *The vast majority of the cortex is given over to sensory processing – only the frontal lobes are dedicated to non-sensual tasks.*

Below: *All incoming sensory information (except smell) goes first to the thalamus. This limbic nucleus acts like a relay station, shunting the data onto appropriate cortical areas for processing.*

The sixth sense

Proprioception is the sense of body awareness telling us the position of our limbs, our posture and equilibrium. It involves the integration of several sensory inputs: touch and pressure sensations from skin, muscles and tendons; visual and motor information from the brain; and data about our balance from the inner ear. Together they amount almost to a sixth sense. Proprioception uses so many different brain areas that it is very rare for it to be lost altogether. Occasionally, though, people suffer brain injuries that so disturb proprioception that they lose all sense of having a body. Certain meditative states involve dissociating the conscious brain from proprioceptive input, inducing a feeling of disembodiment and maybe give the impression of floating or levitation. Out-of-body experiences, in which people report becoming detached from their bodies and floating around in mid-air, may be due to temporary loss of proprioception.

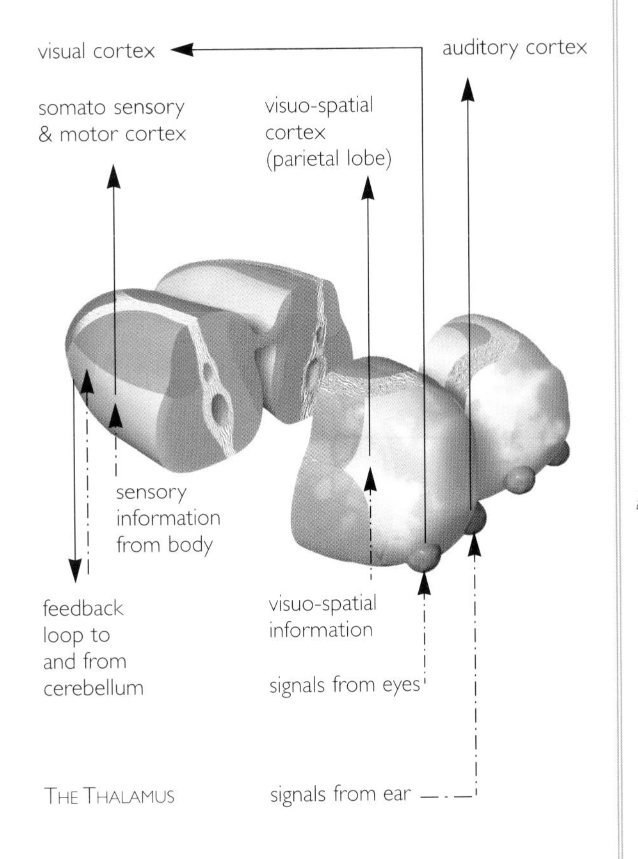

THE THALAMUS

area. For recognition to be complete information is then brought in from memory stores throughout the brain to flesh out the stimulus with the associations that give it meaning. 'My house' becomes 'my home'. Finally, feedback from the limbic area is included so the perception is clothed with emotion. 'My home' is now a place of warmth, love, shelter or loathing as well as the place you hang your hat. The process of recognition is complete.

That is how recognition should take place, but occasionally, in nearly everyone, it falters. When someone who seems to you to be a complete stranger comes up with a beaming smile and a 'How's the family?' either your recognition system or theirs is playing up. Similarly, a disruption in the recognition process accounts for those embarrassing 'know the face – can't quite recall the name' and 'oops, sorry – thought you were someone else' moments. Some people also know the fleeting disorientation of suddenly finding that a much-travelled route seems strange. The opposite – déjà vu – is also common to nearly everyone. But these are momentary glitches – a source of mild embarrassment at worst. For people with serious recognition problems the world is a frightening and alien place.

Altered views

'Sorry – no idea.' The patient shook her head in frustration. In front of her was a large, clear picture of a cat. 'It's a cat,' said the doctor. 'Does that mean anything to you?' 'Not a thing,' was the reply. 'Is a cat, say, an animal?' urged someone else. 'Is a cat an animal?' repeated the patient. 'I only wish I could remember what an animal was.' – *Patient study*[7]

A babbling infant knows what a cat is. But the 69-year-old woman quoted above had no idea. Her world had become furnished with ever-increasing numbers of strange objects – things with odd textured surfaces that ran and jumped and slithered and made odd noises. Psychological investigation revealed she had a condition called agnosia (literally, 'without knowledge') and one of its manifestations in her was the inability to recognize all living things other than people.

Agnosia is a fundamental lack of recognition – a dreadful condition that sometimes affects stroke patients or marks the early stages of dementia. Confronted with a soft, round object a person with agnosia would see it quite clearly but might not know whether to eat it, play ball with it or adopt it as a pet.

As we have seen, recognition is what rolls off the end of a long and complex assembly line. Agnosia is brought about by a fault somewhere along the line, and the point and place at which the fault occurs dictates the type of agnosia that results. Any sense may be affected by agnosia. Visual agnosia – the inability to recognize things that are seen – is the best documented, but there are also people who cannot recognize sounds, smells or bodily sensations.

Cognitive processes can be affected, too. Abstract concepts like 'morality', 'co-operation' and 'revolution' may lose their meaning just like the picture of the cat lost its meaning for the woman who had forgotten animals.

Agnosia is generally divided into two main types:

Apperceptive agnosia results from damage fairly early on in the recognition assembly line, before the perception is properly constructed. If the raw material is not put together correctly, the resulting perception will be so weak or wonky that the brain will not be able to match it with anything it already knows.

Associative agnosia is caused by a fault in the later stages of recognition. Here the perception may be perfect but the memories that are associated with it (and that are essential if it is to have meaning) have either been lost or cannot be retrieved.

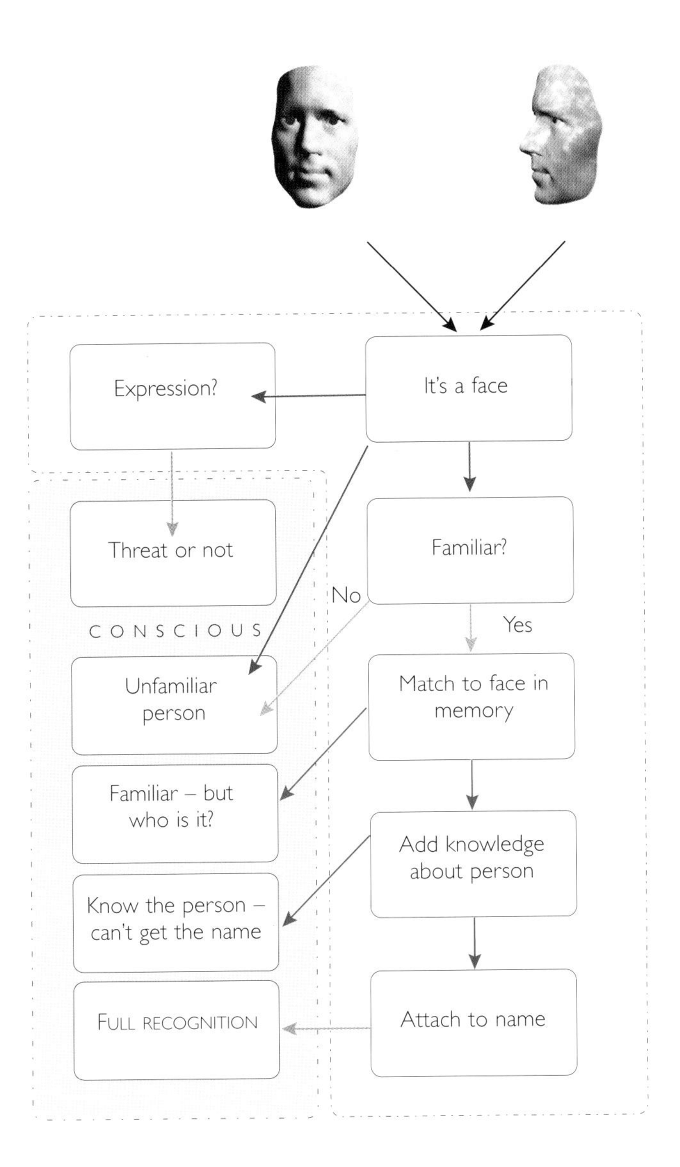

Recognizing someone is a process, most of which is done unconsciously. If the procedure is disrupted the conscious mind cannot achieve full recognition, however well they should 'know' the person.

People with apperceptive agnosia are often affected in one modality only. For example, they may be unable to make anything of objects when they see them but are quite able to recognize them when they are given their names or asked to feel them. If the problem is, say, visual agnosia, they will be unable to copy the object they are looking at except by doing a slavish line-by-line reproduction, and they are usually unable to match objects like for like. Those with associative agnosia, on the other hand, are able to describe the things they are perceiving quite well and visual associative agnosics can copy and match drawings like anyone else.

Loss of recognition may be wide-ranging or extraordinarily specific. One well-studied agnosic, a businessman who had suffered wide-spread brain damage after a stroke, was flummoxed by most things. Shown a carrot he said: 'I have not the glimmerings of an idea [of what it is]. The bottom point seems solid and the other bits are feathery. It doesn't seem logical unless it is some sort of brush.' An onion, he thought, might be 'a necklace of sorts' and a nose he (rather confidently) identified as a soup ladle.[8]

Other people lose the ability to recognize certain categories of things. For example, anything with a proper name may be lost, so a patient may be quite familiar with the notion of a queen or a temple but draw a complete blank if asked about Queen Elizabeth I or the Parthenon.[9] Faces (prosopagnosia) are a specific category and body parts are another. 'Is it a wrist?' asked a patient looking at a picture of an elbow. Then, correcting himself: 'No, of course it's not. It's someone's backside.'

The discovery that a single brain lesion can erase all knowledge of, say, man-made artefacts while another may take away all knowledge of animals suggests that these categories are somehow hard-wired into the human brain – that we all have a set of memory pigeonholes pre-labelled 'proper nouns', 'edible items', 'abstract concepts' and so on.

This seems so unlikely that researchers have expended a lot of time and trouble trying to find an alternative explanation for category-specific recognition deficits. The most hotly debated category is the living/non-living division. Quite a few agnosic patients have come to light who, like the woman at the start of this section, can recognize non-living things but not living. They also tend to fail to recognize food, even when the foodstuff – a block of ice-cream, say – is visually closer to a (non-living) brick than a leopard. Oddly, such agnosics are often particularly bad at recognizing musical instruments, too.[10] Other man-made objects, by contrast, are easily named and so are human body parts.

These patients suggest that the brain, for some reason, puts food, animals and musical instruments into one compartment and man-made objects and human body parts into another. At first sight it seems bizarre. Why musical instruments and animals? Why body parts and artefacts? What does the brain make of these disparate objects so that it lumps them together?

No one as yet knows the answer to that question but, as ever, there are plenty of theories. Some researchers suggest things are lumped together not by whether they are living or non-living but by whether they are familiar or not. If this were true, however, aardvarks would be placed in a different category from cats and there is little evidence to suggest that they are. Others have suggested the distinction is between large or small; homogenous or varying within the category; threatening or benign and so on.

The currently favoured explanation for category-specific recognition failure arises from the idea that the brain sorts and stores things according to the relationship we have with them rather than how they look or what they do.

Our relationships with things – even quite simple objects – are multifaceted. Food, for example, is something to be seen, smelt, handled and purchased, as well as eaten. An animal may be seen, touched, loved, feared, pursued – or eaten. A musical instrument may be heard, manipulated, seen and played. Some of them even get placed in the mouth.

Each facet of these memories of things may be stored in a separate, appropriate area of the brain. Let's call each fragment a 'recognition unit' (RU). A flute, probably has a shape RU in the visual cortex, a word RU in the temporal cortex, a sound RU in the auditory cortex, and a tactile RU – the feeling of something smooth and cylindrical requiring fine hand manipulation – in the somatosensory and premotor cortices.

Each area of the brain is packed with RUs of disparate objects brought together not because the objects they relate to are similar to one another in any overall way, but because they are similar in the particular aspect that concerns that part of the brain. Thus the tactile RU of a flute may nestle up against the tactile RU of a cigarette, and the flute's auditory RU may be next to the RU of a whistling kettle. When we think of a flute all the flute RUs are pulled together to create an aggregate concept, but the facet of the flute with which we are most familiar is likely to be accessed most

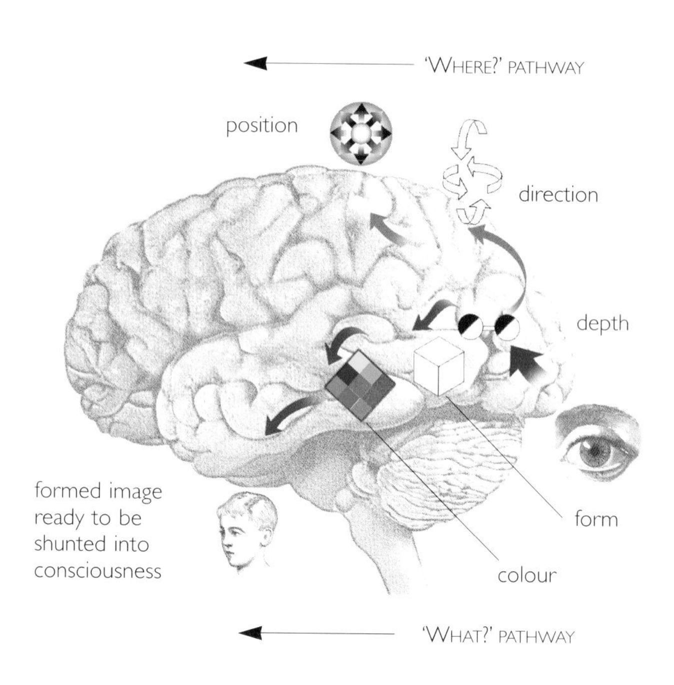

Raw visual material is constructed into recognizable things by unconscious assembly lines. A separate pathway works out where the thing is.

readily. Which one this is will differ from person to person. A flautist would have a strong flute memory in the premotor cortex (handling) and somatosensory (mouth feel) areas. A concert-goer would have a strong auditory memory. And the flute RUs of someone who never listens to music would be mainly visual and verbal.

It may be, then, that the curious categories that seem to emerge during studies of agnosics are essentially the result of an accident of cortical geography. Animals, food and musical instruments may be recognizable to one patient while body parts and tools are not, because – in that particular patient – the first three happen to have their strongest RUs in an undamaged part or parts of the brain, whereas the others do not. The groupings that arise as a result of a system like this would be similar for most people but not universal. For example, animals may be linked to food in one person's mind because, for that person, a picture of a cow triggers the taste (as in a steak) RU of a cow. Vegetarians would presumably make different connections and Hindus who believe cows to be sacred might make yet others. The imaging experiment required to test this thesis has yet to be done.

Face blindness

'When I was about six I told my brother I thought robbers were really dumb to cover their faces with masks when they held up a bank. Why bother, I said, when the rest of their bodies still showed? It took me a long time to realize that faces were special to most people' – *Bill, 50-year-old prosopagnosic.*[11]

Faces are very special to most people. So much so that it seems we have a special brain system dedicated to identifying them. People like Bill, quoted above, have a fault in the system that causes a condition known as prosopagnosia or face blindness. They can see perfectly clearly but a person's face makes no more impression on them than, say, their kneecap. Just as knees – give or take an extra nobble – look alike to normal people, so faces all look alike to Bill, even those he knows well. As he recalls:

'Once, around midday, I met my mother on the sidewalk and I didn't recognize her. We walked towards each other and passed within two feet, on a not-too-busy sidewalk in a neighbourhood shopping district. The only way I know about this is because she told me about it that night. She was not amused.'

Face blindness may be caused by a dysfunction along any point on the cortical recognition pathway. The severity and guise of the condition depend on precisely where the fault lies. If it is early in the process and affects both left and right hemispheres the effect may be catastrophic: one such patient thought a picture of a dog was a man with an unusually hairy beard, while another gave neurologist Oliver Sacks the title for his celebrated case study, *The Man Who Mistook His Wife for a Hat.*

Milder damage to early processing areas may cause faces to be identifiable as faces but to look peculiar – one patient described all faces as 'distorted – a bit like a cubist painting'. Another was only able to tell a person's sex by looking at the hairstyle and said all faces looked like 'strange white, flat ovals with dark discs where the eyes are'. If the problem lies towards the end of the pathway, on the other hand, it may just appear that a person is 'bad with faces'.

The ability to recognize people is so important for normal social functioning that even a mild deficit can make life difficult. A person with a recognition problem risks either offending those they know by failing to recognize them, or embarrassing those they don't know by greeting them heartily – something many face-blind people do to almost everyone 'just to be safe'. Some people with mild prosopagnosia probably don't even know they are unusual – they just find social interaction peculiarly difficult.

THE SEARCHING BRAIN

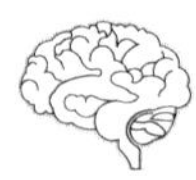

RICHARD L. GREGORY
Professor of Neuropsychology, University of Bristol

The following is an extract from Eye and Brain *(4th ed.) by Richard L. Gregory.*

The regions of the cerebral cortex concerned with thought are comparatively juvenile, and they are self-opinionated by comparison with the ancient regions of the brain giving survival by seeing.

The perceptual system does not always agree with the rational thinking cortex. For the cortex educated by physics, the moon's distance is 390,000 km (240,000 miles); to the visual brain it is a few hundred metres. Though here the intellectual cortical view is the correct one, the visual brain is never informed, and we continue to see the moon lying almost within our grasp. The visual brain has its own logic and preferences, not yet understood by us cortically. Some objects look beautiful, others ugly; but we have no idea for all the theories which have been put forward why this should be so. The answer lies a long way back in the history of the visual part of the brain, and is lost to the new mechanisms which give our intellectual view of the world.

We think of perception as an active process of using information to suggest and test hypotheses. Clearly this involves learning, and ... it [seems] clear that knowledge of non-visual characteristics affects how objects are seen. This is true even of people's faces: a friend or lover looks quite different from other people; a smile is not just a baring of teeth, but an invitation to share a joke ... Hunters can recognize birds in flight at incredible distance by the way each species flies: they have learned to use such subtle differences to distinguish objects which look the same to other people. We find the same with doctors diagnosing X-rays or microscopic slides. There is no doubt that perceptual learning of this sort does take place; but ... we still do not know for certain how far learning is required for the basis of perception.

It is not difficult to guess why the visual system has developed the ability to use non-visual information and to go beyond the immediate evidence of the senses. By building and testing hypotheses, action is directed not only to what is sensed but to what is likely to happen, and it is this that matters. The brain is in large part a probability computer; and our actions are based on predictions to future situations. But we cannot predict with certainty what our or other people's predictions will be – what they will see or how they will behave. This is a price we ... pay for basing behaviour on the probable, or desired, future; but intelligent behaviour is not possible without attempted prediction. Predicting from hypotheses ... from meagre data is the hallmark of perception, intelligence and science. If the brain were unable to fill in gaps and bet on meagre evidence, activity as a whole would come to a halt in the absence of sensory inputs. In fact we may slow down and act with caution in the dark, or in unfamiliar surroundings, but life goes on and we are not powerless to act. Of course we are more likely to make mistakes ... but this is a small price to pay for gaining freedom from immediate stimuli determining behaviour, as in simple animals which are helpless in unfamiliar surroundings. A frog will starve to death surrounded by dead flies; for behaviour ceases when imagination cannot replace absent stimuli.

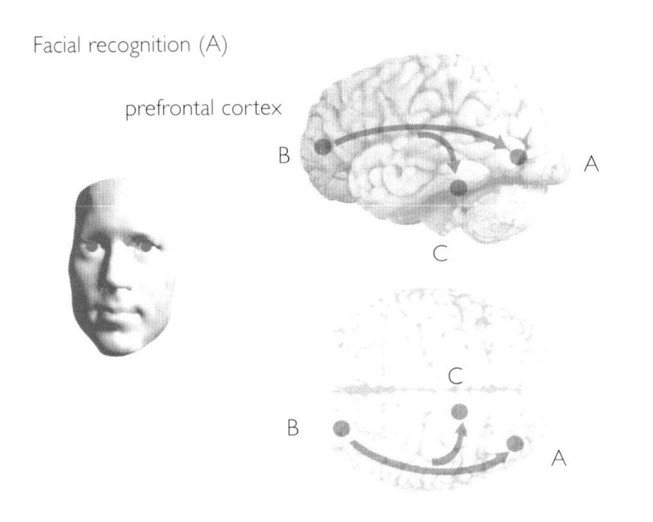

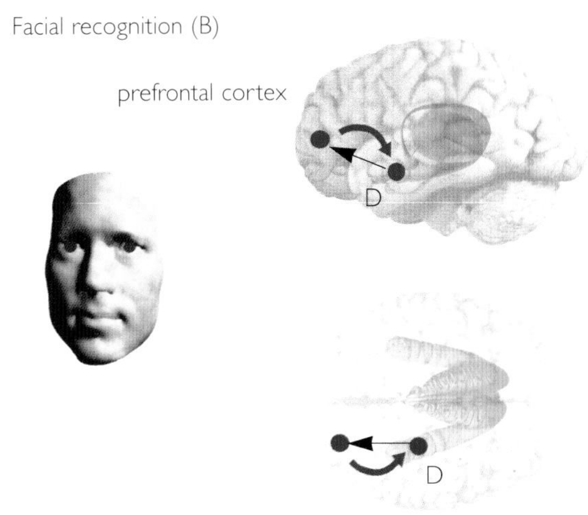

Left: *The face recognition pathway runs from the visual cortex (A) to the prefrontal cortex* (B). *En route it passes through an area which is specifically dedicated to faces* (C). Right: *Information about faces is also shunted to the amygdala* (D), *where it is endowed with emotional meaning. This is then fed back to the prefrontal cortex to give full recognition.*

Those with severe face blindness, on the other hand, can find themselves right out on society's outer edges, perpetually puzzled and lonely. Bill again:

'An ordinary social evening with a group of people sitting round a table is pretty boring for me. It is like it would be for a normal person to sit around all night with only their companions' feet on show. Finding work that is satisfying is difficult because it is so hard to strike up proper relationships with colleagues. I didn't even know how many people worked in the last office I was at because they all looked the same so I never knew if I was seeing the same guy as had been in the room earlier, or another one.'

The brain area that recognizes faces is specific for our own kind: one man, a farmer, became entirely incapable of recognizing people after a brain injury but was able to name every one of his flock of thirty-six sheep without hesitation. The acquired face blindness even seemed to enhance his recognition of the animals – other equally dedicated but non-prosopagnosic sheep farmers were unable to identify more than a few of their sheep without error.[12] Some face-blind people find they can identify faces only when the images are upside down – the opposite of normal.

Prosopagnosia does not usually involve the emotional recognition pathways. This has been demonstrated in studies in which face-blind people have been shown familiar faces while linked up to a machine that measures emotional responses like skin conductivity and heart rate. The subjects typically claim not to recognize the faces, yet when an emotionally significant face is shown their body reacts to it normally.[13] This is called covert recognition – a type of unconscious emotion.

Face-blind people confront endless obstacles: how to negotiate a social occasion without giving offence or embarrassment; how to follow the plot of a play or film when one character is indistinguishable from the next; how to know whom to kiss and with whom to shake hands. But essentially these are practical problems – once they have established a person's identity they are capable of forming normal emotional attachment to them. Awkward though their lives are, they are better off than people who have

I SEE YOUR FACE ...

Familiar faces are stored in the brain in neural circuits – memories – known as Face Recognition Units (FRU). When a new image of a person enters consciousness these FRUs are scanned for a match. If one is found the appropriate FRU is activated, retrieved and attached to the new image. This merging of memory and stimulus is a crucial part of the process of recognition.

New images may come from outside or they may be generated inside the mind – the brain activity is the same in both cases. So imagining a picture of someone you know will activate that person's FRU in just the same way as looking at a picture of them in the outside world will.

FRUs – like other memories – are kept alive by constant use. Each time one is activated it becomes more deeply etched into the brain through the process of long-term potentiation. If an FRU is activated frequently, it will also stay 'warm'. Normally, an FRU needs quite a strong stimulus to be activated – typically the sight of the person it represents, or the sight of something closely connected, or very similar to them. But an FRU that is already warm needs only a weak stimulus to spark it off. In extreme cases – when a FRU has been kept almost permanently turned on for days – almost anything can serve as a visual reminder of the person.

Everyone has experienced this from time to time: when you are in love, or grieving, you see the person who is on your mind everywhere. Then you look again and see it was actually a stranger with only the slightest similarity to the one for whom you mistook them. This is what it means to have someone 'on your mind' – their recognition unit is 'warm' and hair-trigger ready for action.

faults in the *emotional* recognition system – that is, along the pathway that carries facial images into the limbic system where they are clothed in appropriate emotion before being sent back up to the conscious mind.

Profound dysfunction of the emotional recognition system can cause weird sensory experiences. When it occurs in a person who also has a deranged belief system, the result may be pathological.

Fregoli's delusion, for example, is a condition in which sufferers constantly mistake strangers for people who are familiar to them even though they can see the strangers look nothing like the people they believe them to be. The sense of recognition Fregoli patients feel for these strangers is profound – so profound that they find it easier to believe that the strangers are the people they know in disguise than that their own sense of recognition is false.

A typical case was Miss C., a sixty-six-year-old woman who became convinced that her ex-lover and his girlfriend were spying on her. She claimed the couple disguised themselves in wigs, moustaches, dark glasses and hats. Sometimes one of them would pretend to be the gas man, she explained, so they could get into her house on pretence of reading the meter. At night she saw them passing by her window and by day they lurked at street corners. They followed her everywhere, on foot, or in dozens of different cars. Miss C. reported their activities to the police, and sometimes she would approach strangers and demand that they remove their disguises. On the day of her first appointment with a psychiatrist she was hours late because, she explained, she had to do a complicated detour around town to throw the couple off her trail. 'They keep changing their clothes and their hairstyles but I know it's them,' she told the doctor when she finally arrived. 'They should get a medal for acting, they are so good, but I can tell it is them from the way they stand and move about.'

Miss C.'s pursuers disappeared when she was treated with a dopamine-blocking drug. Dopamine excites many subcortical areas, and the strong sense of recognition that was constantly triggered in Miss C.'s brain was probably due to over-stimulation of those parts that normally pick up on familiarity. The drug did not remove her delusions, just the false recognition. She told her doctor:

'They have gone now, the police made them clear their cars away ... I know I've got to change and ignore them if I see them again ... It'll all come out in the open one day ... I am absolutely sure I am not making it up. It's really going on, you know.'[14]

In Capgras delusion the emotional recognition system is under- rather than over-active. Capgras patients can see that people look like themselves, but they do not feel right emotionally. The 'ah-ha!' of recognition is missing and without it Capgras patients cannot believe that their nearest and dearest are really whom they claim to be.

In order to explain the mismatch between appearance and feeling, these patients often come to believe that members of their family have been 'taken over' by aliens. One man was so convinced that his father had been abducted and replaced with a humanoid robot that he ended up slitting the man's throat to look for the wires that animated him.[15] Even animals may be seen as imposters – one woman claimed her cat had been 'switched' for another, because it no longer 'felt right'.

As with those who develop Fregoli's delusion, Capgras patients have cognitive dysfunction in addition to their recognition fault, and it is this – their shaky thinking – that causes them to create weird explanations for their distorted perceptions. The cognitive deficit in such cases can usually be traced to fairly obvious cortical damage: a CAT scan of Miss C.'s brain, for example, showed she had lost quite a large area of cerebral cortex as a result of a stroke.

This raises a question: what would it be like to have an emotional recognition dysfunction in the absence of a cognitive deficit? Say you had an overactive emotional recognition system, so the circuit usually reserved for activation by those who are emotionally important to you is triggered by a much wider range of people. The result of this might be that you would keep spotting people you know, only to discover on closer scrutiny that they look nothing like whom you thought them to be. However, unlike poor Miss C., your intact grasp of reality would prevent you interpreting this as evidence that you are being stalked by people in disguise, so in time you would learn to recognize only those people who trigger an exceptionally strong sense of familiarity.

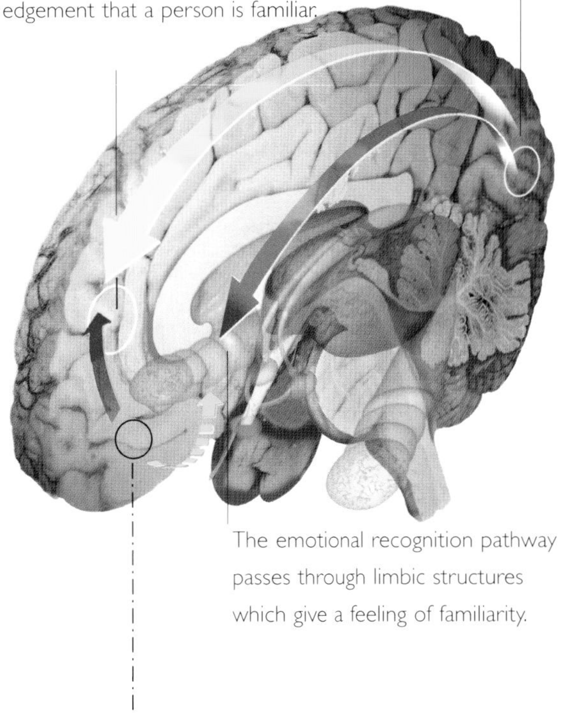

The cortical 'who?' recognition pathway ends in the frontal area with the conscious acknowledgement that a person is familiar.

visual cortex

The emotional recognition pathway passes through limbic structures which give a feeling of familiarity.

These two types of knowing usually come together in consciousness, but if the emotional message does not get through to the conscious brain the person may look familiar but not 'feel' right – a condition known as Capgras' delusion.

You would still feel a twinge of recognition towards others, though, and it seems likely that this would give you a sense of connection, emotional attachment and interest in others that would be greater than that felt by most people. Talking to people would generate a continuous emotional charge; just seeing a friend would give a jolt of pleasure; pictures of familiar people – celebrities, say – would create interest.

If, on the other hand, your emotional recognition system was underactive, you would be slow to recognize those you know and might therefore appear to be aloof. Talking face to face to a person would be only as engaging for you as the content of the conversation – you might find communicating by e-mail just as good. You would probably shun social gatherings, both out of boredom and for fear of accidentally snubbing someone you know. And you may well feel a lack of emotional attachment even to those who are closest to you, as well as detachment from people in general.

These two behavioural profiles match up, of course, with our idea of what it is like to be an extravert or an introvert. So could the degree of activity in the emotional recognition system be a measure of extra/introversion? Could it even dictate which type of person we are?

A recent brain scanning study showed that extraverts have greater activation in the limbic system than introverts, and that it is this, rather than any difference in cortical areas, that determines a person's personality type.[16] Emotional recognition, as we have seen, is a function of the limbic system, so the greater the activation in that area the more active our emotional recognition system is likely to be.

Whatever its role in shaping personality may turn out to be, the emotional recognition system is clearly an important means of forging the emotional connections that are so vital for survival in a highly social species like ours. Without them we cannot truly say we know who's who.

The phantom factory

A short while ago a Dutch eye specialist sent a follow-up survey to patients who had undergone a particular surgical procedure. The questions were compiled, in Dutch, by an English assistant. One of them was intended to ask if the patients saw distortions, and if so, what they were like. However, a slight mistranslation of 'distortion' made it seem that the enquiry was about visual hallucinations.

The doctor was subsequently astonished to receive dozens of accounts of detailed, dramatic phantom sightings – the sort of things that, in certain circles, would be held up as incontrovertible evidence of some Other World. Most of the sightings were clear images of people going about their normal daily business. Sometimes they were complete strangers, sometimes they were familiar. One man kept coming across his wife as he moved from room to room. He found that if he walked straight through the phantom it would disappear. Unfortunately, this led to some painful encounters with the real thing who was alive and well and also living in the house. Eventually, the couple worked out an elaborate identification ritual to avoid collisions.

Some people reported coming across huge buildings in what they knew was empty moorland; others saw whole crowds of people. The images often persisted for hours. A woman reported looking out of the window of her house and seeing a herd of cows grazing in the field opposite. She idly watched the animals all afternoon. It was a cold winter's day, and as night fell she remarked to a companion that the farmer should not really leave the animals out all night. The other person then pointed out that the field was empty, and had been all day.

Many of the patients said they kept their experiences to themselves for fear of ridicule, and expressed relief at being asked about hallucinations because they assumed this indicated that 'seeing things' was a recognized side-effect of the particular eye operation they had undergone. In fact, it is not. Although phantom sightings are much more common among people with damaged sight, the visual system in these patients was generally intact. The error in the questionnaire revealed that hallucinations – at least among middle-aged Dutch people – are far more common than you might expect.

So how do they come about?

The brain does not 'see', 'hear' or 'feel' the outside world. It constructs it in response to stimuli. The stimuli usually come from outside – light waves, for example, bounce off objects and then hit the light-sensitive neurons in the eye. These stimulate the brain to create an image that accords with the information it is receiving.

Sometimes, however, the brain either misreads the incoming information (creating an illusion) or generates its own stimuli, which it then interprets as coming from outside. When this happens there may be no way – other than by deduction – for someone to work out whether what they are sensing is really in the outside world or only in their own mind.

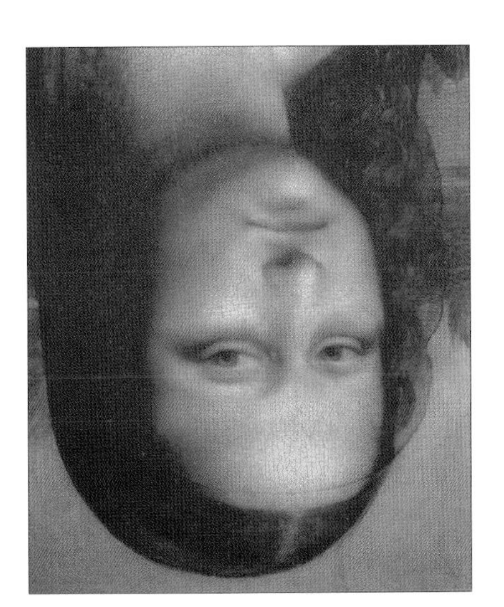

Look at this inverted picture of the Mona Lisa. Looks normal enough, does it not?. Now turn it right side up.

The distortions leap out and hit you only when you turn the picture around because the face recognition system of the brain is designed to work right side up and inverted faces are not processed as 'faces' – just objects. Object processing happens in a different area of the brain which is not equipped to give images as fine scrutiny as the face processing area. Small differences like a droopy mouth are therefore easily overlooked. Incidentally – that little jolt you got when the distorted face registered came from the emotional face recognition area of your brain. If your brain had been scanned at the moment you felt it the area around your amygdala would have been seen to light up.

'Phantom limb' sensation affects 60 per cent of people who undergo amputation.[17] Sometimes it fades after a few months but often it lasts a lifetime. People may simply feel the lost limb in the way that they felt it when it was there. It is quite common for patients with recently amputated legs to cause themselves further injuries by momentarily forgetting their loss and trying to walk on the lost leg. Sometimes a lost limb feels as though it is stuck in an awkward position. One man tends to go through doors sideways because his arm – amputated several years ago – feels to him as though it is sticking out horizontal to his body. A young man who lost his arm in a horrific motorcycle accident cannot sleep on his back because it feels as though his arm is still bent behind him in the position it was forced into when he was spreadeagled across the road. Similarly, a man who lost both legs in a cycling accident still feels them pushing the pedals around from time to time, just as they were before he was hit by a car. It is, he says, quite exhausting.[18]

Hallucinations, imagination, and 'real' seeing are essentially the same thing as far as the brain is concerned. If you look at a scan of a person's brain while they create an internal picture of, say, their bedroom, you will see activity in the same vision and recognition areas that would be activated if they were actually looking at the room. Usually, however, more sensory neurons are activated in response to outside stimuli than in self-generated sensory experience.

You can measure the difference by looking up from this book, memorizing the scene in

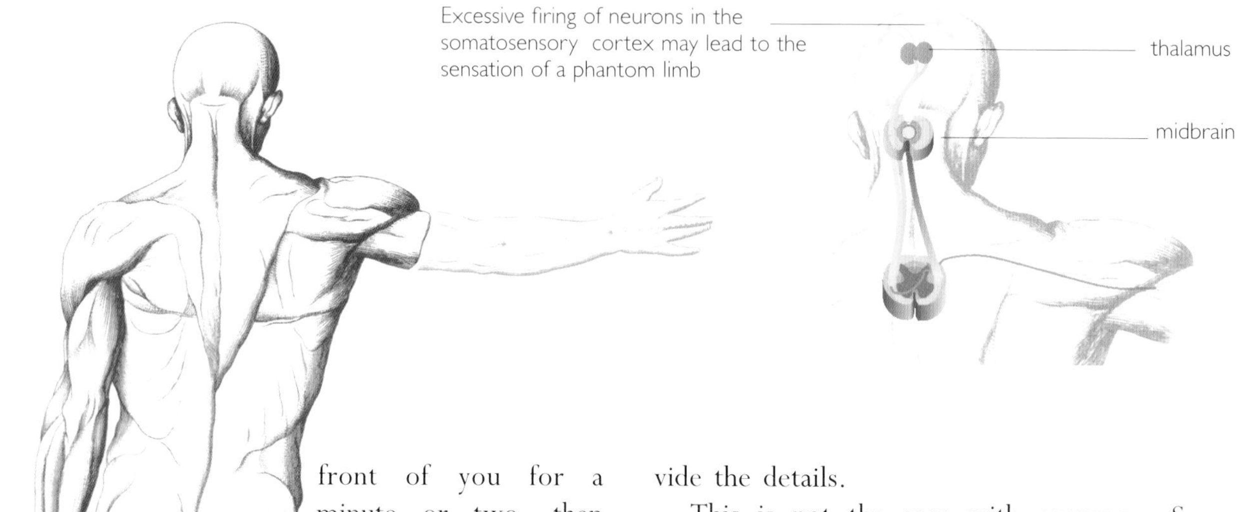

front of you for a minute or two, then closing your eyes and trying to recreate it in your mind's eye. The first impression probably seems very strong but if you home in on details – the titles on the spines of books in the bookcase, for example, or the pattern on a cushion you will probably find they go fuzzy. This is because enough neurons are firing in your visual cortex to give an impression of the scene but not enough to provide the details.

Phantom limbs are commonly felt for a long time after the real limb has been amputated. Studies have shown that the brain cells in the somatosensory cortex which once received signals from the nerves in the amputated part fire excessively when their normal input dries up, and this may give rise to the feeling that ther limb is still there. Another possibility is that the brain constantly generates a ghostly sensory image of the complete body against which incoming signals are matched to ensure that everything is intact and in its right place. The 'sensory image' is created by spontaneous neural activity, and sensation of a phantom limb may arise from this.

This is not the case with everyone. Some people have eidetic – photographic – memory, which allows them to create visualizations that are just as intense as those brought about by the original stimuli. Eidetic memory may be something we all start off with. Studies suggest that up to 50 per cent of five-year-olds[19] have the ability to 'read off' an imagined image as though it was really there. If they look at a picture of a zebra, for example, and then close their eyes, they can count the stripes along its back – the equivalent of reading the spines on the books.

A few adults retain the ability to imagine things with this intensity. Psychiatrist Morton Schatzman reported the case of a woman called Ruth who was 'haunted' by her abusive father. She would wake in bed and find him leering over her, or walk into her living room and find him ensconced in her favourite chair. Sometimes she would lean over to pick up her baby and find her father's face superimposed on the child's. The man was still alive when this happened, but the experience was otherwise very similar to the classic tales told by people who see ghosts, right down to the accompanying feeling of a presence. 'When I'm alone in the house I feel as though there is someone else in the room with me, who wants me to die,' she told Schatzman. 'I feel I am in terrible danger and should run away straight away.'[20]

Ruth was subsequently found to be able to create such intense images that they completely blocked out the real world. In one experiment she was linked to a machine that measured the electrical activity created in her brain by certain stimuli. When she was placed in front of a light her brain at first reacted in the expected way. But when she imagined a person sitting between her and the light her brain no longer responded to the light waves coming from the bulb. The figure she had conjured up actually blocked her vision of the real world. This image differed from the 'hauntings' in one way only: she knew she had produced it herself. Once she recognized that her father's unwelcome visitations were similarly self-generated, they ceased to affect her and eventually disappeared.

Hallucinations can therefore best be understood as exceptionally intense self-generated sensory experiences. People with eidetic memory are likely to experience them more frequently than others. Children who play with seemingly invisible playmates, for example, may actually be seeing these friends as clearly as they see real people.

All types of sensory experience can be generated internally. Tinnitus, for example, may be caused by stimulation of the auditory cortex in the absence of outside stimuli. Some people report hearing a full orchestra playing, just as loudly and clearly as if they were in a concert hall. The Soviet composer Dmitri Shostakovich is said to have been able to hear music by tipping his head to one side, and claims to have come by several melodies by listening in to it. The music appeared after a shell exploded near him during the Second World War, embedding a splinter of metal in his brain. By tipping his head to one side, the splinter presumably came into contact with the auditory cortex and triggered activity in it.[21]

Voices are perhaps the most common type of auditory hallucination. Work with schizophrenics has demonstrated that the voices heard by them are in fact their own. What happens is that they generate speech in one part of their brain and experience it as auditory input in another.[22] This does not usually happen with normal people because the brain generally monitors its own speech production area and signals to the speech recognition area when it is active. This prevents one's own words being mistaken for those of someone else.

Sometimes, however, this automatic distinction between outside voices and one's own breaks down even in normal people. It is quite common for people in mourning to report hearing the voice of the person who has died. And many people hear apparently divine messages when they are in a state of excitement or stress.

Phantom tastes and smells are also well documented. Imaginary smells, for example, are a common feature in early Parkinson's disease, and people who are depressed often report that they can smell themselves or that they have a 'bad taste' in their mouth.

Bodily sensations may also be generated internally. We take mild bodily hallucinations for granted: when we itch, for example, we do not necessarily expect to see something scratching us. Somatosensory hallucinations can be among the most discomfiting of all, however. Phantom limbs, for example, can cause terrible pain and distress for years after the real thing has been amputated.

False messages from the brain centres that keep track of one's own body can also produce profoundly upsetting hallucinations: doppelgangers, for example – apparitions that look precisely like onself – are probably produced by disruption to areas where the body map infringes on the visual association area.

Doppelgangers (properly known as autoscopic delusions) are traditionally seen as harbingers of death and so are regarded as terrible. Yet people who actually experience them usually report being curiously unmoved by the experience. Ms

B., a retired schoolteacher whose case is reported by a doctor in Bristol, first saw her 'other self' when she returned home after her husband's funeral. She opened the door to her bedroom and was confronted by the shadowy shape of a woman facing her. Ms B. reached out with her right hand to switch on the light, and the figure did the same with her left hand, so their hands touched on the light switch. 'My hand immediately felt icy cold, and where it touched me I felt as though all the blood drained out of me,' she told the doctor. Despite this, Ms B. said she did not feel scared, just 'mildly surprised'. Without bothering about the intruder, she went about removing her hat and coat, and noticed that the other woman did precisely the same thing in mirror image. It was only then that she realized she was looking at her double. At that moment she suddenly felt exhausted and frozen, and lay down on the bed. As soon as she closed her eyes she lost sight of the figure – at which point her warmth and strength returned. 'It was as if the life of this astral body flowed back into me,' she explained. Ms B. was subsequently visited almost daily by her double. She found that she did not just see the figure – she felt it as well. Just as normal people are aware of two legs, two arms and so on, she was aware of four. 'It is me,' she explained, 'split and divided.'

Some people take a less benign attitude towards their other halves. Mr F., a 32-year-old engineer, was intensely irritated by his doppelganger. It appeared as a phantom face directly in front of his own and would imitate all his facial expressions. Mr F., like Ms B., acknowledged the face was part of him. He nevertheless spent most of his time in its presence making faces at it and using it as a punchbag. The face – ending as it did at the neck – was unable to hit back.[23]

Some hallucinations may be brought about by a change in attention – a de-focusing on the outside world that allows internally generated stimuli to flood the brain. People with damage to the tegmentum – an area just above the reticular formation, which is part of the attention control mechanism – sometimes report having very elaborate, richly coloured hallucinations of everyday scenes. Sometimes these are familiar while at other times they seem to be new creations. Some are much bigger than life-size, others are tiny. One patient reported watching an entire circus performance, with clowns, jugglers and high-wire artists, all taking place in the palm of her hand.

Although such hallucinations are usually sharp-edged and quite solid-looking, they often seem to lack emotional impact – people typically say they observe them with detached amusement.

A similar matter-of-fact attitude is usually shown by children towards their imaginary friends. It is also characteristic of people who claim to be visionaries. People who say they see ghosts on a regular basis, for example, often seem blandly unconcerned by it. 'They can't hurt you, so why be frightened of them?' is their usual explanation. This suggests that certain hallucinations are created by stimulation of the cortical areas of the brain and do not activate the limbic areas. They may look exactly like real entities but at an unconscious level the brain 'knows' that they do not pose a material threat. Hence – whether or not the person consciously recognizes them to be self-generated – they carry relatively little emotional punch.

Flashbacks, like those experienced in post-traumatic stress disorder, are just the opposite. These, too, take the form of intensely clear and lifelike scenes but their defining quality is terror. Flashbacks (as the name suggests) are not fresh creations but memories. Sometimes the memories are fragmented, sometimes they are precise replays of some traumatic event. They differ from other hallucinations in that they are triggered by memories laid down in the amygdala, and bring with them their full cargo of both sensory and emotional associations. In con-

PATTERNS IN THE MIND

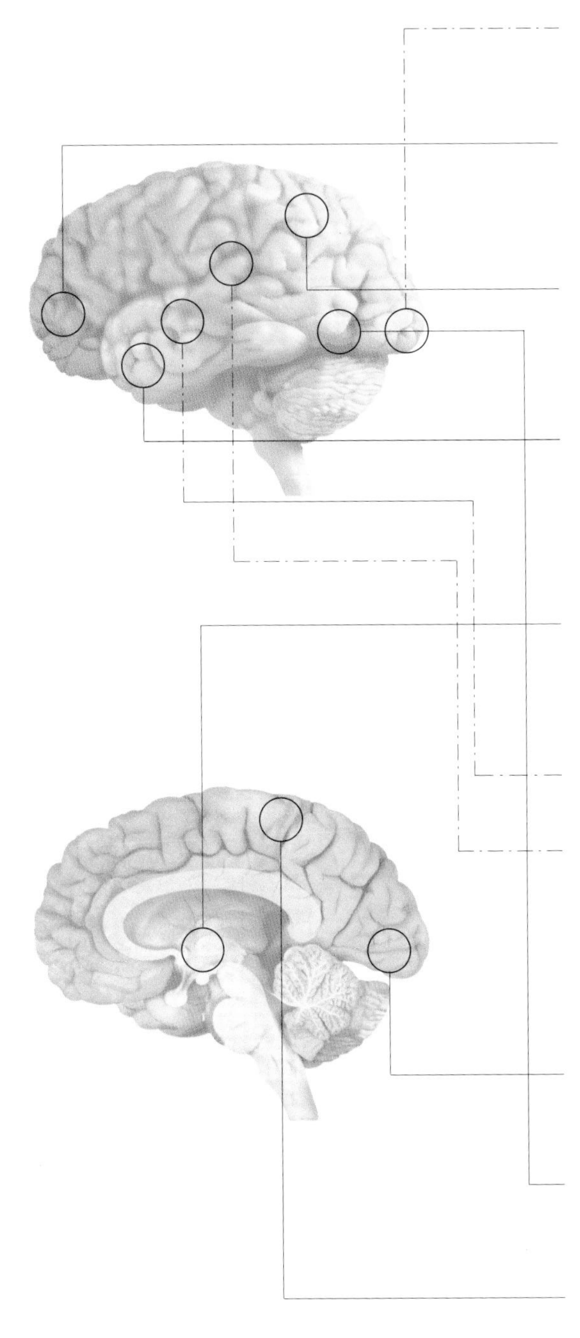

Eyes and V1: 15 per cent of people who lose part of their sight report having hallucinations.

Left frontal lobe: reality testing area – damage to this area may reduce the brain's ability to distinguish between externally and internally generated stimuli.

Occipito-parietal area: damage may make objects seem to appear or disappear due to simultagnosia – an inability to hold two objects in vision at the same time.

Temporal lobe: stimulation in this lobe – by epilepsy or drugs – may produce intense flashbacks and feelings of a 'presence'. Objects may appear strange or change from one form to another.

Temporal/limbic system: stimulation here may produce intense feelings of joy, and a feeling of being in the presence of God. Religious visions may occur.

Auditory cortex/speech areas: stimulation here produces hallucinatory voices.

Superior auditory cortex: sounds and noises, such as hissing, clicks and crashes, may occur if this area is stimulated. The brain may interpret them as meaningful noises. Tinnitus may be caused by activity here in the absence of outside sounds.

Visual shape area (right hemisphere): 'Ghostly outlines' may be triggered by overstimulation of the visual area that detects shape.

Visual face recognition area: images of faces may linger after they have 'really' gone due to overactivity here.

Parietal/sensory cortex margin: doppelgangers – a spectral version of oneself – may be the result of disturbance to this area.

trast to other hallucinations these may be consciously recognized as being false, but they feel more real than reality itself.

Migraine, epilepsy and a huge range of chemicals can create the changes in brain activity associated with hallucinations. Some of them increase or mimic the effect of excitatory neurotransmitters like dopamine, amplifying imagined sensations until they are indistinguishable from those generated by outside stimuli. Others inhibit areas of the brain involved in reality testing. Illicit drugs are often taken specifically for their ability to produce hallucinations, while therapeutic drugs often produce them as a (generally) unwanted side-effect.

Phantom sights and sounds are particularly likely to occur when people are deprived of normal outside sensory stimuli. People who lose all or part of their sight or hearing often find that they experience hallucinations for this reason. It also explains why ghosts are more commonly seen at night. In the absence of competing visual stimuli the brain picks up the shadow in the corner and moulds it into a sinister figure clothed with whatever visual associations – monk's habit, funereal gown – leap up from the memory storage areas.

Why does this happen? The brain evolved to keep constant watch on the outside world, sensing, sorting and shaping every stimulus in order to ensure that no danger creeps up unannounced, or opportunity pass by unnoticed. It needs to keep active so if the usual stream of clamouring external stimuli is turned off, it searches desperately for something to take its place. The slightest sound, sight or sensation is seized upon, amplified and shaped to make something meaningful, and if absolutely nothing comes in from outside, the brain will generate its own excitement. Hallucinations, like dreams, are part of a continuous cabaret that keeps us primed and ready for action. If the stage falls empty, the ghosts will come in to fill it.

Whether you see a goblet or two faces depends on which part of this figure your brain selects as background and which as figure. If you expect to see either a face or a goblet, you will probably see that image first and more persistently.

Illusions

In July 1985 a group of Irish teenage girls claimed to have seen a statue of the Virgin Mary move. The figure, in Ballinspittle, Co. Cork, was wringing her hands as though in grief, said the girls. Other people saw it, too. The story – coming as it did in the midst of the annual 'silly season', when hard news is thin on the ground was picked up and made much of in the newspapers. Within twenty-four hours at least forty other statues of the Virgin were reported to be on the move as well – busily waving, blessing and, in some cases, having a good look around. That summer more than a million people claimed to have seen these miracles. For a short while the Second Coming was confidently predicted to be nigh.[24]

Moving Madonnas are now two-a-penny. Since that major outbreak in 1985, they have popped up with monotonous regularity. This is not surprising because, if you look hard enough at a statue under the right circumstances, you will indeed see it move.

Try this experiment. Block out all the light in the room then place a lighted cigarette in an ashtray at one end of it. Then take a seat at the other end and concentrate hard on the glowing tip. After a while you will see it start to wander. Sometimes it will swoop in one direction, sometimes in the other. It may circle gently, or describe graceful arcs and curves as though spelling letters in a neon thread. The harder you look, the more it moves.

A similar effect is achieved if you stare at any distant object illuminated against a dark background. The phenomenon is known as the autokinetic effect and it occurs when the eye muscles become tired from straining to fixate on a single point. The fatigue would normally cause the eyes to wander away from the fixation point, so, to continue to hold them on the target, the brain sends a stream of 'correction' commands to the eye muscles. These are the same signals it would send, if the muscles were not tired, to instruct them to move. The signals are therefore interpreted by the brain as eye movements. Hence the point of light registers as a moving object, activating neurons in the areas of the brain – mainly V5 – that are responsible for registering motion. Divine intervention is not required.

Illusions like this differ from hallucinations in that they are errors of perception and/or construction, rather than false constructions. A great deal of the perceived world is in fact illusory – the moving pictures of a film or video image are actually a series of stills or pulsating dots, for example – but we generally notice such falsities only when they jump out at us or produce an effect that fails to fit our expectations.

Some illusions are caused by fairly mechanical processes: the halo of brightness that persists after a strong light has been switched off, for example, is due to residual firing of the light-sensitive neurons in the retina. Others, though, give an insight into the way our brains work at the highest level: what may at first appear to be an error of the senses is actually created by an error of cognition. Sometimes the two are combined: the Virgin Mary, for example, may move owing to the physiological effect of staring too hard at a fixed point. But it is a cognitive error that translates the illusory movements into hand-wringing and blessings.

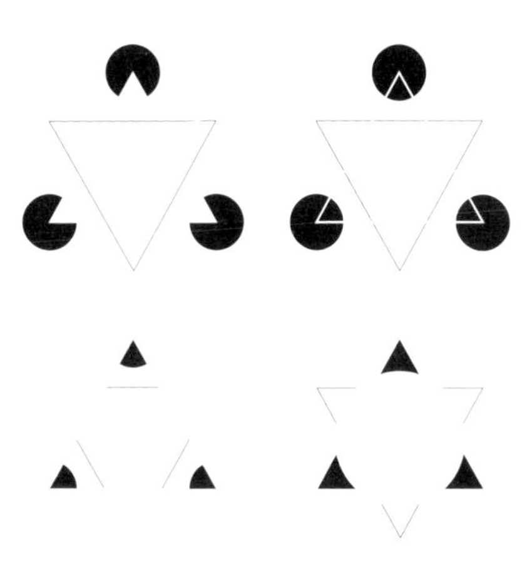

The phantom triangle is constructed in a different way, and in a different area of the brain, from the 'real' one.

Cognitive illusions come about because the brain is full of prejudices: habits of thought, knee-jerk emotional reactions and automatic orderings of perception. These are so deeply ingrained that we are usually unaware that they exist, and when we do become aware of them we think of them as common sense assumptions or intuition. These prejudices are, to some extent, hard-wired into our brains. As we have seen, even infants have strong expectations about how objects should behave. This is why babies are so gripped by games in which things are made to 'disappear'. The concept that material things take up space and do not just dematerialize is clearly very much part of our neural blueprint.

Such pre-programmed theories about the world are useful because they allow us to make quick, practical decisions about how to react to what we perceive. Most of the time they work fine (which is, of course, why they evolved). Sometimes, though, we are caught out by them.

Take, for example, our simple prejudice about object size. If you see two cars ahead, one big and the other tiny, you assume the bigger one is closer. Your experiences of perspective have woven a neural pathway in your brain that

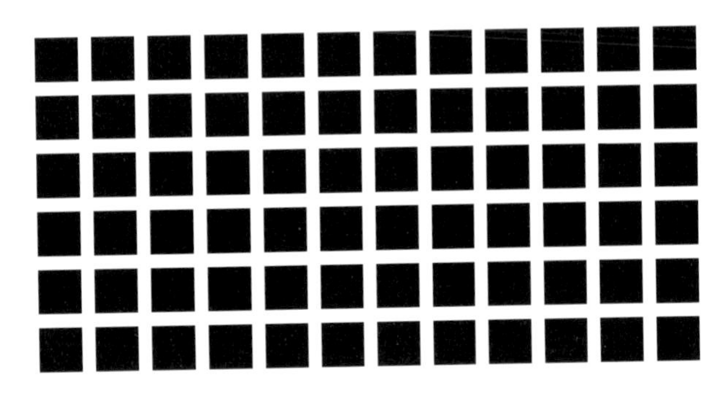

Hermann's Grid
If you look at the grid you see little grey blobs at the intersections of the white lines. This is due to a physiological mechanism affecting the retinal neurons. It is called lateral inhibition. When you look at the intersection of the white lines – the place where the blobs appear – you will see that the blobby area is surrounded on four sides by white but if you shift your gaze to the vertical bands of white between the intersections you will see that there are only two neighbouring white areas. Lateral inhibition causes light-receiving cells in the retina to shut down when the light-receiving cells around them are activated. When you look at the intersection, the cells focused on that area are surrounded on four sides by others which are also firing. When you look at the bands, by contrast, the cells adjacent to those focused on the bands are activated only on two sides.
The cells are therefore more inhibited when they focus on the intersections than they are when they focus on the white bands. The result – dark patches. Magicians take advantage of lateral inhibition to obscure their apparatus. For example, if they want to hide the struts of a floating body they might surround it with brightly shining metallic objects, white cloth, and so on. The parts they want to hide would be black, placed against a black background. The dazzle from the bright objects would cause the cells which would normally spot the darker objects to be inhibited – so the struts would not be seen.

transforms the sight of 'little object/big object' into 'far object/near object'. There is no pre-operative checkpoint at the start of this pathway to question whether the tiny car may in fact be some child's toy cunningly placed to look like a real distant vehicle. Such a test, if it was run each time you encountered two cars of differing size, would take up so much time you would never get around to crossing the road. No, when the stimulus 'big car/little car' gets through the primary vision areas and into the visual association areas the automatic perspective processor trips into action without pause. We are so used to it delivering the right answer that when we find it is wrong – the little car really *is* a child's toy – we receive a jolt.

Similarly, when we see a curved oval object lit from above the brain automatically responds to it as though it is a face. As faces stick out rather than in, we therefore construct a perception of a concave object – even when it is in fact convex. If the lighting is changed so the object is lit from below, the brain no longer sees it as a face (presumably because faces are usually downlit by sunlight) and the object suddenly pops out.

The prejudices that give rise to sensory illusions are generally benign. The illusion of a mirage in a desert may be a cruel trick but it is not one that takes many lives nowadays and magicians may use our various blind spots to deceive but it is usually done by happy collusion with the deceived. As a scientific tool they are very useful, but as individuals there is not much to be gained, in practical terms, by becoming aware of the mechanisms by which we falsify the evidence of our senses.

Prejudices of thought are a different matter. The illusions created by our own cognitive misconstructions make bigger fools of us than any stage magician may contrive. 'Purely' cognitive illusions – those in which the affected perceptions are ideas rather than sensory phenomena – may have arisen for the same reason as sensory illusions: they help us deal quickly and practically with complex challenges. But ideas are more slippery than material objects. If we process them according to rigid prejudices they are very likely to trip us up – and the cost may be high.

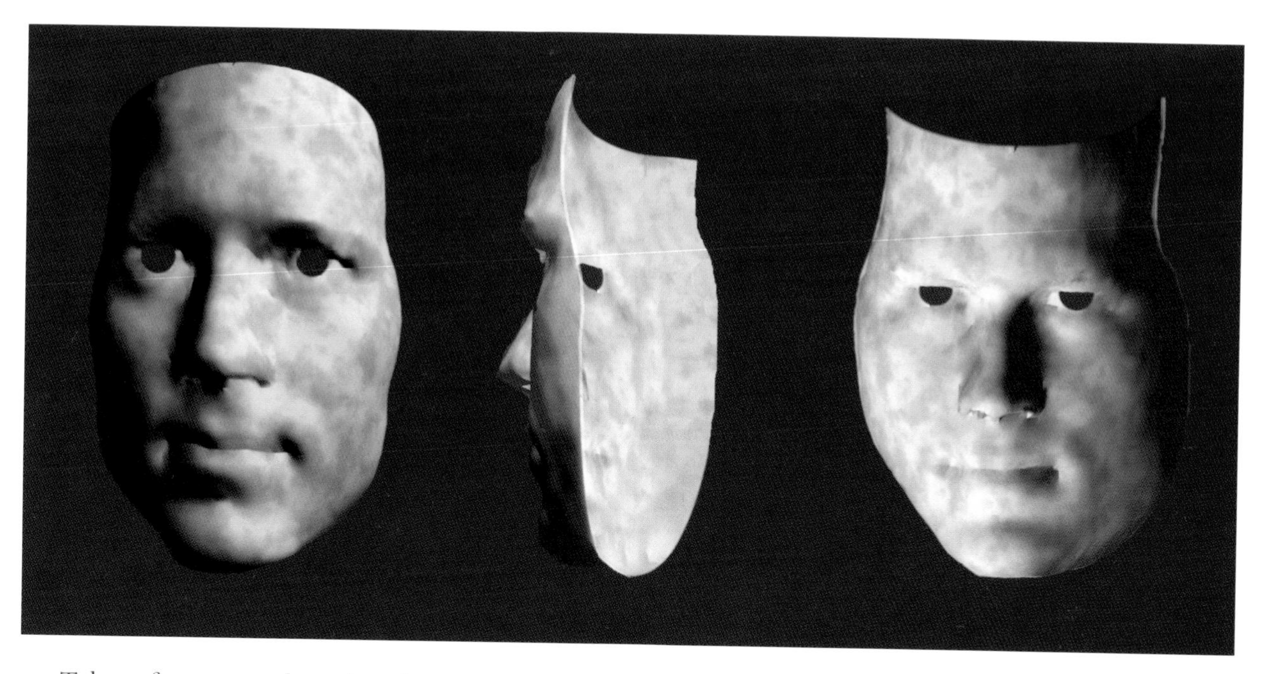

Take, for example, the 'common sense' approach to the following scenario. You are a member of a jury that is trying a man for murder. The prosecution has no evidence against him bar this single item: his DNA matches traces of genetic material left by the murderer on the victim's body. The chances of the DNA found at the scene matching any single person plucked at random are, you are informed, ten million to one. Do you find him guilty?

You probably do. Such evidence has, in the past, sent many suspects to jail, while the juries that convicted them went home to sleep peacefully in their beds, convinced that 'statistically' a guilty verdict was in order.

The apparent unlikeliness of the person in the dock being innocent is, however, illusory. Even if his DNA pattern is seen only once in every ten million in a country with a population of, say, 100 million people, that means it is shared by ten other men. If all ten were in dock, would you have any reason to single out the one that is now before you? The prosecution, remember, has *no other evidence against him*. Logically, then, the chances of his being guilty are one in ten, not (as you might have thought) one in ten million.

A concave mask appears to be convex once enough of it is visible for us to realize it is a face. As it rotates (anticlockwise) the concave side springs out and appears to be convex. This is because the face recognition area 'knows' that faces are convex – and constructs the image accordingly.

This, like many sensory illusions, is one that persists even when we have spotted the error. Our prejudice – in this case the assumption of guilt until proved innocent – is so deeply entrenched that we find ourselves struggling to fit the facts to match it rather than vice versa. 'Ah,' we say, 'but the suspect wouldn't be in the dock if he hadn't given the police cause to suspect him, would he?'

In the past this particular prejudice probably had high survival value. The fact is, in the messy, 'real' world in which we evolved, it was generally safer to assume guilt in anyone who came under suspicion and not to bother with fancy ideas about admissable and inadmissable evidence. Our concern for the rights of the individual, not to mention our ability to analyse statistics, came a whole lot later. They do not therefore come naturally to us. In the example above, furthermore, it is easy to rationalize the assumption of guilt. We know that the legal system puts strong constraints on what evidence may be presented so it is quite possible that there is other evidence against the suspect, even though it is not admissable in court. We also know that the police are unlikely to have picked on an entirely random member of the public to put in the dock. So we make our decision in our time-honoured, prejudicial way, and justify it with a veneer of rationality.

There are real, practical considerations here. For one, things change in the world of ideas and prejudices that were fair enough yesterday may be terribly misguided today. In our example, a computerized database of every person's DNA – not such a distant possibility – would remove one of the major justifications of the assumption of guilt. This is because such a database would make it relatively easy for the police to run a check against almost anyone, not just people they had reason to suspect. So the chances of them effectively placing a 'random member of the public' in the dock are greatly increased.

Cognitive illusions about statistics leave us open to costly errors. Stock markets have been known to collapse under the weight of instinctual irrational selling, and many decisions about how to use public money have been found, on analysis, to be based on misreading of statistics. Cognitive illusions cause us to take extreme precautionary measures against obscure risks and to be reckless about the day-to-day dangers that are actually likely to cause us harm. They also cause us to spend money on silly gambles and to shrink from sensible investments.

Take the way that people select their numbers for the British lottery. They can choose six numbers ranging from 1 to 50. Everyone knows that every number is as likely to come up as any other. Yet most people 'intuitively' – and insistently – choose a nice even spread – typically one number in every ten. A few, determined not to be taken in by that particular illusion, choose 1, 2, 3, 4, 5, 6; or 35, 36, 37, 38, 39, 40. These choices certainly stand the same chance of winning as any other but they both stand *less* chance of attracting the Big Win – the multimillion-pound prize that everyone lusts for. The reason? If two more people select the winning numbers, the prize money is divided between them. So the really big prizes only go to single winners. With so many people looking for 'a nice even spread' the chances of two or more of them coming up with the same particular spread is quite high. Similarly, the idea of putting 1, 2, 3, 4, 5, 6 has occurred to many people. So the best string of numbers to select is slightly bunched, skew-whiff and *unusual* – the string you can least imagine coming up.

What has all this to do with the construction of the brain? Probably almost everything. Such evidence as there is suggests that the brain constructs its ideas – even its most sophisticated notions – in much the same way as it constructs sensory perceptions. Our essential prejudices of thought (not what we think, but how we think it) are created by the layout and connectivity of our neurons, much of it probably laid down before we are born.

There is an important difference between these higher illusion-creating mechanisms and those that make us see sticky-out noses on concave faces. It is that they are much more malleable. Try as you might, you will not get rid of the illusory grey square in Hermann's grid (on page 132). But you can work at the illusion of the guilty man, or the illusion of the 'nice, even, *winning*-type' string of lottery numbers. You can chip away at cognitive illusions. Thinking alters thinking. We can actually change the structure and activity in our brains, just by deciding to. It is the greatest accomplishment of our species.

THE CONVERGENCE OF AUNT MAGGIE

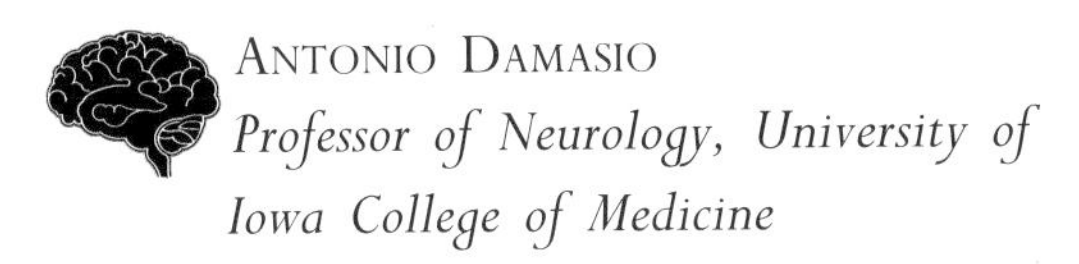

ANTONIO DAMASIO
Professor of Neurology, University of Iowa College of Medicine

The following is an adapted extract from Descartes' Error *by Antonio Damasio.*

Images are not stored as facsimile pictures of things, events, words or sentences. If the brain were like a conventional library, we would run out of shelves just as conventional libraries do. Furthermore, facsimile storage poses difficult problems of retrieval efficiency. Whenever we recall a given object, face or scene, we do not get an exact reproduction but rather an interpretation, a newly reconstructed version of the original, which evolves with our changing age and experience.

And yet we all feel that we can conjure up, in our mind's eye or ear, approximations of images previously experienced. Maybe these mental images are momentary constructions held in consciousness only fleetingly. Although they may appear to be good replicas, they are often inaccurate or incomplete. I suspect that explicit recalled mental images arise from the transient synchronous activation of neural firing patterns largely in the same early sensory cortices where the firing patterns corresponding to perceptual representations once occurred.

But how do we form the topographically organized representation needed to experience recalled images? I believe those representations are constructed momentarily under the command of neural patterns elsewhere in the brain that quite literally order other neural patterns about.

These representations exist as potential patterns of neuron activity in small ensembles of neurons that I call 'convergence zones', which consist of a set of neuron firing dispositions within the ensemble. The dispositions related to recallable images were acquired through learning and can thus be said to constitute a memory. The convergence zones whose dispositional representations can result in images when they fire back to early sensory cortices are located throughout the higher-order association cortices (in occipital, temporal, parietal and frontal regions) and in basal ganglia and limbic structures.

What dispositional representations hold in store in their little commune of synapses is not a picture per se, but a means to reconstitute 'a picture'. If you have a dispositional representation for the face of Aunt Maggie, that representation contains not her face as such, but rather the firing patterns that trigger the momentary reconstruction of an approximate representation of Aunt Maggie's face in the early visual cortices.

The disparate dispositional representations that would need to fire back more or less simultaneously for Aunt Maggie's face to show up in your mind's eye are located in several visual and higher-order association cortices. And the same arrangement would apply in the auditory realm.

There is not just one hidden formula for this reconstruction. Aunt Maggie does not exist in one single site of your brain – she is distributed all over it in the form of many dispositional representations. When you conjure up remembrances of things Maggie she is present only in separate views during the time window in which you construct the meaning of her person.

CHAPTER SIX
CROSSING THE CHASM

Just as each separate brain cell reaches out to make contact with others, so each brain is designed to communicate with its like. Our ability to enter the minds of others, by intuition and by speech, gives human beings a singular advantage over other species. It allows us to create and live in the highly organized hives we call civilization, and as a species we can join in endeavours so grandiose that they alter our environment on a global scale. Language allows us to juggle ideas in a uniquely creative way, and our intuitive knowledge of others' mental machinations makes our relationships complex, subtle and deep.

Language development changed the landscape of the brain radically because it annexed huge cerebral areas once used for movement and sensation. In doing so it created the asymmetry that distinguishes the human brain from that of any other animal.

COMMUNICATION IS NOT a bolt-on extra; for most species it is a matter of survival. Most of the constant information exchange that goes on between living things is unconscious: hormones waft from one creature's gland to another's nose carrying messages about territorial rights and sexual receptivity; reflex pricked ears and swivelled eyes give mute warning of approaching danger within herds; a bee's complex dance, dictated by some mysterious genetic imperative, directs the hive to a cache of pollen.

Once, no doubt, all living things communicated only to the extent that they reacted to changes in others' behaviour or appearance in much the same way as they reacted to other environmental signs. Those who were good at reading these changes must have had a major advantage. If you react to your neighbour's reaction to a rustle in the bushes rather than wait to hear the rustle yourself, you speed up the process of fleeing from a potential predator. Similarly, those who gave big, noticeable reactions must have given a survival benefit to those who hung around and mated with them, so their genes were passed on preferentially. In this way Darwinian selection must constantly have improved communication between individuals until some species became adept at reading the slightest facial expression, body movement or visible physiological change in others of their kind.

Some species, at some point in their development, took communication one step further – they started to do it deliberately. This brought with it all manner of survival benefits. Say you are a dog with a lively litter and you want one of your offspring to stop jumping on your back. You could stop the game by biting the culprit. This would be effective but it might hurt the precious vessel of your genes and thus prejudice its (and your) genes' survival. So instead you snap, giving a deliberate warning that has the same effect but without the risk. Similarly, a horse or a dog or a fish may fake a fighting stance to scare off a would-be aggressor. If it does it convincingly enough it will succeed without involving either of them in potentially dangerous combat.

Hominids, blessed with their free-moving, flexible hands, developed a particularly useful form of deliberate signing: gesture. Gesture is still going strong in humans and can often be more effective than words: try describing a spiral without using your hands, or conveying a Gallic shrug in words. As well as acting as an adjunct to verbal expression it seems to function as a fall-back language when no other is made available. Deaf children who are not taught to sign develop a wide range of gestures spontaneously and they also string the gestures together to form what seem at first to be sentences. However, such children do not develop structured language and without it they do not seem able to make the leap into the world of abstract thought. For that they have to be taught a formal signing language – one that has similar rules and structure to spoken language.

Gesture gave way to 'proper' language about one and a half to two million years ago. The development of language gave humans the tool needed to lift themselves up to a higher level of consciousness. Deliberate communication of the posturing, mimicry sort provided the first step up from the concrete world of the here and now because it created the possibility of an alternative world: a world in which the puppy gets hurt and the audacious horse is attacked. But the range of options that could be conjured

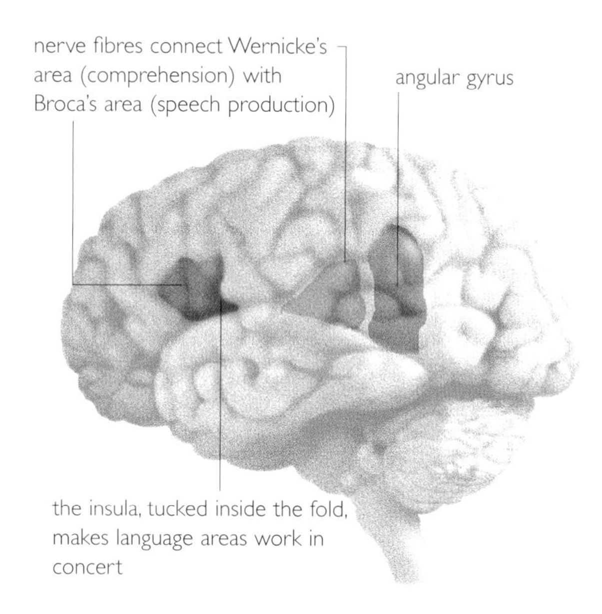

The language areas of the brain are mainly in the left hemisphere, around and above the ear. The main ones are Wernicke's area, which makes spoken language comprehensible, Broca's area, which generates speech (and may contain a 'grammar module') and the angular gyrus which is concerned with meaning.

up in this way were pretty limited. Language, by contrast, opened up a universe of endless possibilities.

Think – what sort of memory would you have without words? What if you needed to remember that bananas, say, were good food? You can't store an actual fruit in your head for later reference so what would be lodged is a sensory impression: the shape and colour, the smooth feel of its skin and the sweet smell of the flesh. Next time you come across a long, smooth, yellow object with a banana-like smell you would match these sensory memories to your bounty and know that it is good.

So far so good, this is (very roughly) how memories are laid down, with or without language, and it works fine in so far as it goes. But what if you wanted to titillate yourself with the idea of a banana? How would you bring those sensory memories to consciousness? Without some symbol, like a name, for an object you have nothing with which to hook the memories of it from storage. They may be brought to mind by sensory reminders – a flash of yellow, perhaps – but voluntary access to them, on demand, would be much harder. Once you can label things, however, you can make your mind into a filing cabinet and pack it with representations of the outside world. You can then pluck out these representations at will and juggle and juxtapose them, creating new ideas in the process. This creates a template within which ideas can be ordered and structured, giving shape and stability to notions that would otherwise remain nebulous. Thus it provides the means to consider abstractions: honesty, fairness, authority and so on.

What it once did for the species, it does for individuals, too. Neurologist Oliver Sacks, in *Seeing Voices*, his book about the deaf community and their language, describes a deaf boy who was reared without access to sign language:

'Joseph saw, categorized, distinguished, used...but he could not, it seemed, go much beyond this, hold abstract ideas in mind, reflect, play, plan. He seemed completely literal – unable to juggle images or hypotheses or possibilities, unable to enter an imaginative or figurative realm.'[1]

Once you can make that jump into the realm of imagination there need be no limit to the mental concepts you create: morality, justice, God... Then, by communicating these ideas to others, you can create social constructs: codes of behaviour, legal systems, religions – and so give your lofty notions practical expression. Such things could never be achieved by grunts and gestures.

The reason for the emergence of language is mysterious, but the brain itself provides some clues. The main language areas are situated in the left hemisphere of the brain, in the temporal (side) and frontal lobes. If you look down on a slice of brain cut horizontally at a certain level, you will see these areas are marked by a distinct, one-sided bulge. The equivalent areas in the right hemisphere are concerned mainly with processing environmental noises and with spatial skills: the rhythm and melody of music impacts here; the 'where' of things in the outside world is registered and fine hand movements – including gestures (but not formal sign language) – are processed. Apart from what seems like a tiny budding in some primates, animals do not have language areas. Instead their brains are more or less symmetrical, and their noises are produced and processed along with environmental noises on both sides.

The region where language developed is also rich in connections to deeper brain structures that process sensory stimuli, and it is one of the places where stored impressions from different senses, particularly touch and hearing, are brought together and reassembled into coherent memories. Assuming that man's immediate ancestors were wired in a similar way to today's primates, it seems that language sprung up in a region where several different and important functions converged. Perhaps the brain of *h. habilis* – the hominid in which language first seems to have blossomed – had already started to expand and was compressed within the skull in such a way that these abutting areas, with their different skills, started to merge. Sound would then become linked to hand gestures and gestures would become associated with the synthesizing of sensual memories into wholes. What better ingredients could there be for creating language?

Once language had taken hold it rapidly annexed large parts of the left hemisphere, growing so fast that it pushed the visual functions back and appropriated most of the locality previously given over to spatial skills. The visuospatial functions were still enormously important, of course, and in the right hemisphere they held their own. This was the beginning of the hemisphere specialization that now makes the human brain uniquely asymmetric in function.

The language vehicle

As far as the brain is concerned the essential thing about language is not that it is a means of communication but that it is a means of communication that follows certain rules. Hence gestures are not dealt with in the language areas and nor are screams, sighs or gasps. Formal sign language is processed in the language department, however, because although it does not have what we think of as the components of language – words – it has a linguistic structure and that is what matters to the brain.

The skill that builds and understands structured language is something quite apart from general intelligence. It does not develop in line with the emergence of important and complex ideas. Rather it seems to be there, potentially at least, in advance of any ideas. It is wired in at such a fundamental level that in normal children a minimal amount of exposure to language at the right time will cause it to flower in all its complexity. Even the most monosyllabic teenage grunt-head is capable of composing grammatically accurate sentences when pushed.

In some people the linguistic vehicle is grander than the ideas it has to convey, and the flimsiest of notions are inflated in order to fill it. Most of us know people who seem to be effortlessly articulate and can talk fluently and expressively without ever actually *saying* anything. At first sight such people may seem socially adept and even gifted, but close acquaintance reveals them to be vacuous.

There is a very extreme form of this condition known as Williams syndrome. Alex is a fairly typical example. He did not babble like a normal baby, nor did he say 'ma-ma' or 'da-da' at the appointed time. At three he was showing signs of mental retardation and he had still not uttered a word.

However, when, at five, Alex finally came out with his first words, they were astonishing. It happened on a hot day in his doctor's waiting room. Alex was restless and toddled over to investigate a portable electric fan that had been brought in to keep the area cool. His mother removed him to a safe distance, but the boy kept returning and sticking his fingers perilously close to the blades. The receptionist, spotting the danger, turned the machine off. Clearly irritated, Alex turned it back on. The receptionist turned it off at the mains. A few minutes later Alex crawled up to the socket and flipped the switch back down. The receptionist responded by pulling out the plug and placing it out of the boy's reach.

Alex was now stymied. Looking severely displeased, he fiddled with the switch, flipped the mains and shook the base of the fan. The other people braced themselves for a wail of anger or some similar infant protest. Instead Alex chose that moment to come out with his maiden speech: 'Jesus Christ!' he said. 'This fan doesn't work!'[2]

From that moment on full-fledged, adult-sounding speech flooded out of Alex's mouth as though it had been stored there, complete and ready-learned, just waiting for the right moment to erupt. By the time he was nine his vocabulary and his grasp of grammar, syntax, modulation and emphasis were as good as any adult's. He has since been supremely confident and outward going, circulating rooms like a seasoned diplomat at a cocktail party. In terms of content, though, his conversation has never become much more consequential than his observation about the fan. Nor, most probably, will it ever.

Williams syndrome is caused by a genetic mutation that produces marked mental retardation (as well as other physical peculiarities) along with extraordinary linguistic skill. Although they often show remarkable intuition and empathy, the average IQ of Williams people is between 50 and 70, around the Down's syndrome mark. Ask a ten-year-old with Williams syndrome to fetch a couple of things from a cupboard and they will get confused and bring the wrong items. They may not be able to tie their shoelaces or add 15 and 20, and if you get them to draw a picture of someone riding a bike they will typically produce an incoherent melange of spokes and wheels, chains and legs. If you ask them to list all the animals they can think of, however, you will get an entrancingly imaginative and fulsome catalogue: one little girl's list included brontosaurus, tyranadon, zebra, ibex, yak, koala, dragon, whale and hippopotamus.[3]

Williams children are unstoppable chatterboxes, endlessly engaging strangers in conversation and taking off on long, expressive narratives interspersed with eccentric exclamations and verbatim inserts:

'...then she said to me: "Oh no! I've left the cake in the oven" and I said: "Heavens to Betsy! That will make for an unfortunate teatime!" and she said: "Well, I think I'll just pop back home and see if I can rescue it before it burns to a crisp," and I said: "Right-ho!"...'

Captivating though the style may be, such stories are nearly always banal, and sometimes they are entirely fabricated. Williams children have no desire to deceive and their confabulations are not designed to bring them any material advantage. It is just that language, for them, is not so much a vehicle for conveying information as a joy to be indulged in for its own sake.

Eye talk

'People talk to each other with their eyes, don't they? What are they saying?' – *Asperger's syndrome subject to a researcher*[4]

Autism is in many ways the mirror image of Williams syndrome. While the chatterboxes seek endlessly to bind non-related and imagined events into cohesive narrative and draw people together in warm community, those with autism see the world as fragmented and alien. They are unable to communicate, either effectively or, in some cases, at all.

There is a spectrum of conditions that can be considered autistic. At one end there are individuals whose only visible function is to make weird, repetitive body movements. At the other there are people who have high IQs, successful professional careers and outwardly normal domestic lives. A few combine extraordinary gifts for drawing, calculating or playing music with a very low IQ. One hallmark of them all, however, is a lack of empathy: autistics do not intuitively understand that other people have minds that may contain an entirely different view of the world to theirs and they cannot 'get into' another person's head.

Autism is strongly heritable and one of the genes that is suspected to be involved lies in a section of chromosome that is also thought to harbour a gene strongly implicated in the development of language.[5] This makes sense: more than anything, autism is a defect of communication – an inability to share feelings, beliefs and knowledge with other people.

The instinctive ability to know what is in another person's mind is known as 'theory of mind'. It is simply illustrated – or rather its lack is illustrated – by a poignant story recounted by the father of an autistic boy. Autistic children do not automatically understand that their desires are subject to constraints; to them their wanting is all-consuming, so they tend to grab what they want when they want it. This particular child had been taught, by painstaking repetition, that he must not just help himself to biscuits. Instead his father trained him to point at what he wanted and then wait for an adult to give it to him. Generally, this worked quite well, and on the whole the child seemed happy living within this and other rules. From time to time, however, the child would throw a massive tantrum for no apparent reason.

One day the father caught sight of his son through a window. The boy was standing alone in a room, pointing at the cupboard where the biscuits were kept. The child could not have known that he was overlooked, and the father decided to stay at the window and watch. After about five minutes of hopeless pointing, the boy started to get upset and after ten he was distraught. Within fifteen minutes he was gripped by a full-scale tantrum. Evidently, the child had been waiting for a biscuit to be given to him. He failed to recognize that this would not happen in the absence of an obliging adult because he had not grasped that the purpose of pointing was to put an idea into someone else's mind. And the reason he failed to grasp that was that he did not have the concept of another person's mind to begin with.

Although we take it for granted, the conceptualizing required to have a theory of mind is complex. It is one of the brain functions that seems least likely to be turned on or off by a single module. Nevertheless, this is just what seems to happen, as the following study by psychologists Uta Frith and Francesca Happé, together with researchers at the Wellcome Department of Cognitive Neurology, seems to show.

A group of normal people read a series of two types of story as they lay in a PET scanner. The following is an example of the first type:

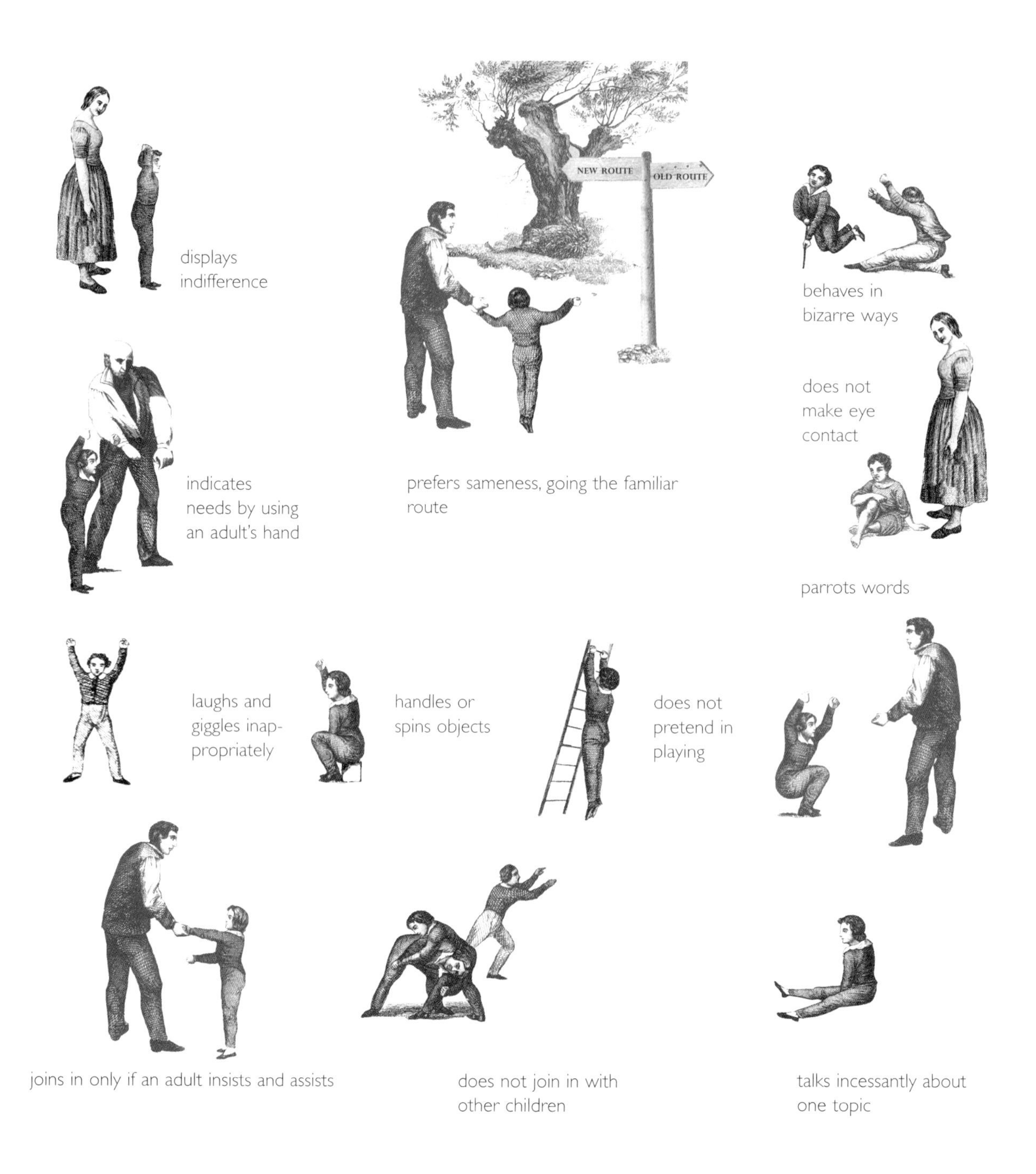

A burglar who has just robbed a shop is making his getaway. As he is running home a policeman on the beat sees him drop his glove. He doesn't know the man is a burglar, he just wants to tell him he dropped his glove. But when the policeman shouts out to the burglar: 'Hey! You! Stop!' the burglar turns round, sees the policeman and gives himself up. He puts his hands up and admits he did the break-in at the local shop.

The following is an example of the second type of story:

A burglar is about to break into a jeweller's shop. He skilfully picks the lock, then crawls under the electronic detector beam. If he breaks this beam, he knows it will set off an alarm. Quietly, he opens the door to the storehouse and sees the gems glittering. As he reaches out, however, he steps on something soft. He hears a screech and something small and furry runs out past him towards the shop door. Immediately the alarm sounds.

After listening to each story the participants were asked a question, and their brains were scanned as they thought about the answer. The question relating to the first story was: 'Why did the burglar give himself up?' The question relating to the second was: 'Why did the alarm go off?' The first question required the participants to have insight into the burglar's mind, while the second required only general knowledge.

The brain scans showed that the normal participants used quite separate regions of their brains to work out each answer. The question that called for a calculation about another person's mental state (in this case the burglar's false belief) lit up a spot in the middle of the prefrontal cortex – one of the most 'evolved' parts of the brain. This did not happen when the subjects worked out the answer to the second question.

The prefrontal area that lit up during the story has wide-ranging connections to many other areas of the brain, in particular those needed to pull in stored information and personal memories in order to 'read between the lines' of a story or 'see behind' the face value of what is presented. These skills are closely related to theory of mind, and are also starkly absent in autism.

A second brain scanning study shows that the reason for this lack may be that the crucial region of brain required for these skills to be activated is not turned on in autistics. In this experiment a group of people with Asperger's syndrome were recruited to listen to the two types of story. Asperger's syndrome is a condition that is characterized by autistic qualities combined with a normal or high IQ. Predictably, these subjects took longer than normal subjects to answer the questions which called on them to see into the burglar's mind, but they got there in the end. The brain activity they used was, however, strikingly different from that of the normal subjects. The prefrontal area identified in the previous scans did not light up at all. Instead a part of the brain below it came into play. This area is known from previous studies to be concerned with general cognitive abilities.[6]

This suggests that the Asperger's subjects worked out what was in the burglar's head by using a brain module that most of us use to work out straightforward causes and effects like those in the second story. They arrived at the answer by working it out like a crossword clue.

Along with their inability to read other people's minds intuitively, people with Asperger's syndrome are extremely poor at reading body

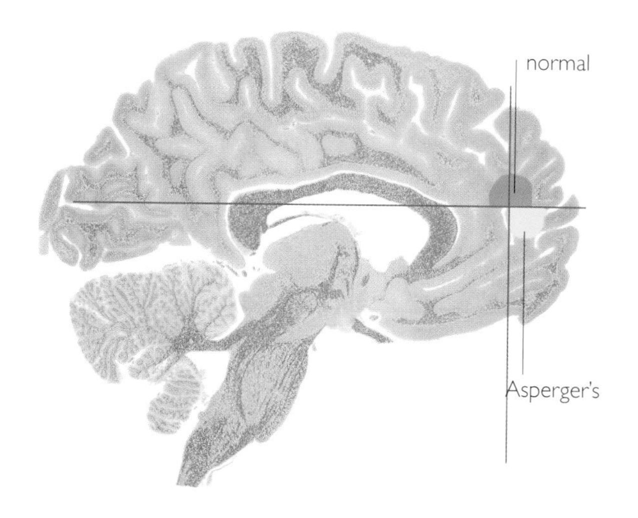

Brain scans show that when normal people read a story which involves inferring someone else's state of mind the left medial prefrontal cortex lights up (hatched area top). But when the same story is read to Asperger's subjects a different area just below it is activated.

language and facial expression. Simon Baron-Cohen and colleagues at the Department of Experimental Psychology at Cambridge recently found an actress with the rare talent of accurately expressing a wide range of mental states. They took photos of her face while she expressed ten basic emotions (such as sadness, happiness and anger) and ten complicated ones (such as scheming, admiration and interest). They then showed both the whole face and separate parts of it – the eyes or the mouth – to groups made up both of normal people and those with Asperger's syndrome. The participants were asked to identify the expressions.[7]

The results revealed what seems to be two tiers of expression reading, each of which uses a different mechanism. For basic emotions the technique seemed to be to read the whole face – seeing the eyes or the mouth alone was less helpful. However, when expressions became complicated, normal people found it just as easy to read the expression by looking at the eyes alone as by looking at the whole face. This suggests that at a certain level of complexity a new communication method comes into play – what Baron-Cohen has called the 'language of the eyes'. People with Asperger's syndrome do not seem to know this language. On basic expressions they could read faces as well as normal people, but when expressions became complicated they could not make them out. They were particularly perplexed when they tried to read the eyes alone.

Once you know what is happening in their brains it is easy to see why people with autistic tendencies are slow on the uptake and exasperating to live with. Asperger's syndrome is thought to affect as many as one in 300 men (nearly all are male) but most of them are unlikely to be formally diagnosed

We use our eyes and our ears to understand the spoken word and when what we see does not match what we hear we often make up a likely compromise. If, for instance, the syllable 'pa' is played to a subject while they watch the speaker's lips move as though to say 'ka', the sound that the subject reports hearing is 'ta'. The perception does not change even when the listener knows what is happening. The illusion is thought to occur because the brain is merging input from two quite separate brain modules in order to construct the perception. Brain scans suggest that this merging is done by the auditory cortex in the right hemisphere.

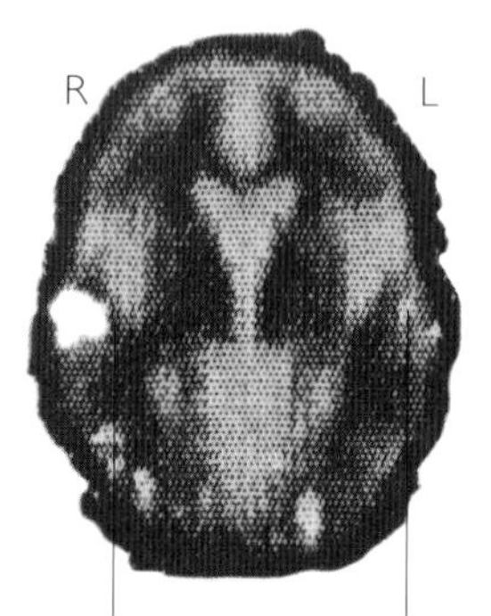

The scan above left (looking up into the brain) shows active areas in the brain of a subject who is reading a person's lips in silence. The lit-up area on the opposite side is the language area, also triggered by lipreading. The scan on the right shows the visual areas which are also activated during lipreading: the visual motion area processes the lip movement while the face area scans the face for expression.

because their general intelligence usually gets them by, despite their lack of intuition. Nevertheless, their behaviour tends to be very odd: they are usually obsessionally wedded to routine, and may indulge in all-consuming hobbies that typically involve collecting or categorizing (train-spotting is, of course, the classic example). Socially, they are a disaster: they do not get jokes based on human foibles and fail to be captivated by gossip. They may drone on for hours about things that are of no interest to others, outstay their welcome or doze off when others are talking. Mercifully for them, they are not given to embarrassment because they are not generally conscious of what other people might be thinking about them.

Some people with autistic tendencies are very high achievers and their oddness may show up only in their preference for being alone, lack of empathy and single-minded pursuit of their own interests.[8] Many very successful academics are thought to fall into this category.

Although most people with subtle forms of autism find it easier to be alone, they often have a strong desire to abide by social convention and some of them get married. Much of the closeness in any partnership stems from the ability of each partner to see into the other's mind, and those who marry people with Asperger's syndrome often feel there is something fundamentally wrong with their relationship, even though they may not realize quite what it is. Psychiatrist John Ratey, in his book *Shadow Syndromes*,[9] quotes Susan, whose husband, Dan, displays typical autistic characteristics:

'It's the inappropriate responses when there is emotional news or bad stuff going on [that is difficult]. Yesterday I found out that I didn't get the job I applied for and I was really upset and crying. Dan said in a little-boy voice: "Oh, better luck next time" and then moved right on to the next topic. Once when I had a devastating

Animal noises, including birdsong, differ from language in that they are largely 'hard-wired' and generated by the unconscious brain.

blow Dan just said: "Oh. Want to go swimming?" I realize he can't help it – but it gets very lonely here [in this marriage] sometimes.'

Music

Music is generally regarded as one of the more elevated endowments of the human world. It seems to be one of the few things we do simply for pleasure – a pure piece of hedonistic icing on a cake of necessities. Yet evidence is accumulating to suggest that our brains are moulded by our genes to create and understand music rather as they are made to form language. Children as young as five months are aware of tiny shifts in musical pitch and by eight months they can remember a melody well enough to show surprise if a single note in a familiar tune is altered.[10]

There is no known mechanism by which purposeless functions come to evolve. Music is therefore likely once to have had some survival benefit, and the most probable one is that it is a prototype

THE LONELY BRAIN

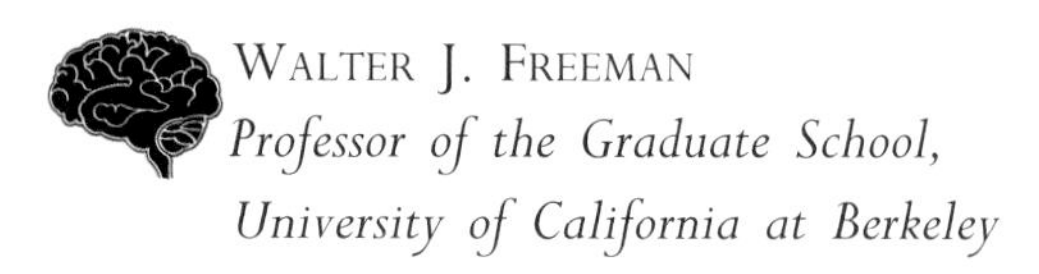

WALTER J. FREEMAN
Professor of the Graduate School, University of California at Berkeley

Nobody can truly feel what another is feeling, though we can empathize. This separation has a good side, in that we have the inalienable right to privacy. The bad side is loneliness and existential dread. And here, paradoxically, is where joy begins to enter.

Our brains don't take in information from the environment and store it like a camera or a tape recorder, for later retrieval. What we remember is continually being changed by new learning, when the connections between nerve cells in brains are modified.

A stimulus excites the sensory receptors, so that they send a message to the brain. That input triggers a reaction, by which the brain constructs a pattern of neural activity. The sensory activity that triggered the construction is then washed away, leaving only the construct. That pattern does not 'represent' the stimulus. It constitutes the meaning of the stimulus for the person receiving it.

The meaning is different for every person, because it depends on their past experience. Since the sensory activity is washed away and only the construction is saved, the only knowledge that each of us has is what we construct within our own brains. We cannot know the world by inserting objects into our brains.

Why not? It is because the world is infinitely complex, and any brain can only know the little that it can create within itself. If everything each of us knows is made inside our brains, how can we know the same things? The answer is simple. We can learn almost the same things. Almost.

Joy comes from surmounting the barrier between us by sharing our feelings and comforts. We cannot ever really cross it, but, like neighbours chatting over a fence, we can be together. However, there is more to this communion than mere talking – there is trust. What is the chemistry of trust?

Answers are found when we look back on our mammalian ancestors. Raising a helpless infant to childhood requires intensive parental care, which comes with bonding between the parents and the infant. Now, how does a carefree child, when it has grown up, become a parent? This change in role requires a catastrophic change in beliefs, attitudes and values to make new parents. We would say they fall in love, with each other and with their offspring. Scientists have learned that, when animals mate and give birth, specialized chemicals are released into their brains that enable their behaviour to change and maternal and paternal patterns of nursing and caring to appear. The most important is a chemical called oxytocin. It doesn't cause joy; it may cause anxiety, as it melts down the patterns of connections among neurons that hold experience, so that new experience can form. We become aware of this meltdown most dramatically as a frightening loss of identity and self-control, when we fall in love for the first time.

Bonding comes not with the meltdown but with the shared activity afterward, in which people learn about each other through co-operation. Trust emerges not just with sex, but also with shared activity in sports, combat and competition, through which people bond by learning to trust each other.

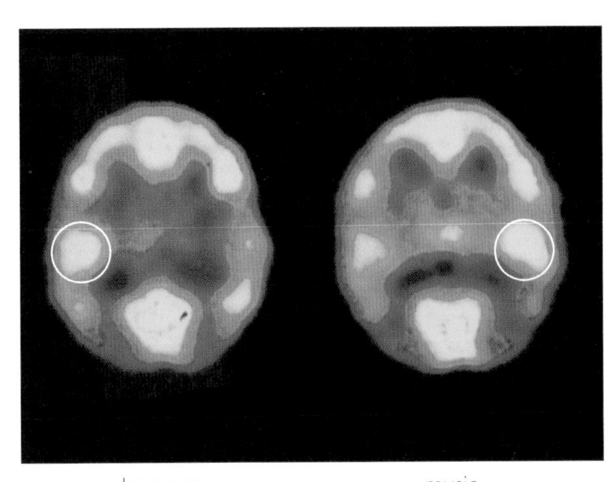

Above: *Words and music are processed in different parts of the brain: the scan on the left shows the left auditory cortex lighting up when a subject listens to words while the scan on the right shows the right auditory cortex springing into life at the sound of music. The language centre of the brain in the left hemisphere is divided into different regions, each of which does a specific task.*

communication system. Support for this idea comes from the fact that music appreciation seems to be wired into some of the dumbest creatures on earth – something that seems rather unlikely if it really were just a cultural frill.

A short while ago Jaak Panksepp, a psychologist in Ohio, recruited a flock of chickens for research purposes. He played various pieces of music to the birds and noted their reactions. Of all the musical delights sampled by the chickens, Pink Floyd's 'The Final Cut' brought the most notable response. The birds ruffled their feathers, shook their heads slowly from side to side and generally gave a strong impression of superannuated hippies at a Woodstock revival festival.[11]

Panksepp concluded that the feather-ruffling was the avian equivalent of the musical 'chill' or 'tingle' experienced by people when they hear certain musical effects. Nearly everyone knows how it feels: an ecstatic sub-orgasmic ripple of tension and release that literally makes the hairs rise up along your spine.

A BBC radio programme called *The Tingle Factor* once asked listeners to nominate their favourite tingle-producing musical excerpts. Certain pieces cropped up time and time again. They tended to be elegiac or sad, familiar and of conventional construction. Mozart, Beethoven and the operatic arias of Puccini and Bizet were among the favourites. Pink Floyd did not figure, but this probably reflected the audience profile of the programme – chickens were not canvassed.

The particular musical moments that are most frequently reported as tingle triggers are sudden shifts of harmony, or sequences that set up an expectation of a particular resolution (the progression from E to F sharp, for instance, sets up the expectation for G sharp to complete the phrase), then delay or subvert it. The emotional pattern in all such moments is relaxation–arousal–tension–relief–relaxation.

As well as sending momentary shivers up the spine, music can tell stories, prompt us to dance, make us happy, aggressive or sad, hurry us along or put us to sleep. The right sort can even dictate what type of wine we buy in a supermarket.[12]

The brain has to do a great deal of construction work to make music from the mere beat of sound waves against a membrane in the ear. The process is similar to that which turns visual stimuli into meaningful images. Each component of the incoming information – pitch, melody, rhythm, location and loudness – is processed separately, then the parts are brought back together and reassembled, along with whatever emotional response they elicit. Glitches in this production line produce the musical equivalent of agnosia – a person may hear a tune clearly, yet be unable to tell if it is 'Jingle Bells' or the '1812 Overture'. Nor can they tell two tunes apart or spot a wrong note. And yet – although they are entirely cloth-eared – such people can nearly always say whether a piece of music is

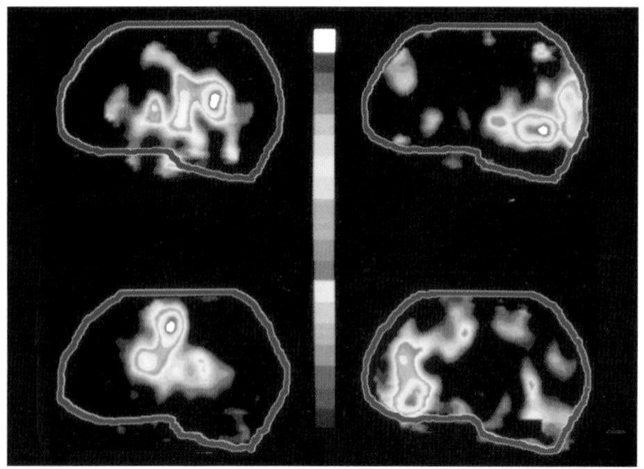

left
right

Above, left: *The brain responds differently to a word according to whether it is (top left) hearing it spoken; (top right) seeing it written down; (bottom left) speaking it or (bottom right) considering which other words it relates to.* Right: *When you look at a word you may see it as a word – that is, a component of language – or it may act as a trigger for the concepts which it represents. In each case a different part of the brain comes into play. The scans, above, show (top) how the language areas are activated when a person is shown a word and asked how many syllables it contains and (bottom) how the same word sparks activity in completely different areas (including those concerned with memory retrieval) when the person is asked to consider what the word means.*

happy or sad. This is because the sound is processed in parallel by the limbic system, which notes only its emotional tone. The feedback from this crude sampling is what informs the tone-deaf person about the happiness or sadness of a piece. The tingle factor probably arises from this primarily unconscious emotional processing. The fact that it is not a function of the conscious brain (even though we may become consciously aware of it) explains why it seems to work for chickens as well as people.

Jaak Panksepp thinks the emotion-tugging effect of certain types of music lies in its similarity to vocal (but not verbal) signals that carry emotional messages between animals. The tension-building sequence with delayed resolution that typically brings about the chilly spine feeling, for example, has features in common with the sounds made by infants – both human and animal – when they are parted from their mothers. In animals these cries have been found to trigger a drop in oxytocin – the brain chemical most closely associated with parental bonding – and they also bring about a drop in the mother's body temperature. When the mother is reunited with her baby, the child responds by 'resolving' the cry – a vocal performance not dissimilar to closing a phrase of music with a satisfyingly final note. At the same time the mother's oxytocin level goes up, and her body becomes warmer. Women have been found to feel the tingle more keenly than men, which fits in neatly with this theory.

Perhaps, then, the tingle is a faint echo of the shiver that helps to motivate a mother to seek out her lost infant. Other emotional frissons we experience when notes take a particular turn may relate to similar signals.

Words

'In trip to spleen is Hardwick. Is the watch in red gone to offer a digital sun to ring machine you will find it warring in another hat. That gas cones the nods aces and that her senses are pipped withers and rum slurp syllabub and dropped currents' – *Author's computer, programmed with Voice Recognition Programme*, 1997

Correction. That should say: 'Interpreting speech is hard work. If you watch the thread of sound on a monitor of a digital sound editing machine you will find that the words run into one another; that gaps come in the oddest places; and that the sentences are peppered with errs and ums, slurred syllables and dropped consonants.'

Which is why, at the time of writing, voice recognition programes for computers still produce gobbledegook like that at the start of this section. Only by training such programes to recognize an individual's voice, and by speaking exceptionally slowly and carefully, can bank-breaking gigabytes of computing power be persuaded to do what most human brains can achieve effortlessly by the time they are three.

Making sense of a stream of spoken words is a formidably complicated business. First, the brain has to recognize that what is coming in is in fact language. This preliminary discrimination may be done in part by the thalamus[13] and completed by the primary auditory cortex. Speech is then shunted to the language areas to be processed, while environmental noises, music and non-verbal messages – grunts, screams, laughs, sighs and exclamations like 'ugh!' – go elsewhere.

Most of what we know about the brain locations concerned with language comes from studying people who have language problems caused by strokes or other types of brain injury. Some of these people have oddly specific symptoms: they may be able to speak fluently, for example, but have no idea what they are saying, or be able to read but not write.

These two scans (taken from the front) are effectively a two-frame movie of a brain at work. Left: *The subject's brain as they read the sentence: 'the star is above the cross'. It shows typical left hemisphere activity associated with language processing.* Right: *This scan was taken seconds later, when the subject was shown a picture of a star and a cross and asked to rotate it in their mind's eye until the star was above the cross. It shows the switch in brain activity from the language areas to the parietal lobes where spatial tasks are processed. At the moment such brain movies are jerky, like early films. When scanning is fast enough to produce smooth pictures it will be possible to chart the procedures our brains use to carry out complicated mental tasks. (Based on: 'Movies of the Brain: Imaging a sequence of cognitive processes' Just M. et al* Neuroimage *Vol. 3 No. 3, June 1996, Part 2)*

The language cortex – situated on the left side in 95 per cent of people – surrounds the auditory cortex in all directions. It is spread over most of the temporal lobes and edges into the parietal and frontal lobes. Two main areas – Wernicke's and Broca's – have been recognized for more than a century, but brain imaging studies have recently suggested that other areas are also involved. These include part of the insula – a hidden expanse of cortex that lies within the great infold, known as the Sylvian fissure, that divides the temporal and frontal lobes.[14] Each main area of language cortex is probably split,

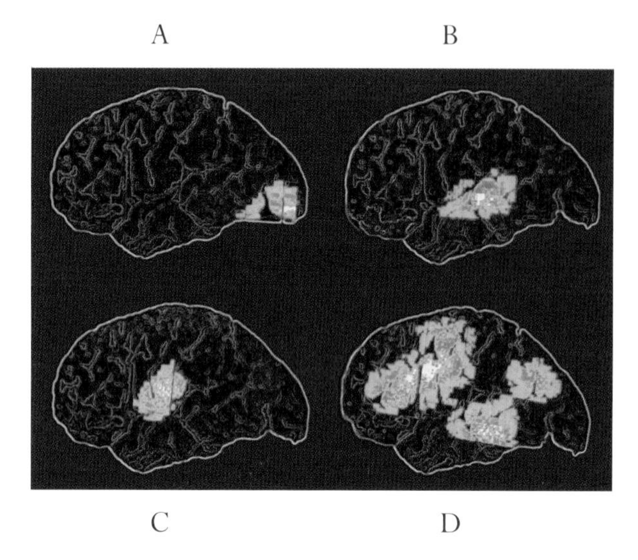

A: reading activates part of the visual cortex
B: listening to speech makes the auditory cortex light up
C: thinking about words makes Broca's area – the articulation centre – light up
D: thinking about words and speaking generates widespread activity

like the sensory cortices, into many different processing regions and sub-regions, but it is only in the last few years that brain imaging studies have started to reveal what they are.

Once speech has been identified, the words are assigned some sort of meaning, and, at the same time the gappy, slurry, tangled ribbon of sound is broken down into its elements – separate words or phrases. The two things are necessarily done together (another example of the brain's parallel processing ability) because without meaning it is almost impossible to make out language construction. Try listening to someone speaking at normal conversational speed in a foreign language and you will be hard pushed to tell when the sentences begin and end, let alone the clauses. Similarly, without awareness of construction it is difficult to make out meaning. there is no punctuation for example in this and already I expect you are finding it hard going yet punctuation is only one aspect of construction if I the words jumbled this like not clue a have would you what of going on about I was so important construction to meaning is. You as see can.

One essential skill required at an early stage is that of discriminating very fast-changing sounds. The difference between two spoken consonants – the 'p' in 'pa' and the 'b' in 'baa', for instance – is only distinguishable for a tiny fraction of a second. If you miss it, there's no knowing whether you are hearing about your father or a sheep. An inability to make these quick distinctions between sounds is thought to be the root cause of specific language impairment (SLI) – a condition in which otherwise bright, attentive children fail to pick up language in the normal way.[15] Neurologists have found a tiny area of tissue – about 1 centimetre square – near to Wernicke's area that lights up only when consonants are heard. When this area is deactivated by electromagnetic inhibition, patients have difficulty understanding words that depend on consonants for identification, whereas they can still make out words in which the crucial distinguishing sounds are vowels. Lack of normal activity in this area may account for SLI.

The cortical area that imposes structure on incoming speech – assuming there is one – has yet to be identified. The eminent linguist Noam Chomsky and his disciple Steven Pinker have argued persuasively for some form of 'language organ' in the brain, but what form it might take and where it might be located are not known. People who lose the ability to structure sentences through brain injury tend to have lesions towards the front of the language cortex[16] and one possibility is that syntax is produced somewhere around Broca's area – the more frontal of the two main language regions. Or it may be something that is dealt with in a subcortical area between the two language areas. Steven Pinker

FINISHED FILES ARE THE RE SULT OF YEARS OF SCIENTIF-IC STUDY COMBINED WITH THE EXPERIENCE OF YEARS

Count the 'F's in this sentence...overpage for answer

has suggested that the language organ may not be a neat module at all – more a smear of neuronal activity resembling road-kill. Alternatively, it may be that there is no physical 'module' for syntax at all – rather that it is produced by some neural interactions or pattern that involve activity in many different areas.

Analysis of word meaning is carried out either in or very close to Wernicke's area – a patch of cortex that is splayed over the top and back of the temporal lobe, edging up to the parietal lobe. Damage to the connections between the primary auditory cortex and Wernicke's area can result in a peculiar language disorder known as word deafness. People with this condition cannot understand spoken words, yet they may still be able to read, write and speak quite normally. 'Voices come, but not words,' says one patient. 'Speech sounds like an undifferentiated continuous humming noise without any rhythm,' says another.[17]

Damage to Wernicke's area itself causes another type of disorder. Wernicke's aphasia patients retain perfectly fluent speech and their grammatical construction is fine; if you listen to them talking from a distance, you'll think there is nothing wrong. Attend to the actual words, though, and you realize that much of what they say is nonsense. Wrong words or non-words are substituted for right ones and much of the content may be made up of meaningless jargon. People with Wernicke's aphasia do not understand their own spoken words either, so they cannot monitor their own speech. For this reason they often seem to be blithely unaware that they are talking nonsense. The communication breakdown makes it difficult to assess Wernicke's patients' general intelligence, but as far as can be gathered their ability to reason is generally unaffected.

The rather surreal drawing below is one of the tools used by psychologists to test for various intellectual impairments. This is an example of the description of the drawing given by a patient with Wernicke's aphasia:

'Well this is...mother is away here working her work out of here to get her better, but when she's looking, the two boys looking in the other part. One their small tile into her time here. She's working another time because she's getting to. So two boys work together and one is sneaking around here making his work and his further funnas [sic] his time he had.'[18]

The jargon generated by Wernicke's aphasics can sound impressive. One patient, trying to explain what he used to do for a living, told a researcher: '[I was] an executive of this, and the complaint was to discuss the tonations as to what type they were...and kept from the different

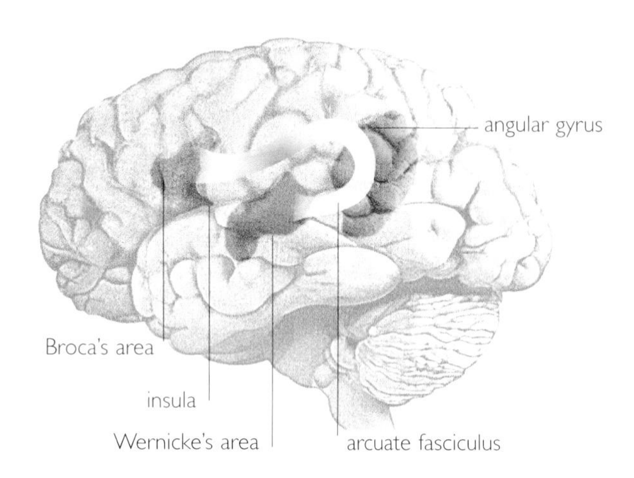

Some types of dyslexia may be due to what is known as a dissociation disorder – a missing or inactive connection between two brain modules. A study in which PET scans were made of dyslexics' and non-dyslexic's brains while they attempted a complex reading task, suggests that the two main language areas, Wernicke's and Broca's, do not work in concert in dyslexics. This appears to be because an important neural link in the vicinity of the insula cortex is not activated during such tasks as it is in others. The result of the 'dead' connection is that the words cannot be understood (a function of Wernicke's area) and articulated (a function of Broca's) at the same time.

tricula to get me from the attribute convenshements.[19]

This type of speech flows effortlessly from the mouths of Wernicke's patients – odd though its content might be, they have no difficulty producing it. This is because speech production is governed by a different part of the brain altogether.

Broca's area is further forward, in the side of the frontal lobe. It abuts the motor cortex, just by the part that controls the jaw, larynx, tongue and lips. This area of cortex seems to hold the programmes that instruct the neighbouring motor cortex to articulate speech. People with damage to this part of the brain can understand what is said to them perfectly well, and they know what they want to say. They just cannot say it. Those words they can get out tend to be concrete nouns or verbs, delivered in staccato 'telegramese'. A Broca's aphasic described the cookie-theft picture: 'cookie jar... fall over... chair... water... empty.' Some people with Broca's aphasia are unable to say anything. Often, though, their economic use of the odd apt word can convey volumes. Professor Christine Temple, in her book *The Brain*, reports visiting a Broca's patient who had a bad head cold. 'Nose,' said the patient, by way of greeting.[20] The meaning, apparently, was crystal clear.

Damage to cortical areas adjacent to these two main ones can cause a wide range of very specific language problems. If the connections between Wernicke's area and Broca's area are damaged, for example, a person may be unable to repeat what is said to them. This is because the incoming words (which are registered in Wernicke's area) cannot be passed on to Broca's area for articulation. Then there are people who seem to be unable to stop repeating what is said to them – the affliction known as echolalia. This may be caused by overactivity of the connec-

FINISHED FILES ARE THE RE
SULT OF YEARS OF SCIENTIF-
IC STUDY COMBINED WITH
THE EXPERIENCE OF YEARS

Did you spot six 'Fs'? Probably not. Most people see only four – they skip the 'fs' in 'of' because the brain processes short familiar words as a single, whole, symbol rather than breaking them down into smaller units as they do with longer or less familiar words. The two types of words are thus probably processed in different brain areas.

tions between the two language areas so that what goes into one is automatically shunted on to the other before other areas of the cortex can step in to inhibit it.

Occasionally, a severe stroke or similar cerebral injury can damage the areas surrounding the language cortex and leave it cut off from the other parts of the brain. Such patients are typically mute and appear not to understand anything that is said to them. Yet they can repeat words and finish well-known phrases. 'Roses are red', for example, will bring the response 'violets are blue'. This shows again how elements of language can exist in isolation from other intellectual functions.

Reading and writing come to children after speech is established, as children have to be taught to read – it does not come so naturally to them. This may be because it is a latecomer in evolutionary terms and, unlike spoken language, we have probably not evolved a specific system with which to read and write. It seems more likely that we produce and understand written language by pressing into use the language system as evolved for speech, together with parts of the object identification and gesturing systems.[21]

As with speech, processing the written word takes place in several different areas. Reading and writing pull on the brain's capacity to see (or touch, in the case of Braille) and to use fine hand movements to manipulate a writing instrument, as well as on the language areas. It is not surprising, then, that the areas dedicated to it are situated around the junctions between the areas given over to these different skills.

Just behind and slightly up from Wernicke's area lies a region of brain where vision, spatial skills and language jostle together on the margins of the occipital, parietal and temporal lobes. It is marked by a bulge called the angular gyrus, which seems to act as a bridge between the visual word recognition system and the rest of the language process. Damage to the angular gyrus itself is likely to disrupt both reading and writing. If the area around it is damaged, by contrast, a person may be left in the peculiar position of being able to write but unable to read silently. One woman, identified by researchers as J.O., is able to understand spoken words and write quite normally, but when she tries to read back silently what she has written she can't make head or tail of it. If she is asked to read the words aloud, however, she can – and can then understand their mean-

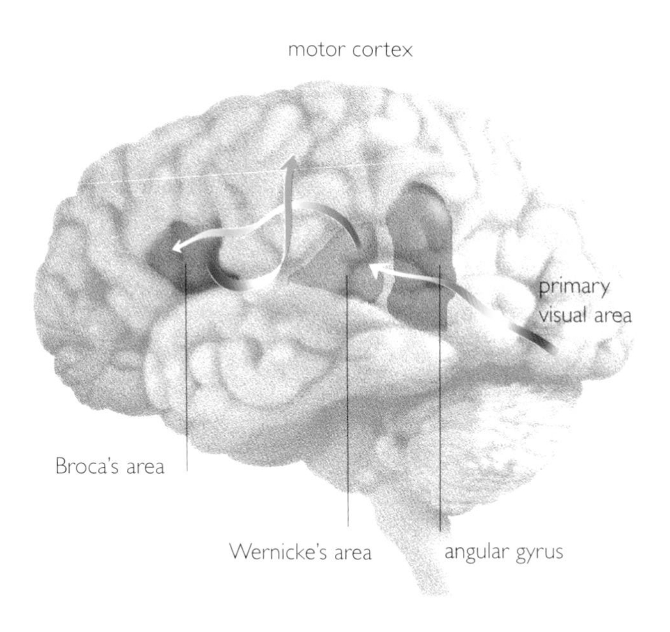

Reading and writing involves more than just the language areas – the visual cortex feeds information in from the page and the motor cortex is required to activate the muscles used for writing. It is therefore important that information flows freely between these areas – if it is blocked or disrupted a form of dyslexia may result.

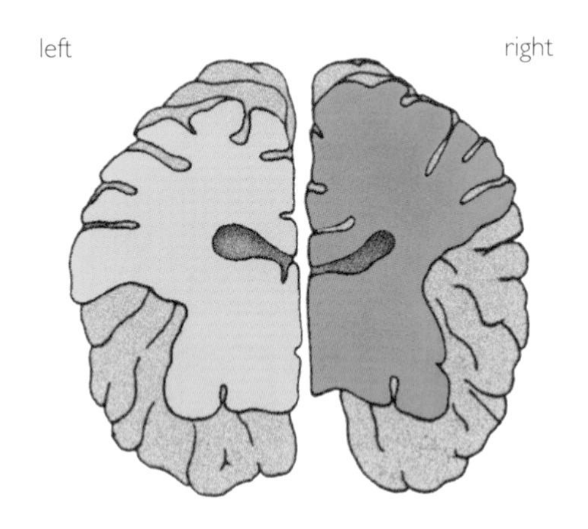

In the human brain the area which processes language, in the left hemisphere, is visibly larger than the equivalent area in the right hemisphere. There is no such asymmetry, generally, in animals' brains – though some studies have detected what looks like the beginning of extra left hemisphere development in some primates.

ing by listening to her own voice.[22] 'I see the words but nothing comes in,' she says. This odd condition seems to come about when the neural path between the visual cortex and the angular gyrus is blocked or broken so visual information about words cannot be matched to their meanings. They can be read, though, because the visual information can still be matched to their sounds – presumably by another route.

The importance of language skills in a literate society can hardly be exaggerated. People are judged on how they speak and nearly all academic teaching is done by means of language. So anyone who is not 100 per cent up to scratch in this area is likely to be seen as deficient across a whole range of skills. This used to happen as a matter of course to people with dyslexia, and it still happens to some extent today.

Dyslexia takes many different forms and probably has many different causes. One form, however, seems to be caused by a particular brain module failing to fire. PET scans of dyslexic people doing word tasks have shown that, unlike in normal people, the language processing areas in dyslexics' brains fail to work in concert so the incoming words get jumbled up and disjointed. The scans, which were done on dyslexic and non-dyslexic volunteers of above-average intelligence, showed that the word tasks caused the language areas in the non-dyslexics to fire in unison, together with a spot in the insula, the deep infold that lies in between them. This spot appeared to act like a bridge between the language areas, orchestrating their activity. In dyslexics, however, the insula did not fire and each language area was activated singly.[23]

The discovery of a distinctive physiological sign for dyslexia should make the condition easier to recognize. Brain scans are not yet practical for general diagnosis, but the discovery of the precise neurological mechanism involved in at least one form of the condition should make it possible to devise functional tests that show up the fault more clearly than the broad-based reading and writing tests used now.

It also raises the possibility that dyslexia may one day be susceptible to a physical cure. As the main problem appears to lie in the link between two brain areas, perhaps it would be possible to insert a tiny artificial bridge – a pacemaker for the brain.

Speech development

Babies are tuned in to speech from birth – or even, perhaps, in utero. If kicks and squirms are taken as reliable markers of approval, they seem to prefer familiar stories to those they have not heard before.[24] Most of them seem to love the sound of the human voice, particularly their mother's, and play an active part in promoting 'conversations' – 'motherspeak' on one side, burbles and grunts on the other – from a few weeks of age.

'Proper' language starts in the second year

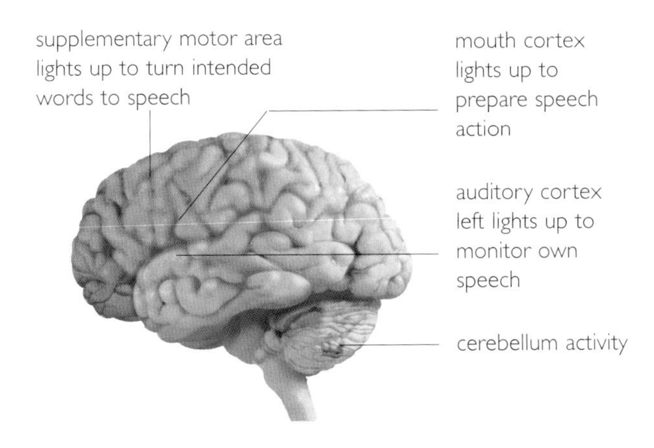

NORMAL
left hemisphere

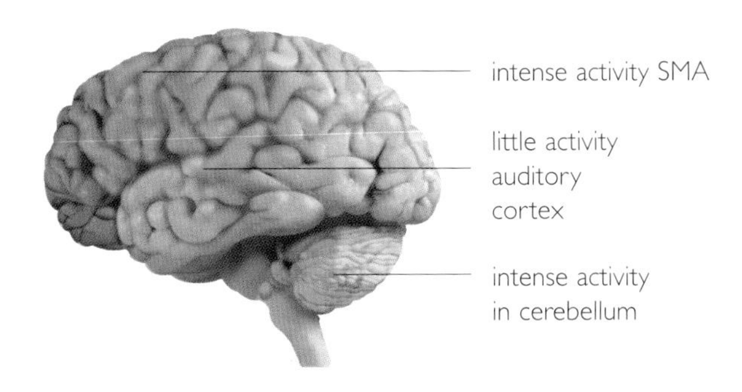

STUTTERER
left hemisphere

little activity in right auditory cortex

cerebellum activity

NORMAL
right hemisphere

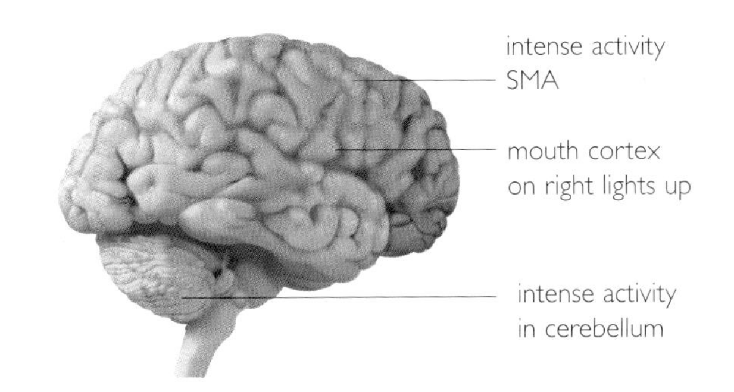

STUTTERER
right hemisphere

When reading aloud, people who stutter show different patterns of brain activity from non-stutterers. In particular, the right hemisphere is more active in stutterers. This suggests that stuttering may be due to competition for dominance between left and right hemispheres. Neither side can decide which is in control, so the two both try to produce words, with disastrous results. Stutterers' brains also show lack of auditory feedback – the loop that usually plays their own voice back to them is underactive. When stutterers read in chorus with other non-stutterers their speech impediment disappears, probably because the auditory feedback is supplied by the other readers.

with the activation of the two major speech areas that occupy separate but neighbouring areas on the side of the brain. One – Wernicke's area – is specialized for language comprehension, while the other – Broca's area – deals with speech articulation. Language initially develops in both hemispheres, but in 95 per cent of cases it shifts by the age of five to lodge in the left hemisphere only and the abandoned speech areas in the right hemisphere are given over to other things, including gesture.

One of the clearest precursors to language is

babbling – the speech-like torrent of sounds that babies typically start generating at about eighteen months. This is followed within a couple of months by a rapidly expanding vocabulary of proper words. This sudden explosion of language coincides with a flourishing of activity in the frontal lobes. Around this same time children seem to develop self-consciousness. They no longer point at their reflection in the mirror as though they see another child, and if a dab of coloured powder is put on their face while they look at their reflection they rub it off – they don't rub the mirror as younger children do.

The simultaneous emergence of speech and self-consciousness may simply reflect the parallel maturation of the two relevant brain areas: language and frontal lobes. Or it may be that the two things are inevitably connected. Language gives the child the tool it needs to form a concept of itself that it can then place outside its own experience and regard in relationship to others. Once it can do that it can start to make plans, and to make plans it needs functioning frontal lobes. So perhaps the language areas, as they stir into life, send wake-up messages forward to the frontal cortex.

Language comes naturally to children – provided they are exposed to it during infancy. But if they are deprived of the sound of speech, their brains may be physically disordered.

In 1970 in Los Angeles a child was discovered who had been locked away, practically from birth, in a featureless room with minimal human contact and nothing to watch, do or play with.[25] When she was found, aged thirteen, she could not hop, skip or extend her limbs, she was unable to focus her eyes beyond the width of her prison, and the only words she knew to utter were 'stopit' and 'nomore'. She was encouraged to learn language and her vocabulary increased dramatically. But grammar – which comes so instinctively to a young child – escaped her.

Brain imaging revealed why: the child had a highly unusual arrangement of cerebral functions. Most notably, when she spoke it was with the right side of her brain, not the more usual left. This is not in itself pathological – about 5 per cent of people (most of whom are left-handed) do this and all it usually signifies is that their language areas have developed in the opposite side of the brain to normal. But in this girl's case something else had happened. Deprived as it was of human speech, her language area had effectively atrophied. When she finally encountered spoken words her brain processed them in the area normally reserved for environmental noises. This area, presumably, had been kept active by the sad little symphony of noises that had reached her during her incarceration: distant birdsong, noises from the bathroom next door, the odd creak of floorboards.

Foreign tongues

We all start off with the potential to speak any language, but if we are only exposed to a single tongue our options soon narrow because the neurological wiring needed to distinguish sounds atrophies if it is not stimulated early in life. Therefore people who learn foreign languages as adults rarely speak them without an accent – Japanese adults cannot reproduce the English 'l' and 'r', for example, because these are sounds not heard in Japanese. Conversely, English speakers cannot manage certain Japanese phonemes.[26]

Second languages are processed in a different section of the language area to the mother tongue.[27] This is why people who suffer very localized stroke damage sometimes lose their ability to speak their native language, while retaining their ability to speak those they learnt as adults.

The importance of gossip

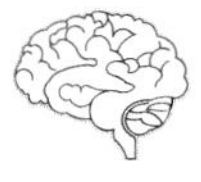

PROFESSOR JOHN MAYNARD SMITH
*Emeritus Professor of Biology,
University of Sussex*

Something very puzzling happened about 50,000 to 100,000 years ago. The fossil evidence is patchy, but it seems that hominids suddenly developed brains that, in terms of size, were much like ours. Yet this apparent growth spurt was not reflected immediately by any great cultural changes. That came 50,000 years later, when a whole variety of artefacts – tools and musical instruments and cave drawings – suddenly came on the scene.

Something must have happened between the physical changes in the brain and the cultural expression of such changes. Most linguists now argue that the something was the development of language. I am sure that our ancestors had been communicating for a long time (half a million years or so) before they became linguistically competent, so perhaps there is something about language itself that led to this acceleration of cultural complexity.

Or could it be the other way around? Could cultural changes have brought about the development of language?

Brain size in primates is closely associated with the size of the social group in which the animal lives. The initial expansion of what was to become the human brain may have been prompted by a growth in numbers that required greater social skills: remembering who various members were, whether they were friends or enemies and so on. One social skill that would have been increasingly required as the social group grew was gossip – the need to swap information about each other. This may have brought about the initial expansion of the frontal cortex and, in its wake, the development of language.

At the moment this is still speculative but I believe that the answer will be found through genetics. If, as some suggest, there is a specific gene or genes for grammatical language, we may be able to identify them and then find out their origin by seeing what related strands of DNA do in our near relatives. The brain is a genetically expensive organ and it would take more than a single happy mutation to bring about something as complex as language – but it may be possible to track those mutations back through evolution.

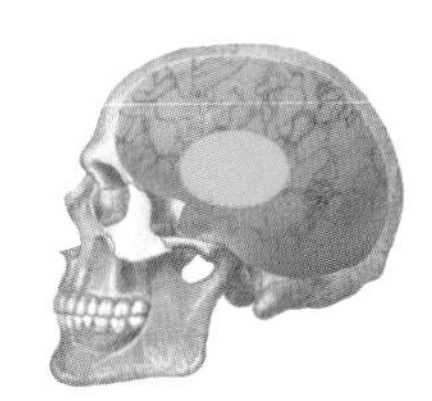

The skull of hominids was pushed into its present shape by the evolving brain

CHAPTER SEVEN

STATES OF MIND

The human brain holds billions of impressions, some fleetingly, some for a lifetime. We call them memories. Just as incoming sensory information is broken down, then rebuilt to form perceptions, so perceptions are broken down again as they pass into memory. Each fragment is sent off to storage in a different part of our vast internal library. But at night, when the body rests, these fragments are brought out from storage, reassembled and replayed. Each run-through etches them deeper into the neural structure until there comes a time when memories and the person who holds them are effectively one and the same.

MEMORY IS MANY DIFFERENT THINGS: it is the picture that comes into your mind when you think of the home you lived in as a child; it is the ability to hop on a bike and pedal away without working out how to do it; the feeling of unease associated with a place where something frightening once happened to you; the retracing of a familiar route; and the knowledge you hold that the Eiffel Tower is in Paris.

Not surprisingly, such a complex, multifaceted aspect of brain functioning is hard to pin down. Each different type of memory is stored and retrieved in a different way, and dozens of brain areas are involved in a complex network of interactions. Little by little, though, the geography of human memory is becoming clear.

To understand memory you have to look at individual cells, because that is where memories are made.

Whichever type of memory you consider it consists of the same essential thing: an association between a group of neurons such that when one fires, they all fire, creating a specific pattern. Thoughts, sensory perceptions, ideas, hallucinations – any brain function (save the random activity of a seizure) is made up of this same thing. One pattern – a group of neighbouring neurons firing together in the auditory cortex, say – brings about the experience of a certain note of music. Another pattern, in a different area, brings about the feeling of fear; another, the experience of blue; another, a particular taste – a hint of tannin, say, in a sip of wine. A memory is a pattern like these. The only difference is that it remains encoded in the brain after the stimulation that originally gave rise to it has ceased. Memories form when a pattern is repeated frequently, or in circumstances that encourage it to be encoded. This is because each time a group of neurons fires together the tendency to do so again is increased. Neurons fire in synchrony by setting off each other like particles in a trail of gunpowder. Unlike gunpowder particles, though, neurons can fire again and again. This firing can be fast or slow. The faster a neuron fires, the greater the electrical charge it punches out and the more likely it is to set off its neighbour. Once the neighbour has been triggered to fire a chemical change takes place on its surface which leaves it more sensitive to stimulation from that same neighbour. This process is called long-term potentiation. If the neighbour cell is not stimulated again it will stay in this state of readiness for hours, maybe days. If the first cell fires again during this period, the neighbour may respond even if the firing rate of cell number one is relatively slow. A second firing makes it even more receptive and so on. Eventually, repeated synchronous firing binds neurons together so that the slightest activity in one will trigger all those that have become associated with it to fire, too. A memory has been formed.

Many things influence whether or not a thought or perception comes to be a memory. Take the taste of tannin. If, when you first taste it, you register it only vaguely as part of a general red winey taste, the association between the neurons that momentarily got together to create the tannin experience will be weak and may in time disappear altogether. If this happens, you will have forgotten the taste, and next time you come across it it will seem as unfamiliar as it did the first time. In practice it is more likely that the tannin neurons will retain some very faint 'special' attraction for one another, so a second tasting will bring a vague sense of recognition.

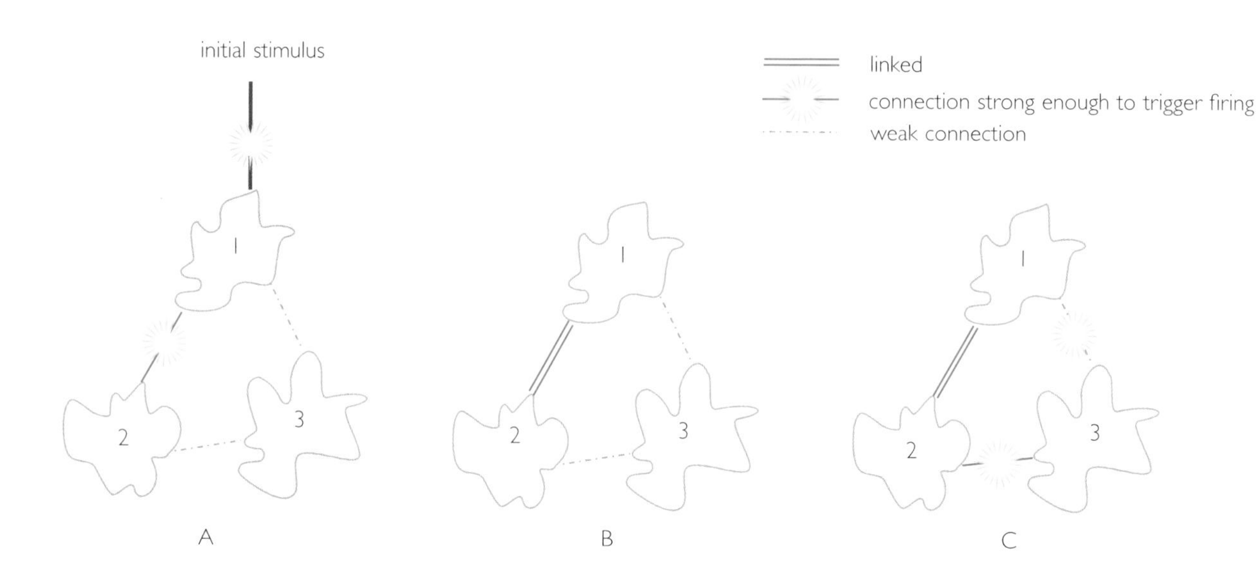

Memories are groups of neurons which fire together in the same pattern each time they are activated. The links between individual neurons, which bind them into a single memory, are formed through a process called long-term potentiation. (LTP) A) *Cell 1 receives a stimulus which causes it to fire. If it fires fast enough it will set off its neighbour, cell no. 2, which will also fire. Cell no. 2 is chemically changed by this – receptors which are normally hidden inside the cell wall are brought to the surface. These make the cell more responsive to its neighbour. Cell no. 2 stays in this 'stand-by' state for hours or perhaps days.* B) *If cell no. 1 fires again during this time, it need do so only weakly in order to trigger a response from cell no. 2. Each time the two cells fire in synchrony the link between them is strengthened. Eventually they are permanently bonded so when one fires the other fires.* C) *When the two cells fire together their combined energy is enough to trigger any neighbouring cell to which they are both weakly attached. If this happens repeatedly the three cells become bound together in a distinctive firing pattern – a memory.*

It would be quite different, though, if – as part of a wine-tasting course, perhaps – you made an effort to distinguish the taste of tannin from all the other flavours in the wine and concentrate on it. In this case the neural pattern created by the tannin will be activated repeatedly, and each time it will be strengthened. Eventually, the association between the neurons will be so strong that they will fire away merrily at the slightest provocation, making the taste familiar and instantly detectable. This will also help to make it likeable. Recognition, particularly of sensory stimuli, is a major part of enjoyment. That is why so many tastes, including tannin, are 'acquired'.

A tannin memory that consists purely of taste is a pretty elementary thing. All it enables you to do is to recognize the taste when you come across it again. If, when you taste it, you link it with its name, however, you will form an association between the neurons that produce the tannin taste pattern and those that produce the tannin word pattern. Your tannin memory will then include both the taste and a label for it. So now when someone says: 'This wine is high in tannin', you will have an idea of how it will taste. You could embellish your tannin memory even more by adding to it the knowledge of its role in wine-making or its chemical structure. The more aspects to it that a memory has the

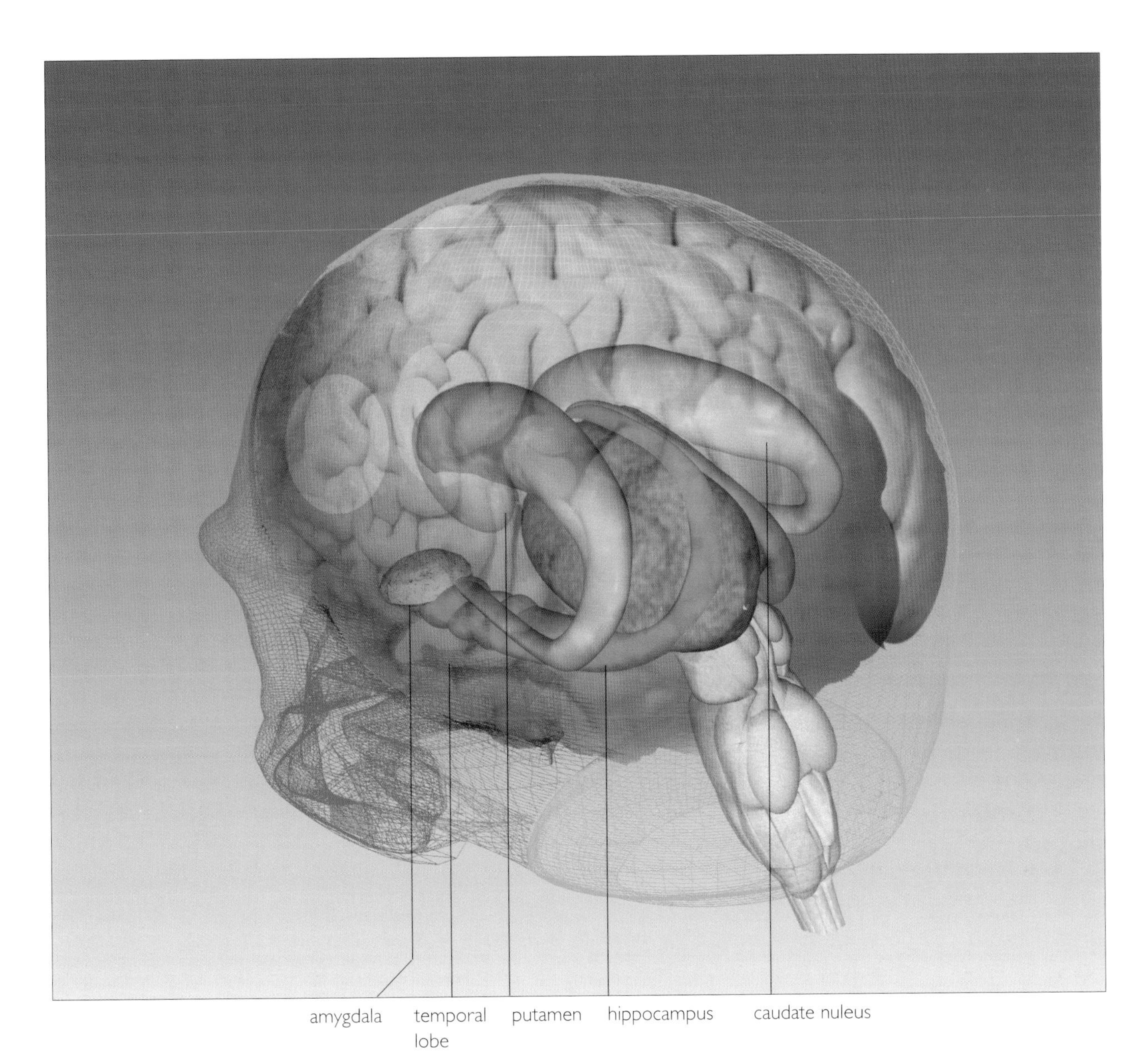

THE HUMAN MEMORY SYSTEM

Many different brain areas are involved in memory. Temporal lobe: long-term memories permanently lodged in the cortex. Putamen: procedural memories, like riding a bike, are stored here. Hippocampus: involved in laying down and retrieving memories, particularly personal ones and those related to finding your way about. Amygdala: unconscious traumatic memories may be stored here. Caudate nucleus: many instincts – which are geneteically encoded memories – stem from here.

more useful it becomes and the easier it is to retrieve because each aspect gives a separate 'handle' by which to yank out the full memory from storage. A multifaceted tannin memory would even equip you to say things like: 'A modest little quaffing wine but firmed up nicely by a subtle hint of tannin' – should you so desire.

Tannin-type memories generally end up in what is known as semantic memory, the store of things that we 'know' independent of our personal relationship to them. When these are first laid down they are, inevitably, part of a bigger construct that includes the personal. The tannin memory will include, for example, where you were when you first tasted it, who you were with, what was said and so on. But unless these personal elements have some particular significance, in time they will fade so that all you are left with is the knowledge of tannin itself. This is true of all the things you 'know': the shape of a mountain, the capital of the USA, the word for the drape of fabric you draw across a window... Once they were all attached to memories of the circumstances in which you learnt them. But the personal marginalia have long since fallen away, leaving the bare but useful facts.

Memories that remain clothed in personal detail are quite different, and the brain deals with them differently. These recollections, known as episodic memories, are usually cradled in a sense of time and space. They include the memory of 'being there', and are personal in a way that your knowledge that the White House is in Washington is not. When we recall them they recreate much of the state of mind we were in when they were laid down.

A state of mind is an all-encompassing perception of the world that binds sensory perception, thoughts, feelings and memories into a seamless whole. To produce it millions of neural brain patterns fire in concert, creating a stream of new 'mega-patterns' – one for every conscious moment. Say you are sitting overlooking the sea, drinking red wine, listening to music and wondering why your offspring are late back from a sailing expedition. The mega-pattern in your head at any particular moment would be woven from the elementary motifs created by the pattern associated with fear; the pattern associated with the taste of wine, the experience of blue and the sound of a musical note. There might also be a pattern that matches your children's faces or your last sight of them; a pattern relating to some previous late-home incident; a pattern relating to life jackets or coastguards; and probably some complex pattern to do with what you will say to them when they finally return. This constellation of neural activity shimmers with constant change as one thought dies away and another comes forward. But so long as your attention is held by the basic theme, the overall pattern, a sort of mega-mega-pattern, will remain recognizable.

Most mega-patterns of this sort never make it to memory – they fire once then fade away. Even mega-mega-patterns, on the whole, leave only a hazy, sepia-washed impression. Yet there are some that stand out in the sludge of our long-term memory like sharp points of light: a

Experiences which are destined to be laid down as long-term memories are shunted down to the hippocampus where they are held in storage for 2–3 years. During this time the hippocampus replays the experiences back up to the cortex, and each rehearsal etches it deeper into the cortex. Eventually the memories are so firmly established in the cortex that the hippocampus is no longer needed for their retrieval.

Much of the hippocampal replay is thought to happen during sleep. Dreams consist partly of a rerun of things that have happened during the day, fired up to the cortex by the hippocampus.

A) *PROCEDURAL MEMORY – the 'how to' sort like riding a bike, are stored in the cerebellum and putamen. Deeply ingrained habits are stored in the caudate nucleus.*

B) *FEAR MEMORIES – phobias and flashbacks – are stored in the amygdala.*

C) *EPISODIC MEMORY – the personal 'filmic' memories which represent our past experience, are encoded by the hippocampus and stored in the cortex. They end up scattered around the cortical areas of the brain. Retrieval, as with semantic memory, then depends on the frontal cortex.*

D) *SEMANTIC MEMORY – facts are registered by the cortex and end up encoded in cortical areas in the temporal lobe. Retrieval is then carried out by the frontal lobes.*

DREAMING UP A MEMORY

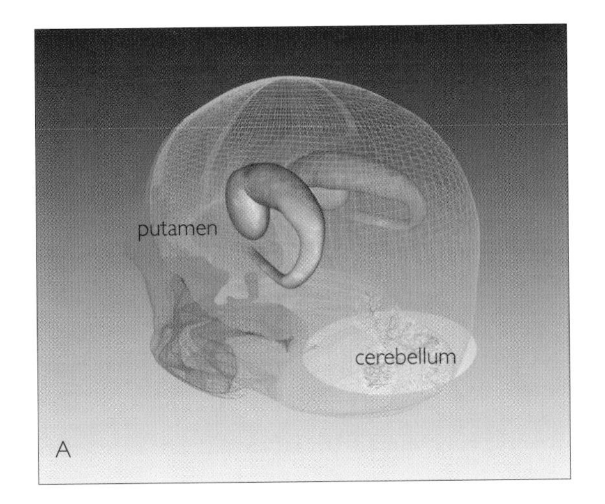

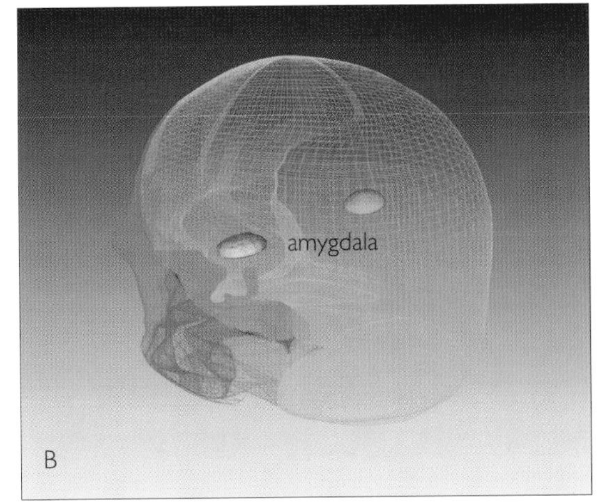

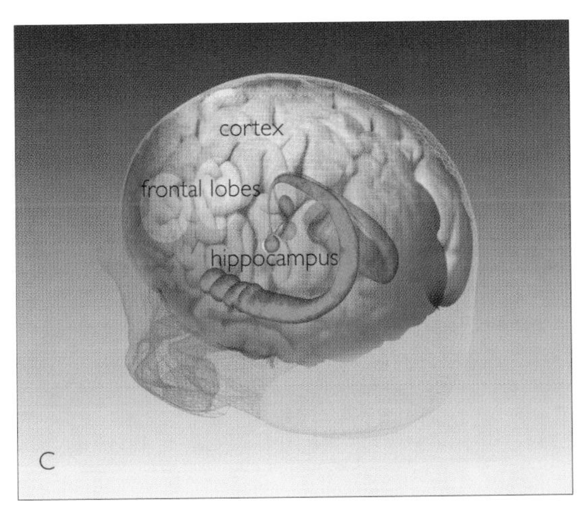

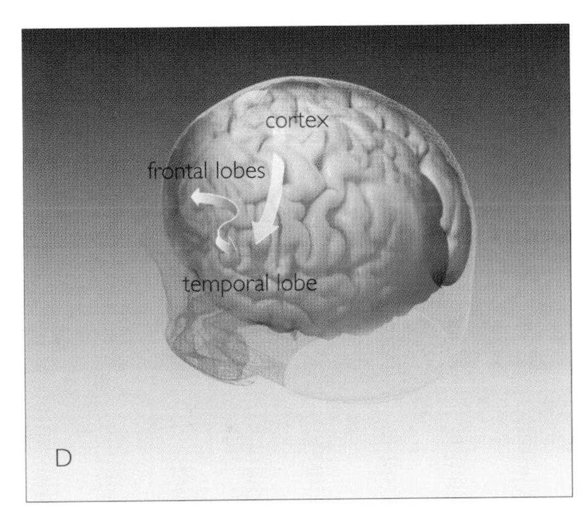

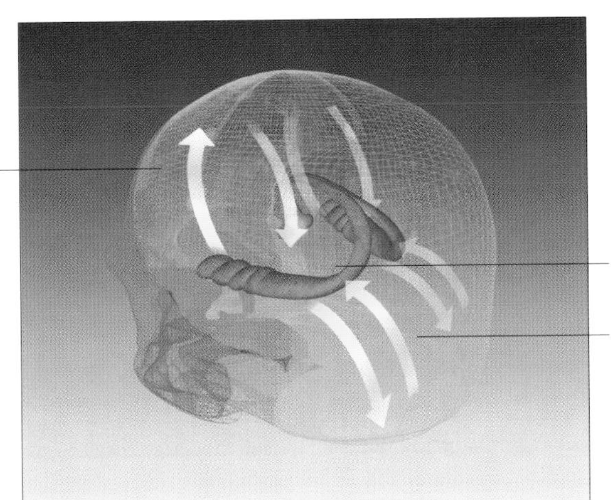

During dream sleep the brain replays recent experiences and etches them deeper into memory

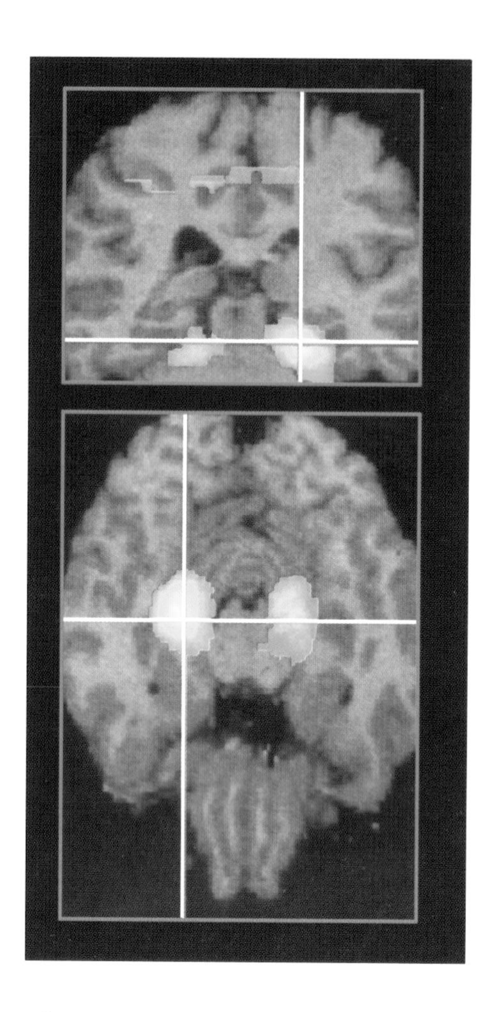

These pictures show activity in the brain of a person watching a film of navigation through a familiar town. The area lit up is the hippocampus.

childhood moment on some beach, feeling the hot sand slip through our fingers; a frozen frame from some distant and otherwise forgotten holiday; the odd, startlingly clear image of a long-dead friend. Why should these brain patterns remain potent while others disappear?

The reason, in most cases, is our old friend emotion. The sort of scenes that stick in our minds are those that, for one reason or another were experienced in a state of emotional excitement. This is because excitement, by definition, is brought about by a surge of excitatory neurotransmitters that increase the firing rate of neurons in certain parts of the brain. This has two effects, both of which have obvious survival value. First, it increases the intensity of perception, producing that 'crystal clear' feeling and sense of slowed time that people typically report when they are in the midst of a crisis. Second, it boosts long-term potentiation, so events that happen in such a state are more likely to be remembered and avoided (if nasty) or sought after (if nice) in future.

The scenario above is a good candidate for long-term memory because it consists of several striking sensory stimuli: the sight of the sea, the sound of music, the taste of wine,each of which provides a different 'handle'. Any one of those might later cause the entire scene to be retrieved, rehearsed and thus strengthened. More importantly, it is marinated in fear. If the incident ended with your children bustling through the door unharmed, it would probably in time become a mere, faint recollection. But if it ended with a knock on the door and a grim-faced police officer telling you of an accident, you would probably carry the memory of that music, the sight of the sea at that moment, the taste of that wine for ever.

Episodes that are destined for long-term memory are not lodged there straight away. The process of laying them down permanently takes up to two years. Until then they are still fragile and may quite easily be wiped out.

Memory consolidation depends on the hippocampus – the funny curved organ that lies beneath the cortex in the temporal lobe. The hippocampus and surrounding areas are connected to almost every part of the neocortex and information about what is happening flows back and forth the whole time. Destruction of the hippocampus can have disastrous effects on memory because without it a person cannot take in anything new. If the organ is removed entirely, they may be unable to hold anything in mind for more than a few moments and be effectively frozen in time. Less severe damage may leave them able to learn new facts but unable to lay down personal memories.[1]

Episodes that are destined for long-term memory appear to be shunted down to the hippocampus from the cortex, where they are registered as neural patterns in much the same way as they are in the cortex. However, because the hippocampus is connected to so many different cortical areas it is able to create a global representation of events. In the waiting-for-news scene outlined above, for instance, the sound of music, taste of wine, sight of the sea and the knowledge of when and where it was all happening would be fed in from separate areas and then bound together to create an 'episode' rather than a collection of impressions or items of knowledge.

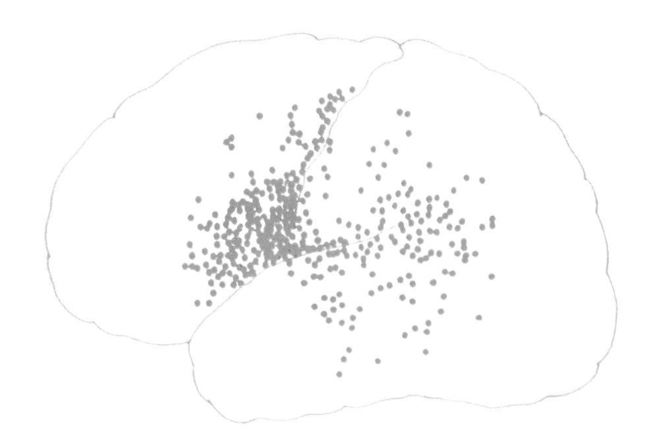

The dots show the places where stimulation elicited snatches of memory during Wilder Penfield's investigations.

These episodes seem to remain in limbo for some time – perhaps as long as two years – before they are finally laid down. During this time they are frequently brought together by the hippocampus and replayed. This happens largely during sleep, and may account for the intrusion of daily events into dreams.[2] Every replay sends messages back up to the cortex where each element of the scene was originally registered. This regeneration of the original neural patterns etches them deeper and deeper into the cortical tissue, protecting them from degradation until eventually the memories are more or less permanently embedded. They also become linked together independently of the hippocampus. This linkage allows any one aspect of the episode to act as a handle for retrieval of the entire thing. If you had experienced the waiting-for-news episode above, for example, you might find that years later the sound of the particular music that was playing then would bring the whole thing flooding back.

Once a semantic or episodic memory is fully encoded in long-term memory the hippocampus is no longer needed in order for it to be retrieved. Bringing a long-known fact to mind activates frontal and temporal cortical areas, while personal episodic recollections have been shown to activate these regions along with several others.[3] The role of each area in memory recollection is still hazy, but it is thought that the temporal lobe activity represents activation of the 'fact' and language storage areas, while

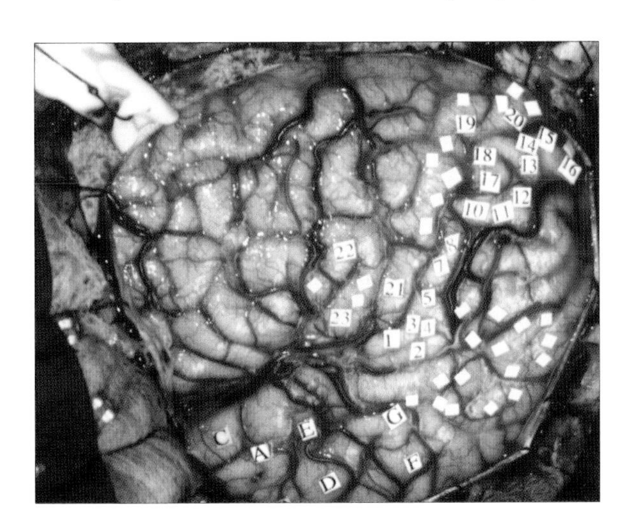

Canadian brain surgeon Wilder Penfield identified the brain regions which produced memories by sticking numbers on the exposed brain during operations.

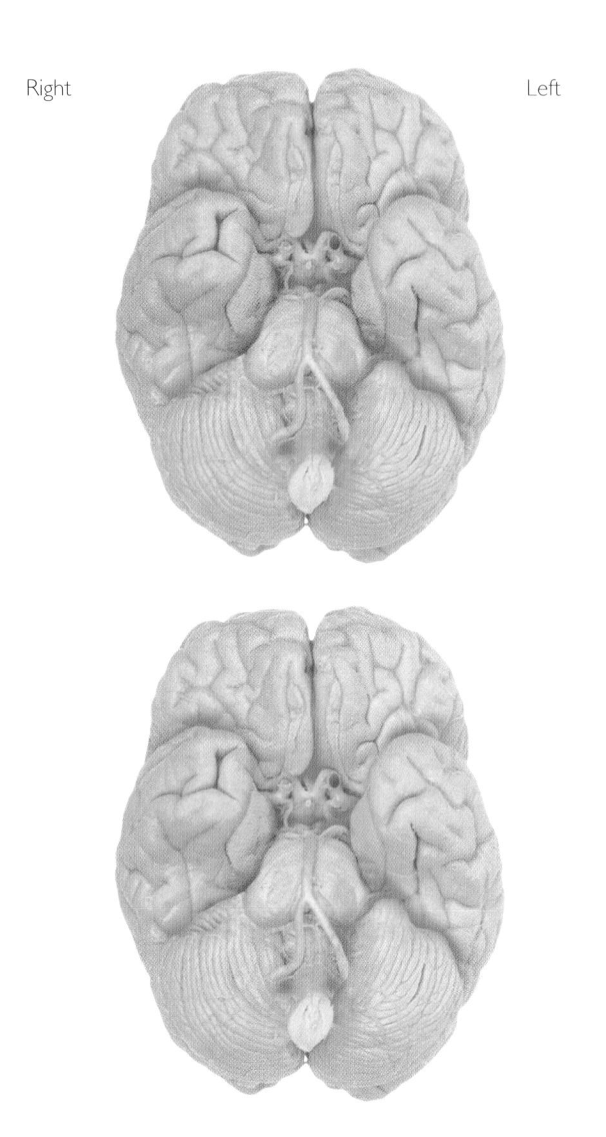

The hippocampus (top) *is activated when people are asked to recall personal or 'episodic' memories. Finding your way around a familiar place also involves the hippocampus – but only on the right side* (below).

the frontal lobe activity draws the memories together and into consciousness. The fact that recollection of episodic memories seems to activate more areas of cortex supports the idea that the various elements of these memories are stored in the cortical areas where they were first registered.

The hippocampus does not relinquish all long-term memories to cortical storage areas, however. Unlike facts and childhood recollections, our memories of space remain encoded in the hippocampal neurons, creating internal maps. This was demonstrated recently in a PET study of London taxi drivers. The drivers were put in the scanner, then asked to imagine the routes that would have to be taken to get to and from various points in the city.

As they travelled the familiar routes in their minds their hippocampi lit up – something that did not happen when they were asked to recall other things, including well-known landmarks.[4]

Although the hippocampus is required to encode and retrieve personal memories, there is evidence to suggest that memories of things that once terrified us may, in part at least, be laid down in the amygdala – the limbic nucleus that registers fear. The emotional effect of flashbacks, like those experienced in post-traumatic stress disorder, are thought to stem from this part of the brain, which is why they have the power to rekindle a physical as well as a psychological replay of the original experience.

How reliable are our various memory mechanisms? One way of judging is to look at what happens when memory goes wrong.

Misleading memories

A course of therapy brought Nadean Cool, a nurse's aide from Wisconsin, more than she could possibly have expected when she entered it. During a decade of treatment Cool came to 'remember' being entrapped in a baby-eating Satanic cult, being raped, having sex with animals and being forced to watch the murder of her eight-year-old friend. She also discovered that she had 120 different personalities – including a duck.[5] When she emerged from what is politely known as therapy, Cool recognized that her 'recovered memories' (not to mention the duck) were therapist-nurtured fantasy. She has since received some $2 million by way of compensation.

Cases like Cool's have generated a raging debate about the nature of memory. Is it possible for a person to recover the knowledge of childhood trauma after years of forgetfulness? Or are such memories necessarily false – the result of clumsy or sinister suggestion? Most of those involved in the issue would dearly like science to provide a simple either/or answer. But the best evidence yet suggests that both recovered and false memories are real phenomena.

False memories are not unusual. In fact, they are the norm. Cool's experience and others like it make headlines only because they are so sensational. Falsities creep into almost every mundane recollection but they generally go unnoticed except for the odd puzzlement ('I could have *sworn* I left my keys on the table') and misunderstanding ('It's not ready yet – I said Thursday, not Tuesday'). They arise because human memory does not lay down a fixed record of objective events in the way that a video recorder encodes a film on tape. Instead it creates and recreates the past, producing, like Chinese whispers, a version of events that may in the end bear little resemblance to what actually happened.

The process begins even as we take in the things that are destined for memory. Most sensory perceptions are not registered consciously and only a few of those that are get retained. Of those, most in turn fade from mind within a few hours. This leaves a tiny distillation of the past to take root in long-term memory. This personal selection of life's highlights is distorted both by selection and by our idiosyncratic way of seeing things. If two people look at a busy scene and are later asked to recall what was happening in it their reports will differ according to what struck each of them at the time as most important or intriguing. Depending on which threads are selected and how they are interpreted, the scene may be remembered as amusing or frightening or just muddled. So memories are not 'pure' recordings of what happens to start with – they are heavily edited before they are laid down.

The process of falsification gets another boost each time a memory is recalled. As we go over things that have happened we add a bit, lose a bit, tweak a fact here, tinker with a quote there and fill in any little bits that may have faded. We may consciously embellish the recollection with a bit of fantasy – the biting comment, perhaps, that we wished we had said but that was actually only thought of later. Then this new, re-edited version is tucked back in storage. Next time it gets an airing it may pop up with the fantasy comment still attached, and this time it will be difficult to distinguish it from the 'genuine' memory. So, by gradual mutation, our memories change.

Given this, it can take very little tampering to create an entirely false memory. Psychologists Elizabeth Loftus and Jacqueline Pickrell at Washington University have shown that false memories can be implanted simply by 'reminding' people of things that never happened. They gave twenty-four volunteers four short accounts of what they claimed were true incidents (supplied by a relative) from the participants' early childhoods. Three of the four were actually true, but the fourth – a tale of getting lost in a shopping mall, crying and eventually getting help from a stranger – was a fabrication. After reading the stories and later being reminded of them, one in four participants was adamant that the false event had really happened.[6]

Although false memories may seem to be true to the person who experiences them, imaging studies suggest that the brain activity involved in recalling a real event differs from that which produces a false recollection. Daniel Schacter at Harvard University took PET scans of the brains of twelve women while they were shown lists of words, some of which they had seen before and some of which were new to them. As each list was shown the women were asked to recall if they had seen it before. Lists that the women had already seen activated the hippocampal and language areas of the brain, while lists that the women *thought* they had seen, but in fact had not, activated these regions plus the orbito-frontal cortex. This, remember, is the 'uh-oh...something funny here!' bit of the brain that was found to light up when things are not quite 'right'. Its activation during false memory recall suggests that – even though the person may not be conscious of it – at some level the brain 'knows' the memory is not correct and continues to generate cerebral question marks. If this finding turns out to be true for long-term memories as well as the recent ones that were involved in Schacter's study, brain imaging may one day have a role in the courtroom or even in therapy to help determine the reality of our imaginings.[7]

People who habitually tell false stories, a condition called confabulation, may be unconsciously trying to 'fill in' otherwise blank spaces in their past. If they have some other condition – mild temporal lobe epilepsy, perhaps, or dementia – the stories may be fantastic: lurid tales of abductions by aliens and so on. But often the falsehoods are quite banal and their very pointlessness underlines the teller's clear belief that they are true. Sometimes the stories are a mixture of truth and falsehood. One such patient, for example, told his doctor: 'I used to work on an assembly line [correct] putting metal rings on the legs of frozen turkeys [correct] at the Hawkeye Packing Plant [incorrect] in the southwest part of town [incorrect].'[8]

Confabulation of this sort is in some ways similar to the compulsive storytelling often seen in people with Williams syndrome. Both are an attempt to bind things together, to make unrelated thoughts into an integrated whole. We all do this to some extent, our brains are constantly seeking to make neat patterns of the information that comes in, and incomplete or fragmented memories (which we all inevitably possess) do not sit easily in our mental filing system. In order to neaten them up the brain may link inappropriate fragments together to create a hotchpotch of half-truths or complete them in a 'likely' way, much as it completes visual stimuli that do not fall into the expected pattern.

The brain also likes events to follow a standard narrative formula: beginning, middle and appropriate conclusion. Studies have shown that when people recall experiences that do not conform to this pattern they will often edit them in retrospect in such a way that they fit the expected structure. In one a group of people who underwent counselling for an anxiety disorder was asked to keep progress diaries. The diaries showed that the course of therapy was bumpy – they would get a bit better, then a bit worse, then a bit better, and at the end many of them reported feeling no better than when they began. When they were asked about the counselling a year or so later, however, nearly all of them reported a straightforward improvement from start to satisfactory resolution.

Habitual confabulation is different from this neatening-up. Some people tell falsehoods as a matter of course – they spin a dense web of falsehood where their past should be. Such people can rarely maintain stable relationships because they are so untrustworthy and they generally lead very difficult lives as a result.

Confabulation is associated with frontal lobe injury. This raises the possibility that the brain's 'lie-detecting' mechanism does not function in such storytellers, so the person does not feel even a faint glimmer of unease when real memories blur into false ones.[9] Such injuries are seen frequently in Korsakoff's syndrome – a condition brought on by alcohol-induced brain damage. Korsakoff's patients suffer severe memory loss, and their confabulation seems to be an attempt to fill in the large blank spaces where real memories should be.

Lost in time

One of the most extraordinary and most thoroughly investigated cases of total amnesia concerns a man, known only as H.M., who has no memory of anything that has happened to him since he underwent a brain operation for epilepsy nearly fifty years ago. H.M. is the Phineas Gage of memory research in that, like Gage, his misfortune has provided researchers with a rare opportunity to see exactly what happens when a clearly defined and usually well-protected area of a person's brain is completely lost. He also demonstrates how the most profound aspects of human personality and experience are rooted in mere flesh.

H.M. had such severe epilepsy as a young man that eventually an operation to remove the areas where his seizures were focused seemed to hold the only prospect of delivering him some sort of normal life. In fact – for reasons that could not at the time have been predicted – the result of the operation was catastrophic.

The area removed from H.M.'s brain included, on both sides, the front two-thirds of the hippocampus, an area of surrounding tissue about 8 x 6 centimetres in size, and the amygdala.[10]

Time effectively stopped for H.M. while he was on the operating table. When he came round from the surgery his memory of the previous two years or so was found to be eradicated. Everything that happened to him until the age of about twenty-five he could remember normally, but there was nothing after that.

This in itself would not have been disastrous

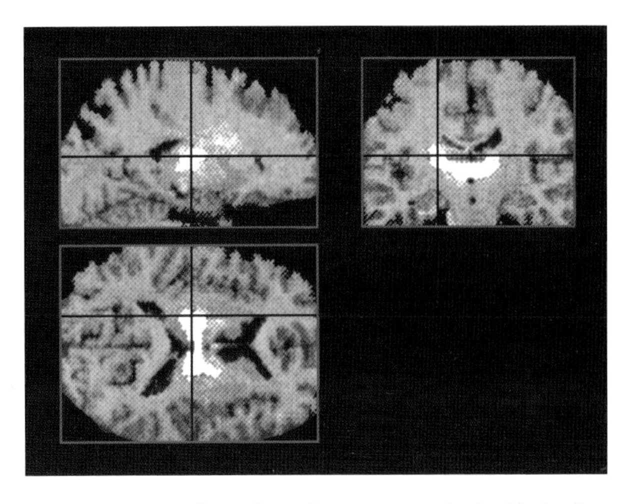

These are scans of people with amnesia – the highlighted areas around the thalamus show abnormal reduction in blood flow. (Based on: 'The Thalamus in Amnesia: Structural and Functional Neuroimaging Studies' Reed L.J. et al Neuroimage, *S 630, Vol. 5, No 4 1997)*

– people who have brain surgery or head injuries frequently have what is known as retrograde amnesia for the period around and before the trauma. As H.M. recovered, however, it became clear that his plight was far more serious than this. He was not just unable to remember the immediate past, he was also unable to lay down anything new. Everything that went into his head stayed put for a few minutes, at most, then faded away...

Try to imagine what this must be like. Normal consciousness feels like a stream – a movement through time. Each moment is made up of a clutch of perceptions, but of themselves these incoming data are meaningless outside of the context of that stream. If you could experience a single moment, entirely uninformed by every moment that has gone before it, you would have no idea what was happening. Our plans, our actions, our thoughts – all these depend on continuity of perception. Even our

Making memories

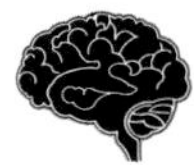

Professor John Morton,
Director, MRC Cognitive Development Unit, London

Very few cases of recovered memories are proved beyond all doubt, though there is no reason to rule out memories being hidden until they are recovered. There is no evidence that event memories can be buried or 'repressed' and then retrieved in pristine form.

Even among academics opinions vary greatly on how easy it is to change memories into complete falsehoods. But it is clear that if forgetting abuse is possible, errors and distortions in recall are inevitable. The presence of an impossible element in someone's story of abuse, then, would not rule out the possibility of an underlying truth and conversely the existence of some verifiable fragments would not guarantee the truth of the rest. The merits of each case must be judged individually within the framework of science.

The process of encoding and representing the world around us and the world inside us is largely automatic. What we end up with is a mental record corresponding to any event, which will not be complete, nor will it be literal. It will be fragmentary and interpretative. When an event is remembered the fragmentary record is retrieved and our cognitive processes operate in order to make sense of it. At this time another memory record is created corresponding to events at the time of recall, including the results of any mental operations, such as problem solving and any accompanying emotions. Other information may be incorporated so when you try to remember the original event you may recall the more recent record with its errors.

For example, volunteers were shown a slide sequence depicting an accident including a scene with a 'Stop' sign. They were later told that the sign had said 'Yield'. In a final test they were shown pairs of slides relating to the original sequence and asked to choose one. Up to 80 per cent indicated that a slide with a 'Yield' sign was what they had originally seen. The usual interpretation would be that the new information was integrated into the subject's representation of the event. But it has been shown that if the pairs of test slides were presented in the same order as the original sequence the misleading effects vanished. What happens is that under the appropriate conditions and with the right cues, you are able to access the record of the original event, and avoid a more recent, erroneous record.

This has two implications: unless the right conditions are set up we may not be able to retrieve particular memories, and secondary records of the intervening activity can be mistaken for primary ones.

Situations when a person may not be able to remember earlier traumas include:

1) When a memory is not part of someone's habitual belief about him/herself, it may not be retrieved without a conscious attempt.

2) When a memory has come to mind whole or in part but was labelled or interpreted as something different.

3) When the memory has not come to mind because the person has not encountered the relevant retrieval cues for many years.

4) When a person's memory has been compartmentalized so that certain events are only recalled in a particular state of mind.

Competing for consciousness

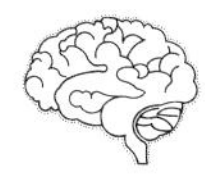

William H. Calvin,
*Theoretical Neurophysiologist,
University of Washington at Seattle*

Jung said that dreaming goes on continuously but you can't see it when you're awake, just as you can't see the stars in the daylight because it is too bright. Mine is a theory for what goes on, hidden by the glare of waking mental operations, that produces our peculiarly human consciousness and versatile intelligence. It suggests that Darwin's evolutionary process could subconsciously operate in the brain, more quickly than in the weeks-long immune response or millennia-long species evolution.

Shuffled memories, no better than the jumble of our night-time dreams, can evolve subconsciously into something better, such as a sentence to speak aloud, improving ideas on the time scale of thought and action. The 'interoffice mail' circuits of the newer parts of cerebral cortex are nicely suited for this job because they're good copying machines, able to clone the firing pattern within a hundred-element hexagonal column. That spatiotemporal pattern is what I call the cerebral code representing an object or idea, the cortical-level equivalent of a gene or meme. Codes are phrase-based translation tables, such as those of bank wires and diplomatic telegrams. A code is a translation table whereby short abstract phrases are elaborated into the 'real thing'. Transposed to a hundred-key piano, my spatiotemporal pattern would be like a short melody – a characteristic tune for each word of your vocabulary and each face you remember. This may not be the most elementary code (some animals actually have a few cells that fire in response to a single, precise element of its environment like the sight of an open hand or a movement left to right) but it's a code that can easily participate in Darwinian competitions. Newly cloned patterns are tacked on to a temporary mosaic, much like a choir recruiting additional singers during the 'Hallelujah Chorus'. But cloning may 'blunder slightly' or overlap several patterns and that variation makes us creative. Like duelling choirs, variant hexagonal mosaics compete with one another for territory in the association cortex, their success biased by memorized environments and sensory inputs, recursively bootstrapping their quality.

Unlike selectionist theories of mind, these mosaics can fully implement all six essential ingredients of Darwin's evolutionary algorithm, repeatedly turning the quality crank as we figure out what to say next or worry about tomorrow. Even the optional ingredients known to speed up evolution (sex, island settings, climate change) have cortical equivalents that help us think up a quick come-back during conversation.

Mosaics also supply 'audit trail' structures needed for Universal Grammar, helping you understand nested phrases such as 'I think I saw him leave to go home' via a tree of linked mosaics, playing together like a symphony of harmonious voices. Even analogies can compete to generate a strata of concepts that are inexpressible except by roundabout, inadequate means – as when we know things of which we cannot speak.

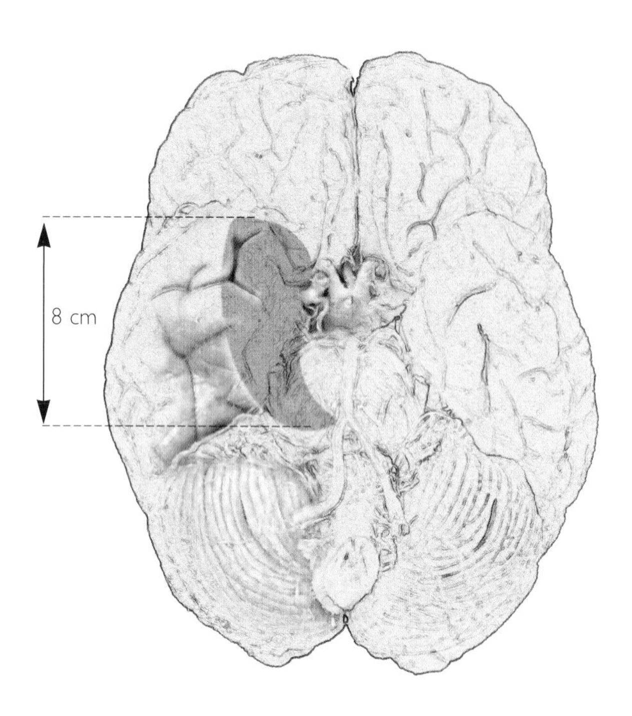

The brain tissue removed from H.M.'s brain included the hippocampus. With it went his past.

identity requires a knowledge of who we were a moment before, and the moment before that.

H.M. does not have the continuity that allows most of us to make meaning of our lives. He is permanently trapped in a single, frozen moment. The stream of his life stopped running when he was twenty-five, so, for him, his identity is suspended there with it. When asked he tells people he is a young man. He talks about his brother and friends, long dead, as though they were still alive. When he is given a mirror to look in his face registers horror as he sees an old man look back at him. The cruelty of inviting him to look at his reflection is mitigated only by the fact that within minutes he has clearly forgotten what he saw.

H.M. is an old man now and is rarely asked to take part in investigations. Over the years, though, he has done exhaustive tests for various psychologists. Some of these researchers spent weeks at a time with him, seeing him each and every day. Yet each session had to begin the same way – with an introduction. H.M. hasn't got to know a person since 1953 – he is always in the company of strangers. He never showed signs of irritation at the constant requests to do the sort of tedious things psychologists learn so much from – tracing a line through mazes, repeating words, naming objects – because to him they were, and remain, all new.

Although each procedure was always unfamiliar to H.M., and his performance on those that require memory function was always uniformly dire, there were some at which he did better each time he tried. Mirror writing, for example – trying to write while monitoring your performance via a mirror – is extremely tricky to do at first, but most people can get quite good at it with practice. H.M., too, became good at it through repeated attempts. But his performance in the later trials surprised him because he had no memory of the previous times he did it. Similarly, he has learned to play new tunes on the piano yet he has no memory of actually being taught them.[11]

The skills that H.M. learnt all involve laying things down in procedural memory – the 'how to' rather than the 'what?' store. He succeeded because this type of memorizing is done in a separate area – a neural loop involving the cerebellum and the subcortical nucleus called the putamen –that was untouched in H.M.'s operation. The 'how-to' mechanism tends to be less subject to degeneration than the hippocampal area and is often preserved in people with severe memory loss. In those with Alzheimer's disease islands of skill – the ability to play golf or to swim the butterfly stroke – may remain when almost all other memory has gone.

The separation of episodic and procedural memory is even more marked in another well-charted amnesic, Clive Wearing, an Oxford musician who suffered severe brain damage after a bout of encephalitis. Wearing came round from his illness, in which he had been comatose, feeling confused and disoriented. But instead of moving on and getting his bearings, Wearing was stuck in that terrifying moment and nothing that happened to him subsequently took root. Yet, despite his complete inability to make new memories, Wearing was able to conduct music as he used to because the flow of the music is programmed into his procedural memory.[12]

Most cases of amnesia are transient, but they can be extremely distressing. The most dramatic is the state of 'fugue', which involves loss of episodic (personal) memory but retention of semantic (facts) memory. This is the good old standby affliction beloved of soap operas in which people say: 'Who am I?' and appear not to know their families. Most cases of fugue are caused by a head trauma like concussion or temporary loss of oxygen to the brain. The causal factor does not have to be catastrophic: physical overexertion, extremes of temperature, even sexual intercourse have all been noted as apparent triggers.[13] One man reported to his doctor an incident in which, in the middle of a family dinner in his own home he suddenly looked around and found himself surrounded by people he did not recognize. At the same time he realized he was unable to recall who, or where, he was. The man did not bother to mention what he was experiencing to the other people at the table, and he told his doctor he was relatively unperturbed by it because 'they seemed rather an agreeable lot and I felt quite happy being with them even though I didn't know who they were'. After a while his memory returned but he visited his doctor just to check things out. A brain scan subsequently revealed evidence of a small stroke in one of the pathways associated with recognition.

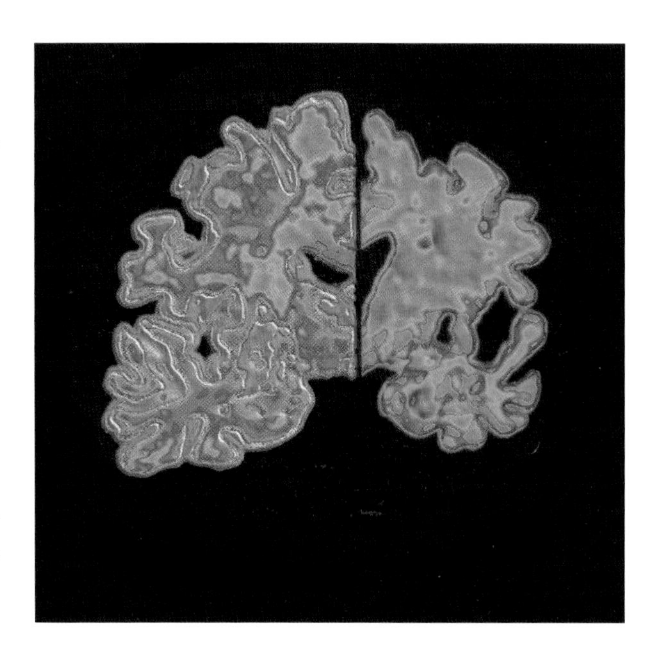

As Alzheimer's disease progresses the brain shrivels and shrinks. Above left shows a slice through a normal brain, above right shows tissue from a patient with severe Alzheimer's disease.

One of the time-honoured fictional plots that are woven around amnesia is the one in which the amnesic person is later found to be having everyone on, usually in an attempt to 'lose' a life they no longer wish to live. It is actually rather difficult for a person to act amnesic convincingly. People who try it usually get caught out because they fake a more complete wipe-out of memory than is in fact likely to occur in the real condition. Unlike H.M. and Clive Wearing, people with fugue usually lay down memories all right and have their entire past stored away. It is just that they are unable to gain access to them. These buried memories may show up without the amnesic person realizing what is happening. A preacher called A. Bourne, for example, had a period of fugue in which he adopted the name A. Brown – a curiously similar name. A. Brown was a zealous churchgoer and on one occasion, during a particularly fervent testimony-swapping session, he spoke of a religious experience he had in fact experienced as A. Bourne – even though he claimed not to remember anything that had happened to him in that identity. Another amnesic patient was reconciled with her family when her doctor suggested that she dial a telephone number at random. The number she chose – without knowing why – turned out to be her mother's.[14]

People who suffer a mental or physical trauma may be left amnesic only for the event and the time around it. Again, they often seem to retain an unconscious memory of what happened. A man who had been the victim of a homosexual rape, for example, became very distressed and even attempted suicide after being shown a Rorschach card that is often interpreted as showing one person attacking another from behind. Another rape victim became very agitated when, without her realizing, she was taken back to the scene of the assault. It happened to be a brick path, and the woman had earlier reported that the words 'bricks' and 'path' kept popping into her mind.[15]

Unconscious memories – often called covert recollection – permeate everything we do. Social psychologist Robert Zajonc showed, for example, that people generally prefer things they have seen before, even if they do not remember seeing them. Similarly, our reactions to people are changed by whether we have seen them before, even if we do not remember them. In one experiment volunteers were shown a number of faces in quick succession – too quick for them to register properly. Later the volunteers were asked to rate another lot of faces – some of which had been shown before and some of which were new – for attractiveness. Although the volunteers were unable to remember seeing the pre-exposed faces they consistently rated them more attractive than new ones. In a sophisticated variant of this experiment some of these same volunteers were asked to join two other people, A and B, to decide the gender of the author of some poems. In fact, the exercise was not about this task at all. Person A, though not person B, was among the faces that the volunteers had already (very briefly) seen. The volunteers did not remember seeing A, but when (by arrangement with the researchers) A and B failed to agree on the gender of one of the poets, leaving the volunteer with the deciding vote, the volunteer inevitably sided with the person they had previously glimpsed.

The unconscious recognition of a stimulus is known to psychologists as priming and the stimulus itself – the quickly glimpsed face – is the prime. As the experiments above show, benign or neutral primes are generally endearing. But nasty primes can make people feel scared or aggressive, without them knowing why.

Covert memories involving fear are presumably stored in the amygdala rather than the cortex and no amount of thinking about them will bring them to mind because cortical activity tends to depress amygdala activation rather than increase it. This is perhaps why traumatic amygdala-based memories tend to pop into consciousness when we relax and allow our minds to wander, as in the psychoanalytical technique of free association. This clearly has profound implications for the debate about the sort of memories experienced by people like Nadean Cool during therapy. Hers were recognized as false, but that does not mean that others are, too. Memories that are laid down in a fragmented way in the cortex but burned indelibly on the amygdala might well remain buried until later life. Indeed, in a recent survey of 129 women who had fully documented (in hospital records) histories of sexual abuse, 16 per cent said they had at some stage forgotten about what happened to them and that the memories had later come back to them.[16] These were often fragmentary, and took the form of flashbacks rather like those experienced in post-traumatic stress disorder. This suggests the memories were located in the amygdala rather than the cortex.

In these cases the cortical, conscious memories of trauma may be difficult to access because they were not laid down in the normal way. Prolonged stress has been shown to affect the hippocampus. Vietnam veterans who suffer from post-traumatic stress disorder (once known as shell-shock) have been found to have 8 per cent less hippocampal tissue than comparable veterans, and adult survivors of childhood abuse were found in one study to have 12 per cent less hippocampal tissue. These people were found to have memory deficits both for the traumas they underwent and for more recent events.[17]

The hippocampal damage seen in trauma victims is thought to be brought about by long-term elevation of stress hormones. A single flush of these hormones, as we have seen, helps to form memories. But if the brain is constantly bathed in them it seems the hippocampus may be damaged with deleterious effects on memory recall and consolidation.

Infant trauma, it has been suggested, may also allow the memory to split itself into different compartments, creating what seem to be several different characters within the same brain: causing Multiple personality disorder (MPD). The first recorded case of MPD was in 1817 but the condition came to public attention with the classic story – made into a successful film – of 'The Three Faces of Eve' in 1957. At that time even those who believed the condition was genuine thought it very rare. Today, however, some clinicians claim that as much as 1 per cent of the population in the USA is affected by the disorder.

MPD (which is also known as Dissociative Identity Disorder) is another issue in which professional (and public) opinion is polarized. Some psychiatrists think it is nonsense, and point out that it is relatively easy to produce the appearance of MPD in patients by suggestion. Others show evidence not only that it exists but also that patients with the condition show distinct alterations in brain organization and neurochemistry that correspond to their changes in identity.[18]

So far there has not been enough laboratory research on MPD to demonstrate if the behaviour seen in MPD it emanates from unusual brain activity, or whether 'ordinary' brain processes, under certain circumstances, can produce the clinical profile seen in these patients. When MPD patients are subjected to brain scanning – as one day they surely will be – some of the secrets of human personality will surely be revealed. We might even be able to see what happens in a person's brain when the resident duck takes over.

How much can memory hold?

The brain has 100 trillion connections joining billions of neurons and each junction has the potential to be part of a memory. So the memory capacity of a human brain is effectively infinite, providing it is stored in the right way.

The human memory is different from a computer's in that it is selective. Items of interest – those that ultimately have some bearing on survival – are retained better than those that are not. So personal and meaningful memories can be held in their billions while dry facts learnt at school may soon fade away.

The brain also works by linkages. If you cannot remember a fact, link it to a meaningful memory and use the latter to hook the former. This is the basis of all mnemonic systems. A few people are able to memorize vast amounts of information – entire telephone directories, for example – by using various mnemonic tricks. Some even seem able to do it without trying.

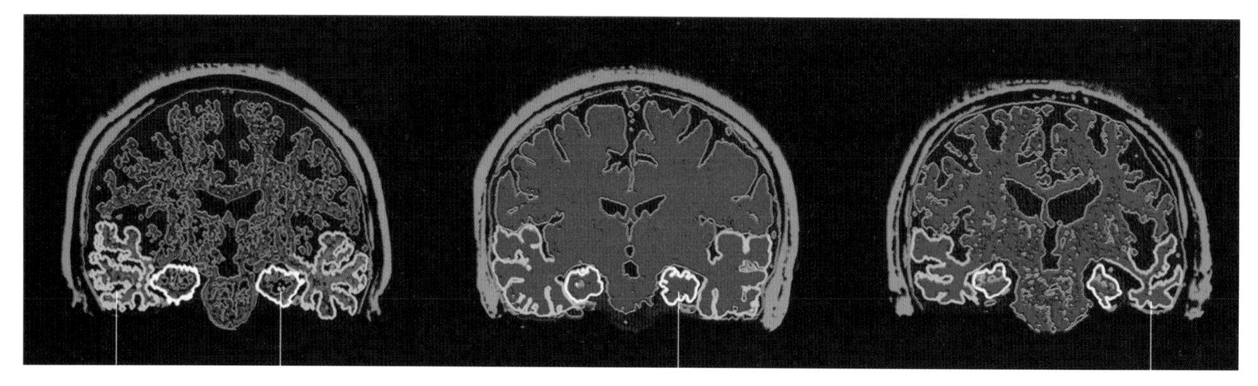

Different types of dementia cause characteristic patterns of memory loss because they attack different parts of the brain. In Alzheimer's disease the first area to go tends to be the hippocampus, where personal memories are stored. The hippocampus is also involved in remembering one's way around, which is why people with Alzheimer's type dementia often get lost. In semantic dementia the temporal lobe is affected first, so people tend to forget general things like the names of objects and what they are for.

The chemistry of learning

Memories are groups of neurons that fire together in the same pattern each time they are activated. The links between individual neurons, which bind them into a single memory, are formed through a process called long-term potentiation (LTP).

Cell 1 receives a stimulus, which causes it to fire. If it fires fast enough, it will set off its neighbour, cell 2, which will also fire. Cell 2 is chemically changed by this – receptors that are normally hidden inside the cell wall are brought to the surface. These make the cell more responsive to its neighbour. Cell 2 stays in this 'standby' state for hours or perhaps days. If cell 1 fires again during this time, it need do so only weakly in order to trigger a response from cell 2. Each time the two cells fire in synchrony the link between them is strengthened. Eventually, they are permanently bonded so when one fires the other fires. When a group of linked neurons fires it triggers a memory.

Losing it

Memory loss is one of the earlier symptoms of dementia, but the type of memory that is lost depends on the form of dementia. **Alzheimer's disease** tends to attack the hippocampus. The hippocampus is essential for retrieving short-term and spatial memories and for laying down most new memories. Eventually, it progresses to all areas of the cortex, therefore affecting other types of memory as well.

Semantic dementia, which involves loss of factual memory rather than loss of personal memory, destroys the cortical area of the temporal lobe first, where semantic memories are thought to be stored.

In the 1920s the celebrated Soviet psychologist Aleksandr Luria studied one such man, identified only as S., in exquisite detail. He later published the case history as *The Mind of a Mnemonist*:

'When I began my study of S. it was with much the same degree of curiosity psychologists generally have at the outset of research, hardly with the hope that the experiments would offer anything of particular note. However, the results of the first tests were enough to change my attitude and to leave me, the experimenter, rather than my subject, both embarrassed and perplexed.

'I gave S. a series of words, then numbers, then letters, reading them to him slowly, or presenting them in written form. He read or listened attentively, then repeated the material exactly as it had been presented. I increased the number of elements in each series, giving him as many as thirty, fifty, or even seventy words or numbers, but this, too, presented no problem for him. He did not need to commit any of the material to memory; if I gave him a series of words or numbers, which I read slowly and distinctly, he would listen attentively, sometimes ask me to stop and enunciate a word more clearly or, if in doubt whether he had heard it, ask me to stop and repeat it. Usually during an experiment he would close his eyes or stare into space, fixing his gaze on one point; when the experiment was over, he would ask that we pause while he went over the material in his mind to see if he had retained it. Thereupon, without another moment's pause, he would reproduce the series that had been read to him.

'The experiment indicated that he could reproduce a series in reverse order – from the end to the beginning, just as simply as from start to finish; that he could readily tell me which word followed another in a series, or reproduce the word that happened to precede the one I would name... It was of no consequence to him whether the series I gave him contained meaningful words or nonsense syllables, numbers or sounds: whether they were presented orally or in writing... As the experimenter I soon found myself in a state verging on utter confusion. An increase in the length of a series led to no noticeable increase in difficulty for S. and I simply had to admit that the capacity of his memory *had no distinct limits...*'

Luria later discovered that S. used synaesthetic linkage, binding items together through multi-sensory imagery. His extraordinary ability was not entirely advantageous, however, because the memories he carried were sometimes a burden. Luria concluded:

'Each attempt [S. made] to move...to some higher awareness proved arduous, for at each step he had to contend with superfluous images and sensations. There is no question that S.'s figurative, synaesthetic thinking had both its high and its low points.'[19]

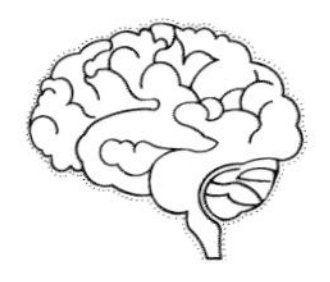

MOLECULES OF MEMORY

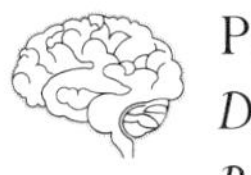

PROFESSOR STEVEN ROSE,
Director, Brain and Behaviour Research Group, Open University

In Alzheimer's disease, plaques of an insoluble protein fragment – beta amyloid – accumulate in the cleft between neurons, blocking communication. This and other abnormalities eventually lead to the death of the cells, especially in regions of the brain known to play a role in the formation of memories, such as the hippocampus. Since the beta amyloid protein fragment is formed by biochemical errors in the breakdown of an essential protein – the amyloid precursor protein or APP – many researchers are now wondering whether APP has anything to do with memory formation.

My group at the Open University has used a simple model of learning to analyse the biochemical cascade involved in making memories. Young chicks peck spontaneously at small bright beads, testing them for edibility. If a bead tastes bitter, the chicks peck only once and thereafter avoid similar beads. Tasting a bitter bead initiates a sequence of biochemical events in particular regions of the chick brain, beginning with the release of the neurotransmitter, glutamate. Within half an hour or so, this activity at the synapses has triggered a complex burst of signalling molecules within the neurons, which turns on specific genes in their nuclei. These genes code for the synthesis of particular proteins that are transported to the synapses and inserted into their membranes, thus remodelling the synapses and altering the pattern of connections. I believe that it is these remodelled synapses that are part of the long-term memory trace.

We have found that in chicks there is a special class of proteins involved in the switch from short-term to long-term memory, called cell adhesion molecules, which are not merely embedded in the synaptic membrane but protrude into the inter-neuronal spaces. Attached to the protein chain are sugar molecules that make the chains sticky and capable of attaching themselves to matching proteins protruding from the opposite membrane. The cell adhesion molecules work like Velcro to hold the neurons together.

To remodel the synapses, two communicating neurons must first unstick the Velcro molecules so that the synapses can move, grow or even divide in two. Later the cells stick the adhesion molecules together again so as to hold the synapses in their new configuration. This requires a wave of Velcro molecules to be produced some four to seven hours after the learning experience – the time of transition to long-term memory. In chicks we have found that blocking the synthesis of the cell adhesion molecules or blocking their sticky ends with antibodies prevents these processes and hence the memory for the bitter bead fades after about six hours.

APP, it turns out, is one of these cell adhesion molecules and blocking its function with, for instance, an antibody also prevents the formation of long-term memory. Another process also thought to be crucial to memory storage is long-term potentiation. This is the persistent increase in the strength of synaptic connections that can be induced in parts of the brain's cortex, especially the hippocampus, by brief episodes of intense activity across the synapses. The question is whether there is a link between long-term potentiation and cell adhesion molecules such as APP.

Blocking the action of APP, or other cell adhesion molecules, in anaesthetized rats with antibodies can reduce the duration of long-term potentiation. So cell adhesion molecules may be necessary for the proper remodelling of synapses during memory formation.

Mice in which the genes encoding various cell adhesion molecules have been inactivated or 'knocked out' show varying learning and memory deficits from subtle to frank, depending which member of the cell adhesion molecule family they lack. But although these mice have impaired long-term potentiation in one major neuronal pathway their ability to learn their way around their environment is normal. By contrast, mice deficient in APP have normal long-term potentiation but find it difficult to learn their way around their environment.

Knockout mice may provide misleading information, however. During development their nervous systems may have compensated for the missing gene by over-expressing other genes with similar functions. Another good model is a mouse strain engineered to express the mutant human APP. As these mice age they show many of the neuropathological signs of Alzheimer's gene. The drugs currently used to treat Alzheimer's boost the function of neurotransmitters otherwise lost during the progress of the disease. But if the transition from short- to long-term memory requires a boost, not of the transmitters but of the cell adhesion molecules, this is the deficit we should be tackling. Already there is evidence that a number of steroid hormones and a class of drugs known as cognitive enhancers may work in this way. Hopefully, understanding the mechanisms involved in Alzheimer's disease and the role played by APP in the formation of memories will eventually enable us not just to treat but to prevent this tragic disease.

CHAPTER EIGHT
Higher Ground

The frontal lobes are where ideas are created; plans constructed; thoughts joined with their associations to form new memories; and fleeting perceptions held in mind until they are dispatched to long-term memory or to oblivion.
This brain region is the home of consciousness – the high-lit land where the products of the brain's subterranean assembly lines emerge for scrutiny. Self-awareness arises here, and emotions are transformed in this place from physical survival systems to subjective feelings.
If we were to draw a 'you are here' sign on our map of the mind, it is to the frontal lobes that he arrow would point. In this our new view of the brain echoes an ancient knowledge – for it is here, too, that mystics have traditionally placed the Third Eye – the gateway to highest point of awareness.

EVERYTHING THE HUMAN BRAIN does is amazing, but not all of it is so special. Computers can calculate better than us; recording machines can rerun the past more accurately; dogs can smell more acutely; birds are better at singing... It is not what we do that makes us so precious (a totally paralysed, mute person is no less a person, after all), it is what goes on inside our heads – the rich and highly developed quality of our consciousness.

You can travel around the brain for a long time without confronting consciousness. Behaviourists, for example, managed to dominate psychology for most of the twentieth century without even admitting that it existed. Now, though, the behaviourists' obsession with rigid objectivity has been replaced by a fervour of interest in subjective experience that is currently engaging some of the greatest philosophers and scientists in the world.

There are two broad schools of thought about consciousness (and by extension mind): one is that it is some transcendent quality beyond understanding. Essentially, this is Cartesian dualism – the idea that the 'spirit' world is separate from the material world in which our brains are rooted. The other is that it is the product of brain activity, a property of the material world that can be explored and will eventually be explained without recourse to the supernatural.

Those who have taken up the challenge of trying to explain consciousness (rather than assuming it to be an impenetrable mystery) are faced with huge questions. Does it have a purpose, or is it just a by-product of neural complexity? Is it a single, continuous stream, or is that feeling of continuity and oneness an illusion? If you could remove the informational content of a living brain and keep it separate from its body – download it, perhaps, on to a floppy disc – would consciousness go with it? And if so, *where* exactly would it be? What bytes of data could you afford to junk when you take Grandad off-line and still be sure he is able to enjoy his virtual existence?

There are no answers yet to these questions but there are some intriguing clues. In particular, the search for the neural basis of consciousness (the philosopher's stone of brain mapping) is providing fascinating insights into its quarry.

A huge volume of evidence suggests that consciousness emerges from the activity of the cerebral cortex and in particular from the frontal lobes. Ask yourself this: 'Where, precisely, do I feel that "I" am centred?' If you are like most people, you will point to a position just above the bridge of your nose. It is right behind here that you will find the prefrontal cortex – the area of the frontal lobe most closely associated with the generation of consciousness. This region is also responsible for our conscious perception of emotion and our ability to attend and focus. Most important of all, it endows the world with meaning and our lives with a sense of purpose. The symptoms of schizophrenia, depression, mania and Attention Deficit Disorder are mainly due to frontal lobe disorder. Our growing knowledge about this area, and the chemicals that make it tick, holds the best hope of restoring people with these disorders to normal life.

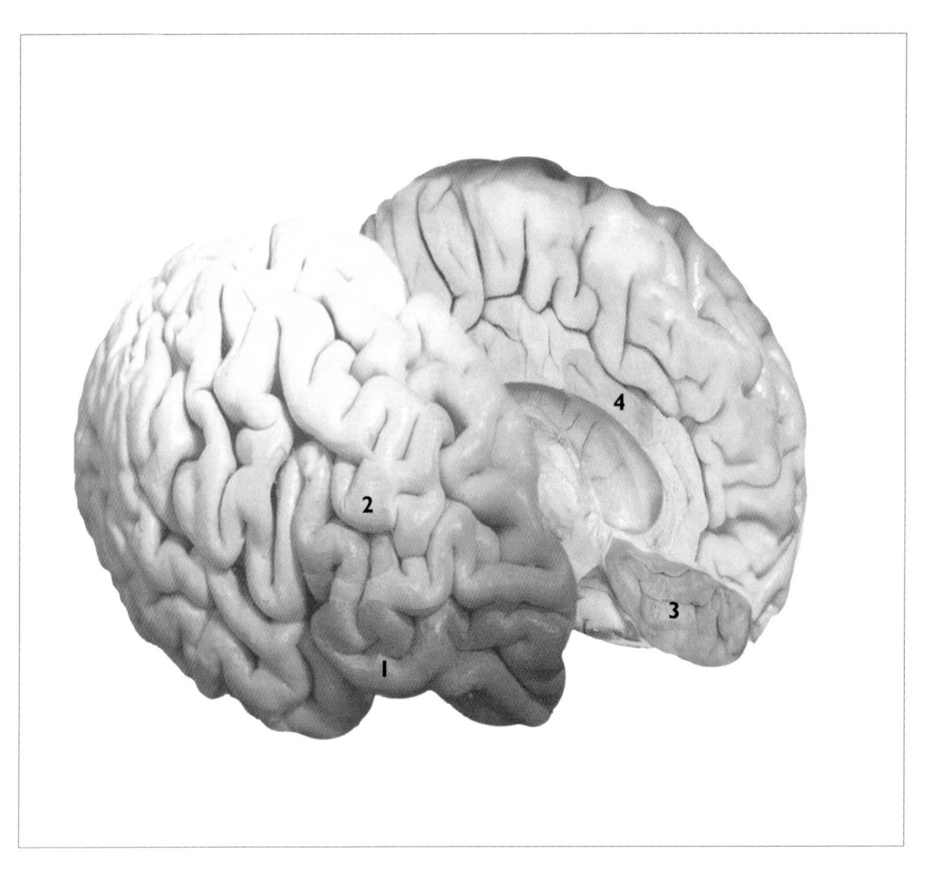

1) Orbito-frontal cortex: this area inhibits inappropriate actions, freeing us from the tyranny of our urges and allowing us to defer immediate reward in favour of long-term advantage.
2) Dorsolateral prefrontal cortex: things are held 'in mind' here, and manipulated to form plans and concepts. This area also seems to choose to do one thing rather than another.
3) Ventromedial cortex: this is where emotions are experienced and meaning bestowed on our perceptions.
4) Anterior cingulate cortex – helps focus attention and 'tune in' to own thoughts.

The frontal cortex is the part that mushroomed most dramatically during our transition from hominid to human and it makes up about 28 per cent of the cortical area of the human brain, a far larger proportion than in any other animal. The back part of the frontal lobe is given over to the parts of the brain that allow us to take physical action. This includes part of the language area (Broca's), which articulates speech, and the motor cortex, which controls movement. Just in front of the motor cortex is a strip called the premotor cortex, or Supplementary Motor Area (SMA). This is where proposed actions are rehearsed before they are actually carried out.

The premotor cortex is an important landmark. It divides the sensing and doing cortex from the area that is given over to man's most impressive achievements – juggling with concepts; planning and predicting the future; selecting thoughts and perceptions for attention and ignoring others; binding perceptions into a unified whole; and, most important, endowing those perceptions with meaning.

Edge forward from the premotor strip and you enter an area known as the prefrontal cortex. This is the only part of the brain that is free from the constant labour of sensory processing. It does not concern itself with the mundane tasks in life such as walking around driving a car, making a cup of coffee or taking in the sensory perceptions from an unremarkable environment. All these can be done adequately without calling on the prefrontal cortex. So long as our mind is in neutral the prefrontal cortex merely ticks over. When something untoward occurs, though, or when we actually think rather than daydream, the prefrontal cortex springs into life and we are jettisoned into full consciousness as though from a tunnel into blazing sunshine.

The frontal lobes are connected by numerous neural pathways to almost all the other cortical areas and also to the limbic region. These paths are two-way – they carry in information from the buried parts of the brain and they also carry signals back. Information must flow in to the frontal lobes in order for them to function, but a heavy input from below can inhibit activity on the surface and vice versa. This means that a sudden flood of emotion may occlude thought, while an arduous cognitive task may dampen emotion. This is why terror can (momentarily at least) wipe a brain clean of thought, while young men who wish to prolong sexual congress are traditionally advised to practice calculus.

The ebb and flow of neural traffic is mediated by the neurotransmitters dopamine, serotonin and adrenaline, and any disturbance to these chemicals, or damage to the tissue that is sensitive to them, can have catastrophic effects on the way we think, feel and behave.

High-quality consciousness requires a fair amount of frontal lobe activity, and this is fuelled by information from many other areas. If the mechanisms that switch on consciousness are faulty, or if the information coming in from the unconscious back rooms of the brain is distorted or incomplete, consciousness will be reduced or partial.

Awareness, perception, self-awareness, attention, reflection – these are all separate components of consciousness and the quality of our experience varies according to which and how many of them are present. It is a little like a printing process in which an image is built up by laying down its component colours one on top of the other. The first pass of the ink-jet leaves a vague monochromatic impression, the second gives it greater definition, the third brings a richer and clearer view and so on. At some point the smears of ink become the picture, and at the end of the process, although the image may be formed from just five colours, it can contain thousands of hues. Of course, the picture is the result of many processes that precede its printing: an original painting, text or photograph has to exist first and the original has to be set up for printing either as a digital code (in a computerized system) or as an etching on a metal plate. The equivalent mental processing that precedes consciousness involves many of the neural systems in every brain region. However, just as a final printed image may be built up on the page from just a handful of component colours, so it seems that the penetrating, multifaceted consciousness that normal healthy humans enjoy may be created by the activity of relatively few brain parts.

The paramount importance to consciousness of the frontal lobes was discovered from brain injury studies many decades ago, and cases like that of Phineas Gage further suggested that specific parts of this region are responsible for specific qualities like self-awareness, personal responsibility, purposefulness and meaning. Even so, until functional brain imaging came along it was difficult to imagine that such amorphous concepts could be pinned down to precise clumps of nerve cells. That, though, seems to be the case, and in its short history functional brain imaging has proved spectacularly successful at teasing out the physical loci of our most elevated states of mind.

Although consciousness emerges from the cortex it requires an entire brain to feed it. The brainstem, midbrain and thalamus are essential because they are part of a system that directs and controls conscious attention by shunting neurotransmitters to various parts of the cortex. Activity in these areas alone is sometimes seen in people in deep comas and although it is not sufficient to give rise to consciousness it can

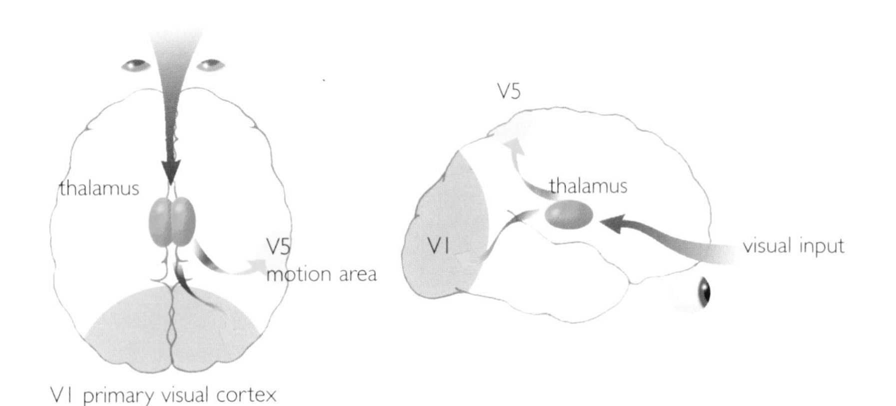

Most visual input goes to V1, the primary visual cortex at the back of the brain where sight is consciously experienced. A smaller neural pathway leads direct to V5, however, and this is how some blind people may nevertheless be aware of movement, so-called 'blindsight'.

produce an eerie simulacrum of conscious behaviour. The patient's eyes may lock on to a moving target, for example, so they may seem to be watching those who pass by. They might clutch at things, too, and grimace if pricked with a pin. These actions are purely reflex but nevertheless deeply disturbing for those who see them.*

What has to happen in the brain for the first smear of consciousness to be laid? A clue to this comes from studies of a curious condition called blindsight, which is a godsend to researchers because it allows them to study what happens on the very edge of consciousness.

Blindsight is the ability of people to see things without being conscious that they are seeing them. People experiencing it are fully conscious of all but the blindsight itself, so the fully conscious bit of them can report on how it feels to be aware-yet-not-aware of one element of their brain functioning. The effect is almost as though a fully conscious brain is in telepathic contact with a less conscious one.

Blindsight first came to light on the battlefields of the First World War when blinded soldiers were seen to duck bullets even though they had no idea they were doing so. It has since been methodically investigated in several subjects. Some researchers claim to have produced blindsight, by clever tricks, in fully sighted people, but it is easiest to detect in people with a form of blindness caused by damage to the primary visual cortex (V1), which is essential for normal sight. The neurons on this part of the cortex are arranged so that they each respond only to their own part of the visual field. If some neurons are killed by injury, the area of visual field that they formerly registered becomes (or seems to become) a blind spot.

Oxford University's Larry Weiskrantz, the first scientist to investigate blindsight experimentally, found that people with this sort of injury could often point accurately at a target that was moved across their blind area even though they had no conscious awareness of it. Later he found they could often tell the shape

* There are those who argue that the emergence of consciousness from the cortex cannot be known for certain, and they are right. The evidence is essentially negative in that no one without a functioning cortex has ever reported or displayed behaviour consistent with consciousness. The assumption that the cortex is essential for experience informs many ethical decisions in current medicine, including turning off life-support machines of brain-damaged patients and taking organs from bodies that are still breathing. To discover it is wrong would remove one of the few lodestones in a notoriously boggy area of morality, and shed an unwelcome new light on past practices. Happily, nothing has emerged from the new imaging studies to suggest that a rethink is in order.

Attention Deficit Hyperactive Disorder is a condition marked by lack of concentration, short attention span and physical restlessness. It is usually diagnosed in children, many of whom are so disruptive that normal play and schooling is impossible for them. The condition is often blamed on bad parenting, or a 'bad attidtude', but brain imaging studies show clearly that children with this disorder have an underlying neurological dysfunction which almost certainly accounts for their behaviour. Essentially the problem is caused by a brain that has yet to come fully 'on-line'. The limbic system is working at full steam in these children but the cortical areas which focus attention, control impulses and integrate stimuli have yet to become fully active. Imaging studies show this to be so: the brains of children with ADHD show marked lack of activity in several right hemisphere regions. They include the anterior cingulate – an area associated with fixing attention on a given stimulus; and the prefrontal cortex, an area concerned with controlling impulses and planning actions. An area in the upper auditory cortex has also been found to be hypoactive in such children. This region is thought to be concerned with integrating stimuli from several different sources, and it is possible that ADHD occurs partly because lack of activity in this part of the brain prevents the child from grasping 'the big picture'. Instead the world seems fragmented, with one stimulus after another vying for attention. Adults with attention disorder show a similar pattern.

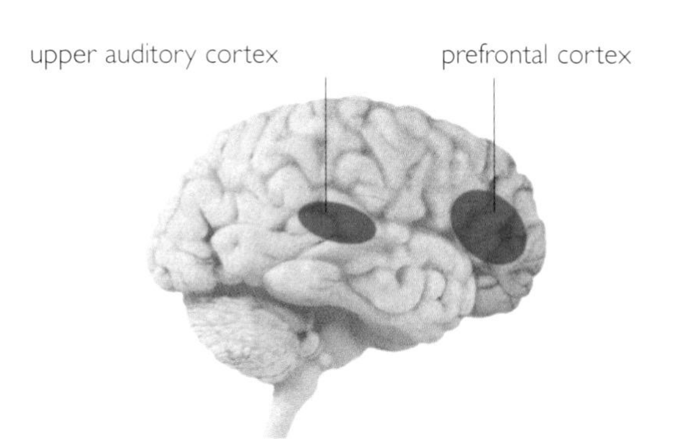

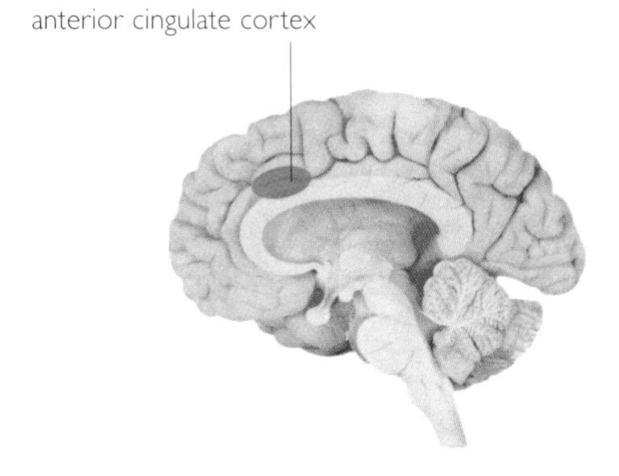

Amphetamine-type drugs which raise the level of excitatory neurotransmitters in the cortex reduce Attention Deficit. The cortical activity they produce inhibits the limbic system, substituting thought for action and producing more controlled and focussed behaviour.

Drug treatment for ADHD stimulates these underactive areas, causing the brain to concentrate and focus.

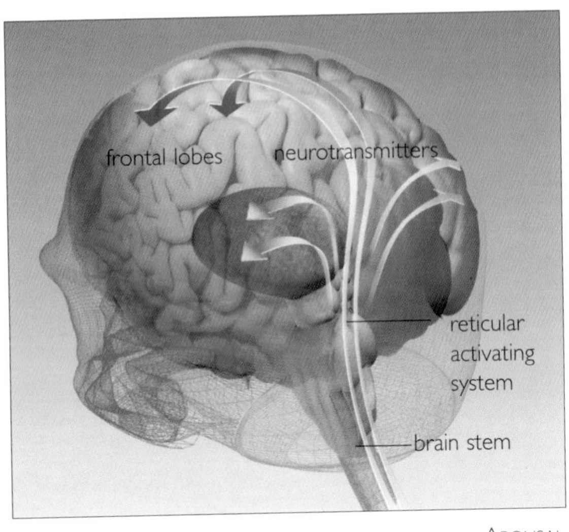

AROUSAL

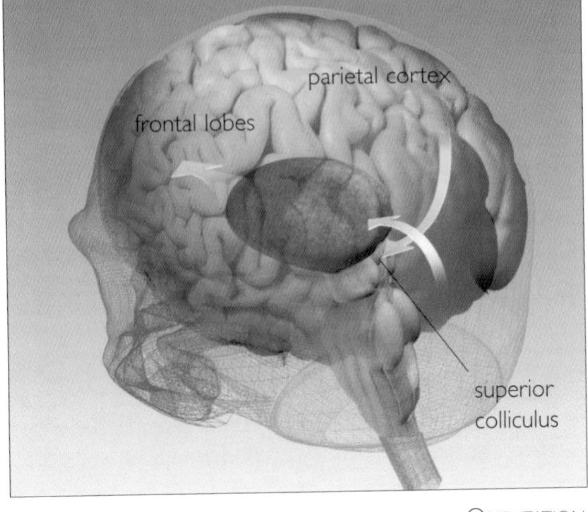

ORIENTATION

ATTENTION

The brain started as the body's alarm system, and alertness can be thought of as a special mechanism to ensure that the brain is at its most efficient when danger is about.

If the brain picks up a stimulus that may be a threat – a rustle in the bushes, say – the reticular activating system releases a rush of adrenaline throughout the brain. This closes down all unnecessary activity, so an alert brain shows up on a brain scanner as very quiet. It also inhibits body activity: the heart rate slows and breathing becomes shallow and quiet.

While the brain waits on alert for something to which to react, activity is maintained in the superior colliculus, the lateral pulvinar (a part of the thalamus) and the parietal cortex. These areas are concerned with orienting and focusing. Once a cue comes the appropriate area of the brain springs into activity and shows a greater level of activity than a brain that was not previously alert.

Attention is necessary for thinking, and possibly for consciousness. The brain constantly scans the environment for stimuli. This is done largely by automatic mechanisms in the brainstem. Even people in Persistent Vegetative State show scanning eye movements, which are part of this system.

Attention requires three elements: arousal, orientation, focus.

Arousal is dependent on a group of nuclei in the midbrain – the top of the brainstem – called the Reticular Activating System. The core of the brainstem is made up of neurons that have unusually long dendrites stretching up and down. Some of these travel right up to the cortex. Some of them are responsible for consciousness; concussion often results from disturbance of the system, and damage to them may result in permanent coma. Others control the sleep/wake cycle. A third group is responsible for controlling the level of activity in the brain. When they are stimulated they release a flood of neurotransmitters, which sets neurons firing throughout the

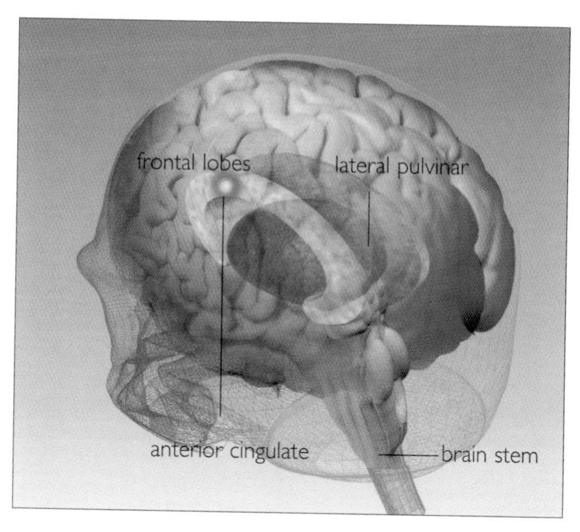

Focus

brain. The ones known to be particularly involved in activating the prefrontal lobe are dopamine and noradrenaline. Stimulation of this group of reticular neurons also creates alpha brainwaves – oscillations of electrical activity at 20–40 Hertz – which are associated with alertness.

Orientation is done by neurons in the superior colliculus and parietal cortex. The superior colliculus turns the eyes to the new stimulus, while the parietal cortex disengages attention from the current stimulus. Damage to the superior colliculus may cause occulomotor apraxia – in which eyes can't lock on to a new target. This makes a person functionally blind.

Damage to the parietal cortex may make a person unable to disengage from a stimulus.

Focus is brought about by the lateral pulvinar – a part of the thalamus – which operates rather like a spotlight, turning to shine on the stimulus. Once it is locked on, it shunts information about the target to the frontal lobes, which then lock on and maintain attention.

and orientation of a target, too. This is the conversation that took place after an experiment in which the subject successfully identified the position of a symbol placed in his blind spot at every attempt. Weiskrantz speaks first:

'Did you know how well you were doing?'

'No, I didn't because I couldn't see anything. I couldn't see a darn thing.'

'Can you say how you guessed? What it was that allowed you to say whether it was horizontal or vertical?'

'No, I could not because I couldn't see anything. I just don't know.'

'So you really didn't know if you were getting them right?'

'No.'[1]

Blindsight is probably brought about by the remnants, in humans, of a primitive visual system that once – in our deep evolutionary past – shunted all light-borne information to the subcortical areas of the brain that focus attention and trigger appropriate physical reactions to stimuli. It was entirely concerned with the practical business of organizing the animal's behaviour and as such it did not register anything that was not likely to be of immediate physical importance. It was probably similar to the system now observed in, say, a lizard. Anything outside a lizard's own little bit of space is beyond the concerns of a lizard's visual system. It probably cannot register a fly in a distant corner, only a fly within tongue's reach. And stationary objects do not have the same impact on it as moving objects because stationary objects, by and large, do not signify either food or attack. The lizard-level visual system is purely for survival – it is not there to help the lizard appreciate Picasso. And so it once was with us.

As the cerebral cortex evolved, however, there came a point when it became advantageous

THE WORKING MEMORY

Memory used to be regarded as a simple library with a long-term store – childhood memories and so on – and a short-term store – a temporary holder in which information is retained for as long as it is needed, then discarded. As experimental techniques became refined, however, it has become clear that there is no rigid dividing line between a memory and a thought. A new term has therefore crept into use to describe how we juggle perceptions, memories and concepts: working memory.

Professor Alan Baddeley, of Bristol University, has developed a model of working memory based on three parts:

* The Central Executive – co-ordinates information from a number of sources, directs the ability to focus and switch attention, organizes incoming material and the retrieval of old memories and combines information arriving via two temporary storage systems:

to shunt light messages to this new brain as well as to the existing 'visual' system. The cortex – being more complex and flexible than the old brain beneath it – was able to bring about more sophisticated and effective reactions to visual stimuli. Gradually, as the cortex evolved more and better tricks, the brain remoulded itself to take advantage of them. More light-borne information flowed to the cortex, where it stimulated the growth of more grey matter, which in turn thought up more imaginative reactions, which encouraged more light to flow to it and so on. Meanwhile, the old subcortical system became increasingly redundant. Blindsight, however, reveals what this ancient system can do when it is not eclipsed by the cortical visual system. The ability that some skilled tennis and cricket players have to hit a speeding ball before its existence can possibly be registered by the cortex may also be due to blindsight.

On the face of it blindsight seems no more significant for consciousness than the reflex actions seen in people in vegetative comas. Yet there is evidence that it is a little closer to consciousness than that. When pressed to describe the experience of blindsight some subjects admit to a vague awareness. To quote one: 'I sort of feel that something is there...when it moves it feels as though something is coming towards me – like a billiard cue being aimed at me.' Furthermore, subjects who took part in blindsight experiments got better at them as time went on.[2] This suggests that, although blindsight may itself be unconscious, it may mark the beginning of a route that leads to consciousness.

The difference between blindsight and the automatic reactions is reflected by a difference in brain activity. Reflex actions do not involve cortical activity – clutching and reacting to pinpricks, for example, is sometimes observed in babies with anencephaly, where no cortex exists.[3] By contrast, fMRI scans show that a subsection of the visual cortex called V5 – the area that registers

* Visuo-spatial sketch pad holds images
* Phonological loop holds acoustic and speech-based information.

Brain imaging studies at the Wellcome Department of Cognitive Neurology have found that the three parts are echoed precisely in the activity seen when people carry out cognitive tasks.

A) PHONOLOGICAL LOOP (left)

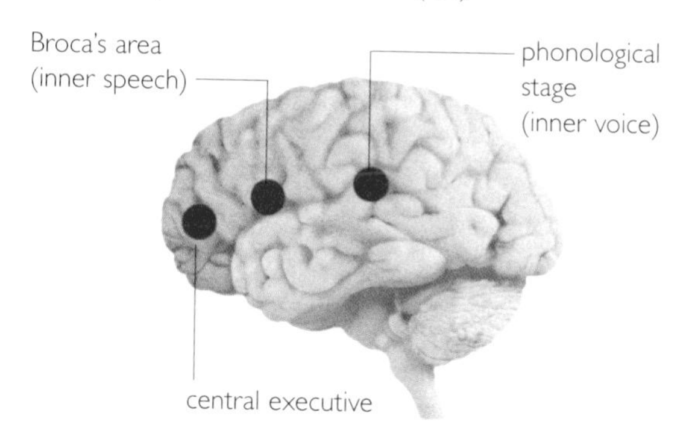

B) VISUO-SPATIAL SKETCH PAD (right)

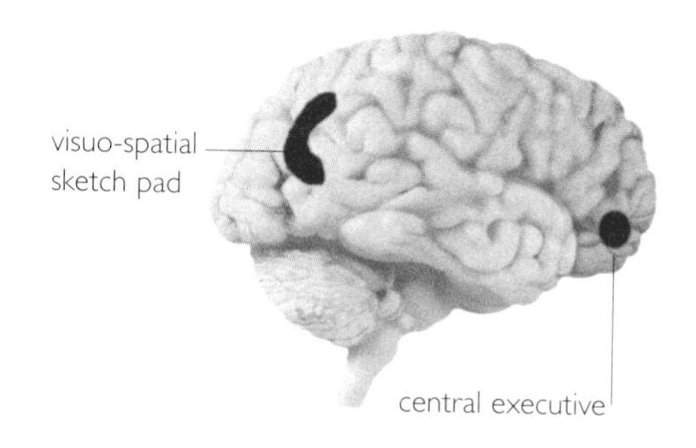

SHORT-TERM MEMORY

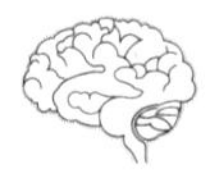

ALAN BADDELEY
Professor of Psychology, University of Bristol

My colleague Graham Hitch and I developed this model in order to account for the results of an experiment in which we tried to interfere with the operation of short-term memory in normal people. We asked students to learn word lists, comprehend prose or do reasoning tests while their short-term memory was occupied with remembering and repeating back telephone numbers. We found that performance was dented but not catastrophically impaired by this simulated 'short-term memory loss'. We suggest that this is because the phonological loop of these patients is impaired by our telephone-number task.

We believe that the phonological loop can be split into two components: a memory store that holds a fast-decaying (one to two seconds) speech-based trace, and a rehearsal system that repeats the trace and registers it in the memory store via non-vocalized speech. Hence a visually presented set of letters can be remembered by saying it to oneself. But because the memory trace fades while the rehearsal is going on we can typically only remember as many words as we can say in two seconds.

Adult patients with pure phonological loop impairment can get by as long as they do not attempt to learn new words. Recent research on a group of eight-year-old children with language deficits found that, while their non-verbal intelligence was appropriate to their age, their language was delayed by two years and their capacity to repeat back an unfamiliar nonsense word by four years. Since this latter task is closely linked with vocabulary development and is a good predictor of future language and reading skills it suggests that the phonological loop has evolved as part of the mechanism for acquiring language.

Less is known about the more complex visuo-spatial sketch pad although the four active regions so far identified by functional imaging are thought to represent 'what', 'where', executive control and possibly image rehearsal.

Working memory allows us to use our memory systems flexibly. It enables us to hold on to information by rehearsing it in our minds, to relate that information to older knowledge and to plan our future actions.

movement – lights up during blindsight even though V1 – the primary sensory area that is essential for normal sight – is not active.[4] So blindsight, perhaps, is not entirely unconscious. It brushes the cerebral cortex, and in doing so it may send tiny whispers of that activity up to consciousness via some little-used neural byway, producing the first glimmer of awareness.

This still falls far short of anything we understand as consciousness. To crank that faint awareness up to a full-blown perception a sensory stimulus has to be registered by an area of primary sensory cortex (V1, for example, in the case of vision). These areas kick off the unconscious assembly lines that begin with the raw material of sensory stimuli and end by delivering to the brain's frontal surface well-processed mental constructions. For the perception to be laden with emotional as well as sensory content a parallel processing line must run from the limbic system (especially the amygdala) to the frontal lobe.

To create full consciousness it is not enough for perceptions merely to 'pop up' in the front of the brain. The pattern of brain activity in which this alone happens – high in the back and sides, low in the front – is seen in people who, one way or another, are not quite 'there'. It is similar to the brain activity recorded in sleep and in Attention Deficit Disorder. It is also seen in the type of schizophrenia that is characterized by withdrawal and inertia.[5] At its most extreme, low frontal activity may result in catatonia, a state in which people fail completely to respond to their surroundings. One woman who spent months lying in bed without speaking or voluntarily moving later explained how she had felt. She had been aware, she said, of what was going on around her, but none of it prompted any thought. 'I couldn't say anything,' she said, 'because nothing came to mind.'

To bring such a semi-slumbering brain to full, thinking, feeling awareness more activity is required in the frontal lobes. Let us see where it must take place to bring about each component of that state.

The Mechanics of Thought

Consider thinking. Thinking is not just a generic term for the collection of skills housed in the brain. It *involves* many of them: recollection and imagining in particular. But it includes something that is not a part of any other function: self-awareness. This aspect of thinking is captured in the word that is often used to describe it: reflecting.

A FAILURE OF WILL

Several brain states reduce or remove free will. People with hysterical paralysis, for example, are unable to overcome their inability to move even though the affected body parts and their connections to the brain may appear to be intact. Once you look inside the brain, however, it is possible to see what is happening. A woman with hysterical paralysis in one leg underwent PET scanning while she tried, but failed, to move the affected one. The scans showed that the woman's frontal lobes lit up when she tried to move the paralysed leg, but the motor areas actually needed to do it remained 'dead'. The normal, automatic knock-on effect from the planning area in the frontal lobe to the premotor cortex that executes movement was disengaged, and as this mechanism is not subject to conscious volition no amount of trying on the woman's part would shift the limb.[6] Hysteria may have significant survival value: it may, for example, be related to the 'play dead' mechanism that occurs in some small mammals when they are in jeopardy.

Although it sounds such a little thing, this self-awareness is of momentous importance. Ultimately, it makes the difference between merely *being* – a passive automaton – and *doing* – a creature of volition.

Thought processing is in one way like the later stages of sensory processing: just as the various parts of an image – location, colour, shape, size and so on – are brought together and integrated into a whole, so we bring together various memories and imaginings and put them together into a new concept. The big difference is that whereas sensory construction is unconscious, thought processing is done consciously. As the frontal cortex carries out its task it monitors what it is doing. So while an image simply 'arrives' in consciousness, a concept carries with it the knowledge of how it came to be.

As we have seen, memories and imaginings are created by the same neural activity that occurs when things really happen. So if you simply imagined or recollected experiences you would not be able to tell if they were happening inside or outside your head. You would therefore be unable conceptually to 'do' anything with what you had brought to mind. You would merely experience it, as though in a trance.

Thinking requires a degree of attention – that is, a focusing of activity in which irrelevant stimuli are ignored. There are two types of attention: the automatic engagement of the senses that occurs when your eye is 'caught' by a flash of movement; and the deliberate turning of the mind to a subject. Attention is created by a flood of neurotransmitters that turns important areas on and unimportant ones off. In the case of sensory engagement, the areas turned on are those that are needed to scan the outer environment and the areas turned off are those that monitor information coming from the body and other parts of the brain. Hence the sight of a stunning person of the opposite sex may

AUTISM – THE UNCENTRED MIND

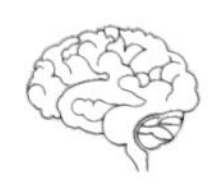

UTA FRITH,
Professor of Cognitive Development, University College London

A lack of 'theory of mind' is well able to explain the poor social understanding in people with autism (for example, being baffled that people deceive each other). However, autism also has characteristic non-social features. Examples are: being obsessively interested in the light fittings in coaches of certain passenger trains; insisting on taking exactly the same route to the supermarket and following a set routine at bedtime; reacting with evident

distress to the touch of a metal button; being inextricably drawn to listen to every aeroplane that flies overhead. These bizarre behaviours, too, demand an explanation.

Not all of these features are handicapping. For instance, many people with autism have special talents, such as prodigious memories, extraordinary calculation abilities, superb drawing skill and absolute pitch. Two cognitive theories tackle these largely unexplored features of autism.

One theory has built on the many parallels with patients who have suffered frontal lobe injury. Indeed, a range of tests of executive function (such as the Wisconsin Card Sorting Task) have been applied to individuals with autism, who find them very difficult. Executive function is an umbrella term covering cognitive processes of the highest order, including flexibility and stopping behaviour that has become inappropriate. Problems in these functions explain the rigidity of people with autism in everyday life (for example, insistence on the same route; obsessively following a bedtime ritual) but cannot explain the typical islets of ability in autism and the feats of superior performance.

The idea that people with autism make relatively less use of context and pay preferential attention to parts rather than wholes goes some way towards explaining the assets seen in autism, as well as some of the deficits. This theory of weak central coherence originated from the attempt to explain why children with autism were particularly good at doing jigsaw puzzles (even when they are upside down) and recalling random word strings. They remember unconnected material with surprising ease, while normally there is a huge advantage in favour of coherent material. Most people attempt to get the gist rather than the surface form of a message. They strive after meaning. This drive for meaning, which appears to be a general organizing principle of the mind, may be weak in autism. If so, we can explain why a child with autism will pay attention to an earring rather than to the person wearing it, and why the child can fail to understand a message but can repeat it verbatim.

Experiments have demonstrated the problems that result from focusing on local rather than global meaning. For example, people with autism might complete the sentence 'The sea tastes of salt and...' with the word 'pepper', and pronounce the word 'lead' like the metal when reading the sentence 'The lead guitarist arrived late'. On the other hand, there are also situations where their weak central coherence leads to superior performance. For example, hidden figures can be found in puzzles more easily if the global meaning is disregarded.

Problems in theory of mind explain why people with autism are unable to form friendships. Problems in executive functions explain why they have difficulty planning their everyday life and cannot live independently despite high ability: without help they might fail to throw rubbish away, or to ask for a doctor when they need one. Weak central coherence can explain why they can have exceedingly narrow interests and mass encyclopaedic knowledge about their pet subjects. It may also explain why they develop absolute pitch and overreact to certain perceptions that are normally experienced within a meaningful context from which they derive their value.

SLOW-WAVE SLEEP The entire brain oscillates in a gentle rhythm quite unlike the fragmented oscillations of normal consciousness. Brain scans show less activity in the limbic system. HYPNOSIS Brain scans show increased activity during hypnosis, particularly in the motor and sensory areas suggesting heightened mental imagery. Increased blood flow in the right anterior cingulate cortex suggests that attention is focused on internal events. The brain activation seen in this state is quite different from that seen in normal waking or sleeping. SCHIZOPHRENIA Lack of activity in the frontal lobes is a feature of states of mind in which consciousness is disturbed or decreased. In chronic schizophrenia the dorsolateral prefrontal cortex is especially hypoactive. This might account for the state's common reduction in planned or spontaneous behaviour and social withdrawal. The anterior cingulate cortex – thought to distinguish between external and internal stimuli – is also underactive, which may be one reason schizophrenics confuse their own thoughts with outside voices. DREAMING Vivid visual dreams light up the visual cortex; nightmares trigger activity in the amygdala and the hippocampus flares up from time to time to replay recent events. The areas which seem to be most commonly active are the pathways carrying alerting signals from the brainstem and the auditory cortex; supplementary motor area and visual association areas – all of which produce the 'virtual reality' effect of dreaming. Activity is decreased in the dorsolateral prefrontal cortex, the area of waking thought and reality testing. MEDITATION Scans of people in a self-induced state of 'passive attention' have been shown to 'turn off' areas of the brain normally associated with seeking stimuli, including the parietal, anterior and premotor cortexes.

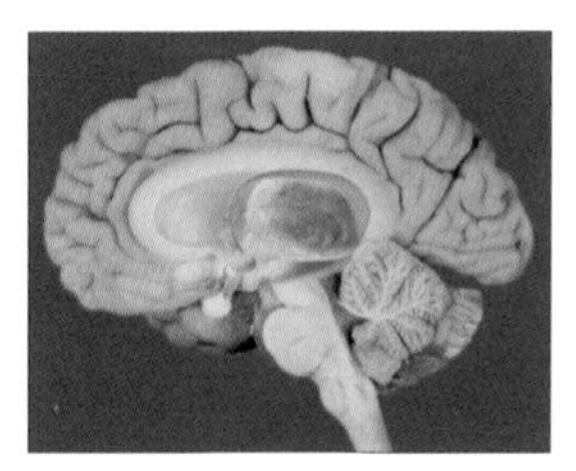

slow wave sleep

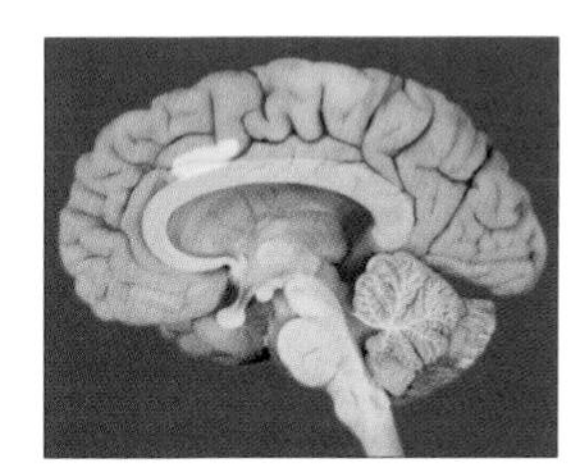
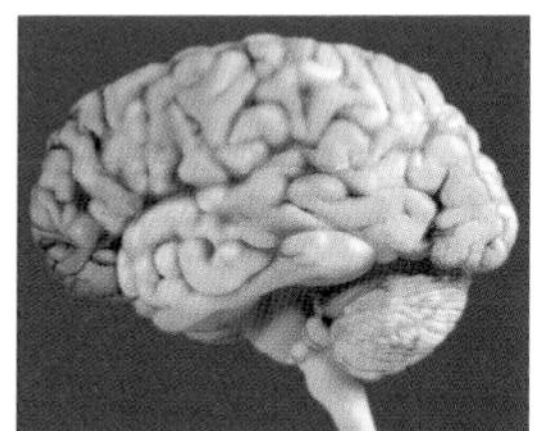

hypnosis

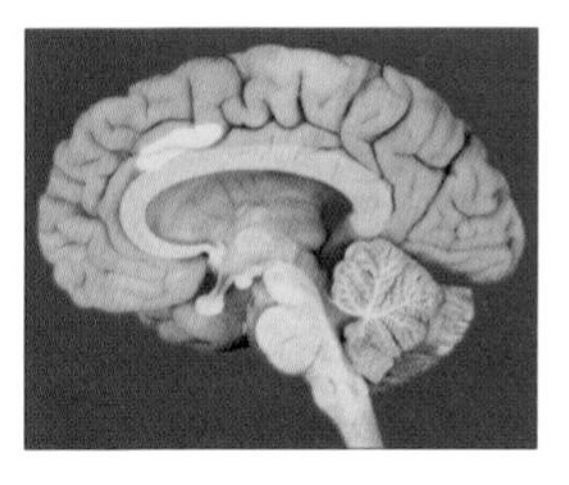
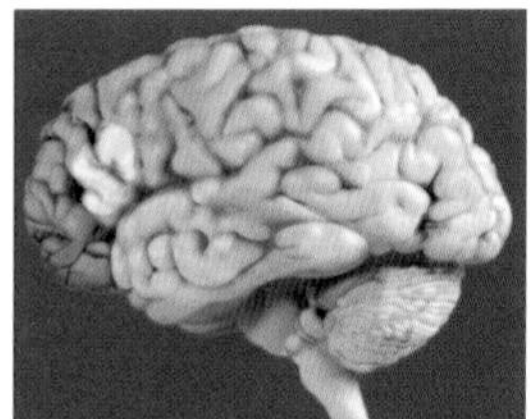

schizophrenia

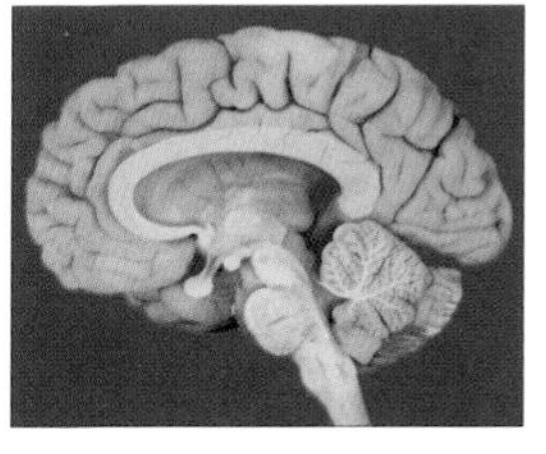
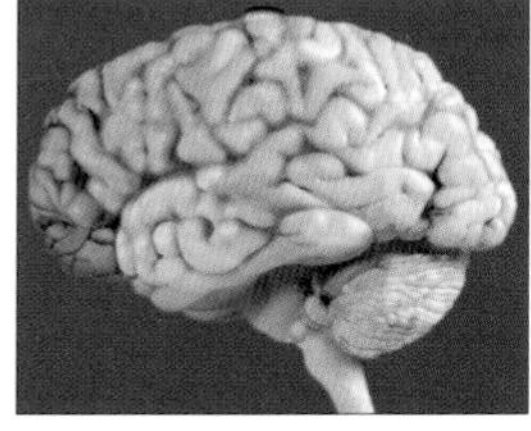

dreaming

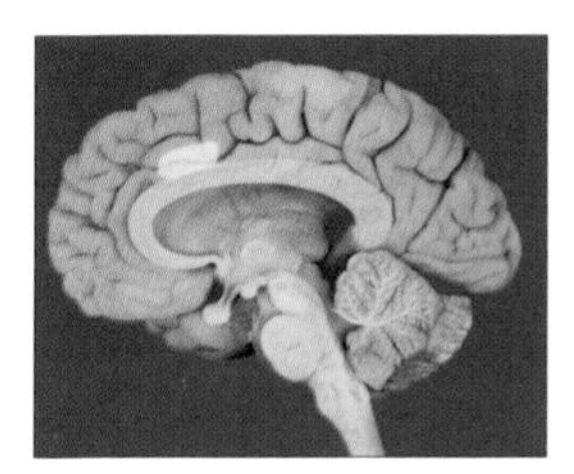
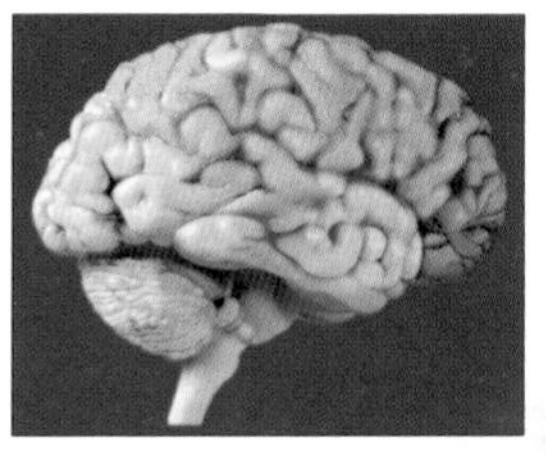

meditation

momentarily banish all reason, disrupt your ability to put one foot in front of another and dispel the pain in your neck. It may also cause you to drop your jaw and pop your eyes as though straining to pick up every minute molecule and light ray that might enhance your perception. If, by contrast, you deliberately think about such a person, your external sensory apparatus will be turned down – one reason why a person deeply involved in a fantasy may fail to hear the call to dinner. A person grappling with a tough reasoning problem may similarly ignore background incidents because attention also requires a certain amount of spare brain capacity.[7]

Many brain regions are involved in directing and controlling attention. One that is especially concerned with holding internally generated stimuli in focus is the anterior cingulate cortex, a region on the inside front edge of the longitudinal fissure, the deep chasm that runs from the front of the brain to the back. This region is sensitive to information from the body and seems to play a part in labelling stimuli as coming from outside or in. It lights up fiercely when a person feels pain, and it also becomes active when we are conscious of emotion.[8] Indeed, the picture generated by a brain experiencing physical pain is in some ways similar to one feeling emotional pain – one reason, perhaps, why the words we use to describe the two states are so often the same.

The nuts and bolts of thinking – holding ideas in mind and manipulating them – takes place in a region of cortex on the dorsolateral (upper side) prefrontal cortex. This is also the location of the closely related activity called working memory. Planning takes place in this area, and it is here that choices are made between various possible actions. Some studies suggest that each type of information has its own special temporary storage niche. An area in the upper reaches of the right hemisphere prefrontal lobe, for example, has been found to light up when a person holds information about objects that are temporarily out of sight. Another spot nearby seems to hold the memory of how many times you have done a thing before. This may be part of a sort of metamemory – the ability to 'know what you know' and to recognize when a particular activity has been 'done to death' – both of which are skills that often seem to be missing in people with frontal lobe damage.[9]

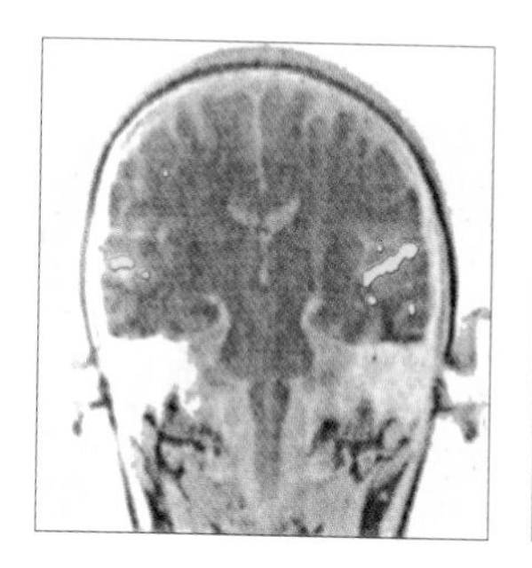
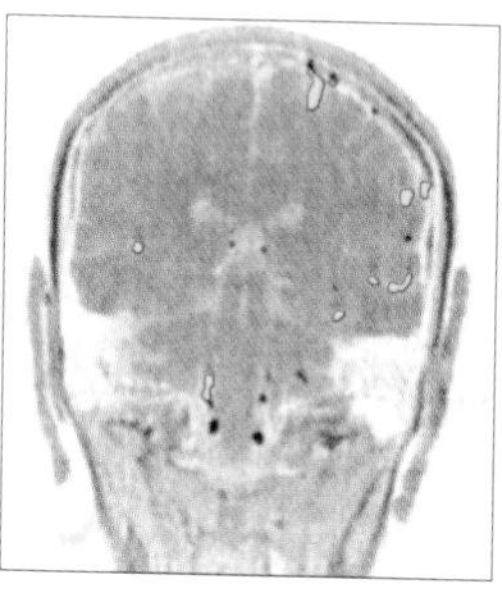

The same stimulus activates different brain areas according to whether the subject is paying attention to it or not. The scans above show, left: the brain of a person who is hearing speech, but concentrating on their breathing. The auditory cortex is responding, but few other areas of the brain are active. Right: Now the subject is actively listening to the words – note how many other areas of the brain have come alive.

Damage to this part of the prefrontal lobe undermines a person's ability to monitor their own performance and to learn from their mistakes. It may also wreck their working memory, making them absent-minded and unable to do tasks like adding figures or doing two or three things in succession. Long-term memory may not be affected, however – it is the ability to juggle memories, not to recall them, that is likely to be worst hit.

One way in which this sort of damage may show up is by bringing about a sort of mental stasis. People in this condition have slow and blunted thought processes. They seem to be

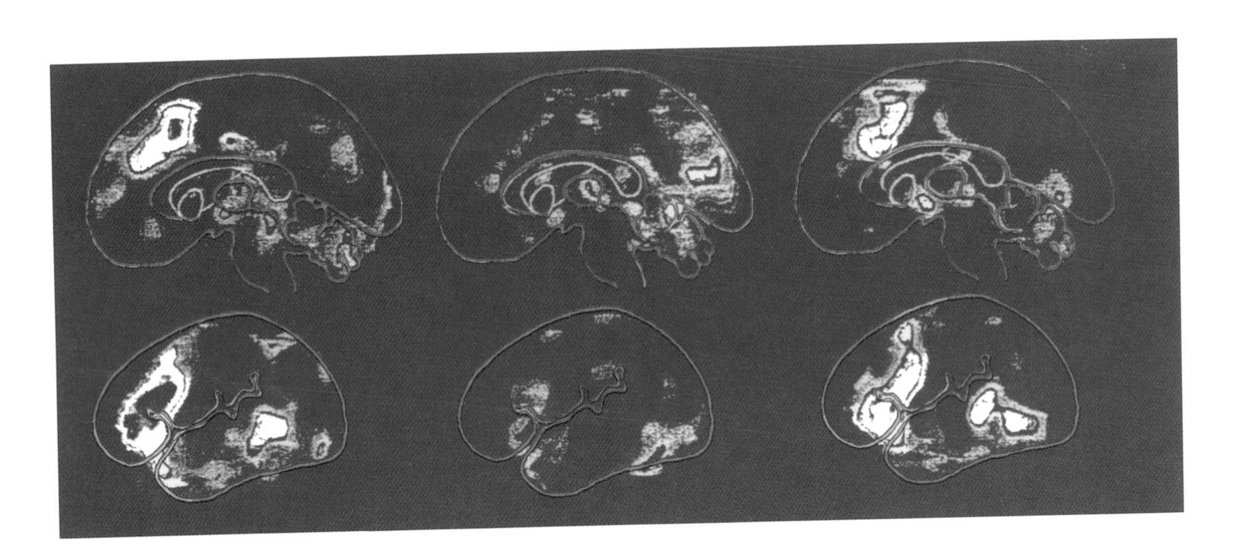

Novel actions that require choice involve more brain activity than routine ones. The scans on the left show the brain of a person choosing which words to say. The areas that are lit up are those concerned with making decisions and focusing attention. In the scans in the middle, the person has practised until the task has become routine and these areas remained switched off. On the right the person is choosing new words, so the activity returns.

rooted in mind-ruts, unable to move in a fresh direction, even when old ways repeatedly fail to bring reward.

A test called the Wisconsin Card Sorting Task shows up this state very clearly. It involves getting participants to sort differently marked and coloured cards into categories. They are first asked to start sorting but are not told how to do it – whether by colour, shape or into equal piles. When they start putting things in piles, using whatever method they choose, the investigator at first feeds them approval. But after a while the investigator starts saying they are doing it wrong, even though the sorter is still sorting the same way. Normal people soon abandon the original sorting method and try another, at which point the investigator again rewards them for a while, and then starts saying they are doing it wrong. The normal person then shifts to yet another tack, showing an overall strategy that maximizes approval. People with frontal lobe injury often fail to do this. Once they have received approval for sorting in one way they keep doing it that way even when they are told it is no longer correct. It does not take much imagination to see how the mindset that causes this will create problems for them in their daily life.

The ability to make a plan of action is useless without the ability to carry it through. One of the things that contributed to the downfall of Phineas Gage after his accident was that he made a dozen plans a day and was unable to follow any of them through. The essential requirement for following through a plan is to put aside things that are immediately attractive in favour of those that further long-term strategy. This ability seems to be located in the orbito-frontal cortex – the region that lies behind the bridge of the nose and continues beneath the bottom curve of the brain, running backwards towards its core.

As we have seen, the basic drives, urges and desires that motivate behaviour come from the unconscious brain and are essentially reflexive: automatic responses to environmental stimuli. If you see food, for example, and your hypothalamus is registering hunger, your unconscious brain prompts you to eat it.

In practice we suppress most of these urges in favour of more complicated (and ultimately beneficial) behaviour. We don't eat when we see food – we wait until we have bought it, or until it has been served on a plate and put before us on a table. If we are trying to lose weight, we may resist the temptation to eat it altogether. That way we attain long-range aims: to stay out of prison; to maintain a civilized lifestyle; to get into last year's jeans.

Children find it more difficult to resist their impulses, partly because they have yet to learn that self-control is generally a useful strategy, and also because the prefrontal lobes are very slow to mature. Until the prefrontal lobes are fully working – which may not be until a person is in their twenties – the limbic system is the stronger force. It is therefore accurate to say that a child does not have as much free will as an adult.

The orbito-frontal cortex has rich neural connections to the unconscious brain where drives and emotions are generated. The down signals from the cortex inhibit reflex clutching and grabbing, and if you take away that control – as happens sometimes in frontal lobe injury – the unconscious retakes the body. This is seen in a bizarre condition called magnetic apraxia. Patients with this disorder automatically scan the environment for anything that catches their attention. When something does, they reach out and grasp it. Sometimes they are unable to let it go.

Orbito-frontal cortex seems, then, to be the area of the brain that bestows a quality we may refer to as free will.

Yet even this heady endowment does not make consciousness complete. Its most important component is not the ability to plan, or choose, or follow through a strategy despite the insistent urgings of our unconscious brain to chase each passing shadow. Rather it is the intuitive sense of meaning that binds our perceptions into a seamless whole and makes sense of our existence.

Can that, too, be pinpointed?

Astonishingly, it seems that it can. Meaningfulness is inextricably bound up with emotion. Depression is marked by wide-ranging symptoms but the cardinal feature of it is the draining of meaning from life. People in a severe state of depression fail to see life as a unified pattern and start instead to see it as a fragmented, incomprehensible sequence of pointless events. Social bonds are severed, normal activities seem purposeless – everything seems to be falling apart. By contrast, those in a state of mania see life as a gloriously ordered, integrated whole. Everything seems to be connected to everything else and the smallest events seem bathed in meaning. A person in this state is euphoric, full of energy and flowing with love. They are also in a state of high creativity – the connections they see between things, which are often invisible or overlooked by others, are often used by them to make new concepts.

The area of the brain that is most noticeably affected in both depression and mania is an area on the lower part of the internal surface of the prefrontal cortex – the ventromedial or subgenual cortex.[10] This, as we have seen, is the brain's emotional control centre. It is exceptionally active during bouts of mania, and inactive (along with other prefrontal areas) during depression. The connections between this region and the limbic system beneath it are very dense, closely binding the conscious mind with the unconscious, and this configurement is probably what gives it its special status: it is, if you like, the part that best incorporates the whole of our being, making sense of our perceptions and binding them into a meaningful whole.

NEGLECT – A PARTIAL VIEW

The patient had been left half-paralysed by a stroke, but seemed not to know it. This is a (condensed) conversation that took place between him and his doctor:

Doctor: Would you clap both your hands together, please?
(Patient lifts right hand and makes clapping motion in air, then puts it back on bed. Smiles, apparently satisfied.)
Doctor: That was just your right hand. Could you raise your left hand and do it again with them both, please?
Patient: My left hand? Oh. It is a little stiff today. My arthritis.
Doctor: Could you try to lift it though, please?
(Pause. The patient does not move.)
Doctor (repeats): Could you try to lift your left hand, please?
Patient: I did. Didn't you see it?
Doctor: I didn't. Do you think you moved it?
Patient: Of course I did. You can't have been looking.
Doctor: Would you lift it again for me, please?
(Patient does not move.)
Doctor: Are you moving it now?
Patient: Of course I am.
Doctor (indicating left hand lying on bed): What is that then?
Patient (looking): Oh, that. That is not my hand. It must belong to someone else.

The weird refusal to face facts shown by that patient is a well-recognized condition called

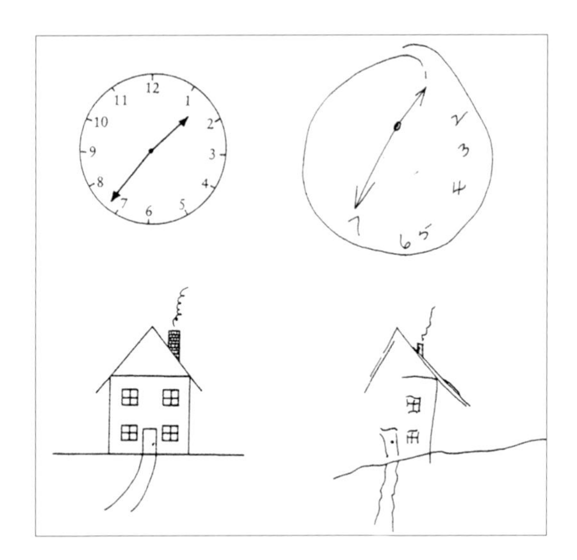

People with certain types of brain injury fail to acknowledge half of the world – everything to one side (usually the left) of their centre line of vision is ignored. The pictures above right show attempts by such a patient to reproduce the pictures on the left.

anosognosia – a word meaning 'lack of knowledge of illness'. It is caused by damage to an area of the brain that is concerned with paying attention to one's own body. Anosognosia is fairly common among patients who have strokes that cause left-sided paralysis. This is because the area that is affected in anosognosia nestles very close to the motor cortex, and strokes (or other injuries) that affect the right motor cortex (therefore the left body side) often affect the area that is associated with anosognosia, too. Sometimes this peculiar dissociation from half of the body comes about when there is no

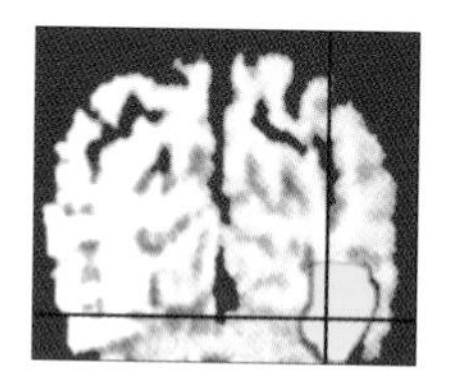
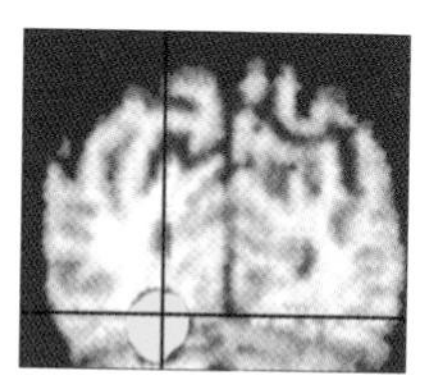

Brain activity switches from one hemisphere to the other as subjects turn their attention from side to side. The scans show the changes in a subject's brain as they look (left) at a visual stimulus on their left (right hemisphere lights up) and then (right) to one on their right.

paralysis. Patients just behave for all the world as though everything to one side of a vertical mid-body line has ceased to exist. They forget to move the limbs on that side. If they walk, they drag the neglected leg behind them. They do not comb one side of their hair. Sometimes – to the extent that it is practically possible – they even forget to put clothes on half their body. The condition is one form of an intriguing condition known as neglect.

Neglect may just be for the left half of the body or it may extend to involve everything in one half of the visual field. Again, the left side is usually the half that is blanked off.

Patients with this form of neglect do not appear to see or be aware of anything to their left. They leave food on the left side of their plate, ignore people who approach them from the left and turn only to the right. If they are asked to draw a clock, they will typically draw a vague right-handed hemisphere, with all the numbers crowded on to one side.

The one-sidedness usually continues even in the person's imagination. If they are asked to close their eyes and imagine walking down a familiar street they will describe, from memory, the buildings on the right but make no mention of those on the other side. The only way to get them to describe the other side of the street is to get them to turn around in their mind's eye and walk back up the other way.[11]

To all intents and purposes people with neglect are 'blind' to half the world. But this is no ordinary blindness. The parts of the brain that deal exclusively with visual input (the primary visual areas) are invariably intact and can be seen on scans to be processing incoming images in the normal way. The blindness is at a higher level in the brain – the level at which sensory input becomes a concept rather than just a stimulus.

People with neglect do not think: 'I cannot see to my left.' Their left simply does not exist in their minds as something to think about at all. Whereas a person with 'normal' half-hemisphere blindness would compensate by turning their head and body towards the blind side in order to bring it into sight, a person with neglect never feels the need to do this. When they read they tend to start each line from the middle of the page and continue to do this even when the text is plainly nonsensical as a result. It just does not occur to them that there is anything 'leftward' to look at.

Neglect is best understood as a defect of attention – an inability of the brain to be conscious of some part of the outside world. What you are not conscious of you cannot miss – hence the patients' airy disregard for their condition.

We all have a touch of visual neglect. There is a blind spot in the normal visual field that corresponds with the area of the retina where the optic nerve leaves the eye. There are no light-sensitive neurons in this spot, so there is no way that light falling on it can be registered by the brain. The disc of blankness this creates is quite big – big enough to obliterate five or six degrees of the visual field. When we look at things with both eyes each eye 'covers' for the other's blind spot. But if you shut one eye your field of vision has a blind area right in the middle.

You can demonstrate this by looking, with one eye only, at one of the crosses below. If you move the book towards your nose at some point the other cross will disappear. This does not, however, produce any conscious awareness of blindness. The impression is that the visual field is complete – an intact page, but one that does not have a cross on it.

\+ \+

Magicians sometimes use their knowledge of the blind spot to trick people right under their noses. In fact, it is only 'under their noses' that such disappearing tricks work because at greater distances the blind spot takes up too little of the visual field, and is too well covered by the other eye, to provide effective camouflage. They are also experts at directing attention away from things they do not want you to see, creating a temporary form of neglect.

Some cases of neglect may arise from damage to the parietal lobe, where it is thought we carry an internal map both of our own body and of the outside world. It is a sort of conceptual amputation. Others seem to be caused by a failure of attention and are asociated with damage to the frontal lobe, the cingulate cortex (the area within the deep fissure between the two hemispheres) and parts of the basal ganglia concerned with controlling movement. The particular aspect of attention that seems to be affected in neglect is the automatic turning towards a stimulus, known as orienting. Like many unconscious processes, orienting is mainly carried out by the right hemisphere of the brain. One way in which this hemisphere may specialize in orienting is by being equipped to turn attention either to the left or to the right. Therefore left-hemisphere damage generally leaves left–right orienting intact. The left hemisphere, by contrast, is thought to orient exclusively to the right, so right-brain damage leaves a person without any left-orienting facility. This is thought to be why people with right-sided brain lesions are more likely to develop neglect than those with left-hemisphere damage.

The most extreme form of anosognosia is a refusal to acknowledge total blindness. This condition is known as Anton's delusion, and people with it blunder about, apparently contained in a visual world entirely of their own construction.

At the other end of the spectrum there are subtle forms of neglect that apply to us all. The stereotypical absent-minded professor who fails to notice he is wearing non-matching socks; the workaholic husband who is staggered to return home one day to find his wife has left him; the feckless debtor who appears not to see the mounting piles of red bills – all these are examples of conceptual neglect that may have a neurological basis. Just as a person who sees shapes more than colours may do so because they have fewer colour-sensitive neurons and more shape-sensitive ones, so the professor probably has fewer neurons in the area of brain associated with self-grooming and more in the area concerned with abstract problem solving. The workaholic may also have a neural deficit. Or perhaps he is short of the neurotransmitter (oxytocin, probably) that would stimulate the area concerned with home and family. The person who runs into debt may have less activity than others, and perhaps fewer brain cells, in their frontal lobes. Their behaviour stems directly from the way they see things.

If prompted, people can usually turn their attention to these areas – just like people with mild left-hemisphere neglect can be forced to attend to the blank side by continual challenge of the 'clap your hands' type. 'Look at it my way' is a more literal figure of speech than it at first appears to be. But unless prompted time and again, it is unlikely that the professor will think about his socks for longer than it takes to change them or that the workaholic will attend to his wife for much longer than it takes to get her back. Most of us make little effort consciously to change our view of the world, and as time goes by we allow our areas of neglect to solidify. Our view of the world may not be entirely internally generated, like people with Anton's delusion. But we are all just a little like them.

Problems up front

Between them the various regions of the prefrontal cortex produce those qualities that we think of as most essentially human: the ability to plan, to feel emotion, to control our impulses, to make choices, and to endow our world with meaning. What, then, of those whose frontal lobes do not function as they should?

The story of Phineas Gage echoes through history because it was the first well-reported case to raise the awkward possibility that morality, free will and responsibility for one's actions were rooted literally in flesh and could be removed without removing the whole person. Since then plenty of other Gages have been discovered, though none has come about their injuries in quite such a sensational way. Most of them, in fact, are victims of common or garden cerebral accidents like strokes. There are also many cases of people who never gain the 'higher' qualities of mind because of developmental brain damage.

J.P. is a classic case. As a young boy he had a normal IQ and could, when he wanted to, do most things – including school work – as well as any other boy of his age. His social behaviour, however, was monstrous. He lied, cheated and stole. Once he borrowed a glove, defecated into it, then returned it to its owner. He seemed to have no grasp of the fundamental concepts of sportsmanship. As he grew up his behaviour took him predictably in and out of various prisons and psychiatric hospitals, picking up the usual labels – schizophrenia, mania, psychopath – along the way.

When he was about twenty J.P. came to the attention of neurologists S.S. Ackerly and Arthur Benton. They noted he had a complete absence of anxiety, incapacity for insight and an inability to learn from punishment. They described him as having 'an unawareness of his total life situation involving todays and tomorrows'.

Brain imaging – using a now obsolete technique in which air was injected into the brain to show up any cavities and an exploratory operation revealed something very much amiss in J.P.'s brain. His left frontal lobe was severely shrunken – and the right lobe was missing altogether. Ackerly and Benton followed J.P. for thirty years. Their final report, written when J.P. was fifty, described him as 'the same uncomplicated, straightforward, outrageously boastful little boy he was at twenty'. They concluded that their patient was 'a very simplified human organism with only rudimentary mechanisms for social adjustment'. Their final words were: 'J.P. has been a stranger in this world, without knowing it.'[12]

Catastrophic frontal lobe injury like that found in J.P. is rare, but prefrontal dysfunction is common to many conditions. Depression and mania, as we have seen, are marked by changes of activity in this area and so is schizophrenia. Indeed, scans of depressed people and withdrawn schizophrenics have many similarities as do those of schizophrenics and people with acute mania.[13] Mania – in behavioural terms – is the 'sunny' side of paranoia – a state in which everything seems to be linked together in some grand scheme. The paranoid delusions of schizophrenia also revolve around mysterious linkages, but the grand scheme is usually sinister rather than glorious. Autism, too, is associated with abnormal prefrontal activity. More controversially frontal lobe dysfunction has been linked to aggressive and violent behaviour. Imaging studies of prisoners have found that a significant number have abnormalities in these lobes – and psychiatrist Itzhak Fried, of the University of California Los Angeles Medical School, has proposed that prefrontal dysfunction might explain the behaviour of men who turn into monsters. As he wrote in *The Lancet*:

'Repeatedly throughout history, groups of individuals, usually young men, have violently attacked other members of society often with the approval of and encouragement from, those in authority. The victims are usually defenceless and are no direct threat to the attackers. Some of the notable manifestations of the phenomenon in this century are the killing of Armenians by Turks in 1915–1916, of European Jews during World War II, of Cambodians during the Pol Pot regime and the ethnic killings in Rwanda in the 1990s. Civil strife, extreme conditions, and ethnic conflict have often had a role in these events, much as poverty and lack of hygiene lead to outbreaks of infectious disease. Yet these events would not have happened without a distinct transformation in the behaviour of individuals.'[14]

Fried suggests that the transformation, which he dubs Syndrome E, is brought about by spasms of overactivity in the orbito-frontal and medial prefrontal cortices. This creates heavy neural down traffic from the cortex, which inhibits the amygdala and prevents emotion from rising to consciousness. In this state people with Syndrome E – typically young men, according to Fried – can carry out horrific acts of violence without being assailed by normal feelings of fear and disgust. After the spasm of hyperarousal the prefrontal cortex falls back, exhausted, into a state of underarousal that precludes normal reflection and self-awareness and thus prevents those with the syndrome from acknowledging the awfulness of what they have done.

The involvement of prefrontal dysfunction in monstrously aberrant behaviour like that cited by Fried is as yet unproven. It is, however, known to cause a certain type of compulsiveness, which can lead to antisocial acts. French neurologist François L'Hermitte has investigated a number of people with damaged or partially

COULD A COMPUTER EVER UNDERSTAND?

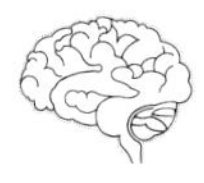

SIR ROGER PENROSE
Rouse Ball Professor of Mathematics, University of Oxford

Few would claim that the machines we use today have much, or even any, understanding – but there are many who argue that it is only a matter of time before computers, or computer-controlled robots, will possess genuine intelligence and therefore really know what they are doing. Indeed, those who hold to what has come to be known as 'strong Artificial Intelligence' (AI) think that machines will eventually possess all the attributes – consciousness, self-awareness, the ability to reflect and so on – that we now think of as essentially human. If they are right, it would mean that understanding, along with all those other human qualities, is something that can be achieved by working things out rather than as a result of some other process or phenomenon.

It seems clear to me that understanding is something that requires awareness – to be fully aware of a situation is the first step to understanding it. A lot of number-crunching may create the appearance of understanding and, conversely, true understanding may circumvent a lot of computation – but the two things are no substitute for one another, rather they are complementary.

I do not think that non-biological machines can ever cross the chasm between computation and understanding. To explain understanding I believe we have to step outside the conventional framework of our present day material world and look to a new physical picture that incorporates the quantum universe – a state whose mathematical structure is largely unknown. This does not mean that understanding has no connection with the brain – in fact, I think there is a specific component of brain tissue that gives rise to it.

The human body contains structures called microtubules – tiny tubes that are especially prevalent in nerve cells. Those in brain cells could, I propose, give rise to a stable quantum state that would bind the activity of brain cells throughout the cerebrum and in doing so give rise to consciousness. Such a state could not be replicated in a computer.

The arguments that underlie my proposal are complicated and some are admittedly speculative. Beneath the technicality I have a strong feeling that it is obvious that the conscious mind cannot work like a computer. That feeling, which comes more easily to children than adults, is surely something that a computer could never have.

CONSCIOUSNESS – NOT A THING BUT A PROCESS

FRANCIS CRICK
Salk Institute of Biological Studies, San Diego

Francis Crick was awarded the Nobel Prize in 1962 for his discovery, with James Watson, of the molecular structure of DNA. Since then he has turned to another huge scientific challenge.

The explanation of consciousness is one of the major unsolved problems of modern science. Indeed, the overwhelming question in neurobiology today is the relation between the mind and the brain. In the past the mind (or soul) was regarded as something separate from the brain but interacting with it in some way. But most neuroscientists now believe that all aspects of the mind, including its most puzzling attribute, consciousness or awareness, are likely to be explainable in a more materialistic way as the behaviour of large sets of interacting neurons. As William James, the father of American psychology, said a century ago: consciousness is not a thing but a process.

Until recently, however, most cognitive scientists and neuroscientists felt that consciousness was either too philosophical or just too elusive to study experimentally. But in my opinion, such timidity is ridiculous. I believe that the only sensible approach is to press the experimental attack until we are confronted with dilemmas that call for new ways of thinking.

The major question that neuroscience must answer is as follows: What are the differences between the active neuronal processes in our heads that correlate with consciousness and those that don't? Are the neurons involved of any particular type? What – if anything – is special about their connections and firing?

Although, in the long run, an all-embracing theory taking in emotion, imagination, dreams, mystical experiences and so on will be necessary, my work assumes that all the different aspects of consciousness employ a basic common mechanism (or perhaps a few such mechanisms). I hope that understanding the mechanism for one aspect will go most of the way to helping us understand them. So my colleague Christof Koch and I, thinking it wise to begin with the aspect of consciousness likely to yield most easily, selected the mammalian visual system because firstly humans are very visual animals and secondly because so much work has already been done on it.

I hold that the biological usefulness of visual consciousness in humans is to produce the best current interpretation of the visual scene in light of past experience (either our own, or that of our ancestors embodied in our genes), and to make this interpretation directly available for a sufficient time to the parts of the brain that contemplate and plan voluntary motor output such as movement or speech.

But there actually seem to be two systems: the rapid acting 'on-line' or unconscious system and the slower, conscious 'seeing system'. To be aware of an object or even an event the brain has to construct a multilevel (for example, lines, eyes, faces), explicit, symbolic interpretation of part of the visual scene. A representation of an object or event will usually consist of representations of many of the relevant aspects of it, which are likely to be distributed over different parts of the visual system. Much neural activity is needed for the brain to construct a representation, most of which is probably unconscious.

The term 'visual consciousness' almost certainly covers a variety of processes. When one is actually looking at a visual scene the experience is very vivid, whereas the visual images produced by trying to remember the same scene are much less vivid or detailed. I am concerned here mainly with the normal, vivid, experience. Some form of very short-term memory seems almost essential for consciousness but this memory may be very transient, lasting for only a fraction of a second. Psychophysical evidence for short-term memory suggests that if we do not pay attention to some aspect of the visual scene, our memory of it is very transient and can be overwritten by subsequent visual stimulus.

Although working memory expands the time frame of consciousness it is not obvious that it is essential. Rather it seems to be a mechanism for bringing an item or a small sequence of items into vivid consciousness, by speech or silent speech. In a similar way, episodic memory, enabled by the hippocampal system, is not essential for consciousness but a person is severely handicapped without it.

Consciousness, then, is enriched by visual attention, though attention is not essential for visual consciousness to occur. Attention is either caused by sensory input or by the planning parts of the brain. Visual attention can be directed to a location in the visual field or to one or more moving objects. The exact neural mechanisms that achieve this are still being debated. But in order to interpret visual input the brain must arrive at a coalition of neurons whose firing represents the best interpretation of the visual scene, often in competition with other possible but less likely interpretations.

missing frontal lobes and all of them, he reports, have one thing in common: when confronted by some cue that suggests they should do something, they seem to be compelled to go ahead and do it. In many of them this shows up as a compulsion to steal: if a purse is left lying unguarded, or a car unlocked, it is, to them, just *demanding* to be taken.

L'Hermitte christened this type of compulsion 'environment dependency syndrome'. In an imaginative series of experiments with two frontal lobotomy patients, he demonstrated just how extreme this automatic obeisance to outside stimuli can be.

In one experiment L'Hermitte invited two of the patients to his home and, without explanation, ushered the first one, a man, into a bedroom. The bed was made up with the top sheet turned down ready for use as in a hotel room. When the patient saw this he immediately got undressed (including removing his toupee), got into bed and prepared to go to sleep, even though it was the middle of the day.

When the patient had been gently persuaded to get up again, L'Hermitte ushered a second patient, a woman, into the room. When she saw the (by now rumpled) bed she, too, went straight to it. But instead of getting in she started to make it. In neither case had the patients' actions been suggested by any word of encouragement or instruction. The actions seemed to be the result of pre-programming. The fact that the first patient got in the bed while the second patient made it presumably reflected something about male/female role playing rather than a fundamental difference in information processing!

In another study, L'Hermitte took one of the patients to another room and said, just before he opened the door, the single word 'museum'. As soon as the patient entered he started to examine the paintings on the wall, just as if he were in a museum. Halfway down the wall there was an obvious gap, where a picture was missing.

Nearby L'Hermitte had placed a hammer, nails and, leaning on the wall beneath, the missing picture. The patient, again without any conversation or instruction, picked up the hammer, banged a nail in the wall and hung the picture.[15]

Although people with frontal lobe damage respond to cues in idiosyncratic or pre-conditioned ways, some reactions are very common. Men with frontal damage (particularly to the part of the brain just above the eyes) often exhibit sexual inhibition or aggression in response to what they see as sexual cues. A recent, rather touching, case of this occurred in a man who was injured in a road accident and was subsequently unable to stop himself proposing to women. He asked one woman to marry him within three days of meeting her and another woman received a proposal during their first telephone call. The judge who heard the man's subsequent claim for damages described him as having changed from a loveable and gentle man to a 'sexual, over-forceful pest who will not take no for an answer'. He also noted that the man had become totally irresponsible with money – buying things like a child, then casting them aside. The £2-million compensation he was awarded is held for him by a state-appointed guardian.[16]

Clearly, these modern-day Phineas Gages do not choose their fate – it is foisted on them by accident or illness. By any normal meaning of the word they cannot be said to possess free will.

So can – or should – they be held responsible for their actions?

The debate (mad or bad?) is ancient and well rehearsed. Does this new science of brain exploration add anything new to it? Surely, yes.

At the moment our legal and moral code is founded on the assumption that each of us contains an independent 'I' – the ghost in the machine that controls our actions. This notion is essentially the same as the dualism first formalized by Descartes. It has endured largely because it *feels* right – how else could mere flesh and blood produce experiences like love, meaning, passion and reverence?

So long as our feelings and actions emerged as though by magic from the black box of our brains it was inevitable that the intuitive explanation of mind would prevail, and as a working hypothesis it has done us proud for centuries. But now that the black box has been opened dualism is fast becoming difficult to sustain. As the studies in this book show, when we look inside the brain we see that our actions follow from our perceptions and our perceptions are constructed by brain activity. In turn, that activity is dictated by a neuronal structure that is formed by the interplay of our genes and the environment. There is no sign of some Cartesian antennae tuned into another world.

Many people baulk at the idea that our actions are entirely mechanistic and some envisage a doomsday scenario if it ever caught on. If people could not be held responsible for their behaviour, they argue, we would all abandon any attempt at responsibility and fall into passive fatalism, acting carelessly on every urge without constraint.

One answer to this is that yes, perhaps we would – if we *could*. But the machine does not work that way. As we have seen, some illusions are programmed so firmly into our brains that the mere knowledge that they are false does not stop us from seeing them. Free will is one such illusion. We may accept rationally that we are machines but we will continue to feel and to act as though the essential part of us is free of mechanistic imperatives.

The illusion of free will is deeply ingrained precisely because it prevents us from falling into a suicidally fatalistic state of mind – it is one of the brain's most powerful aids to survival. Like so many of our survival mechanisms, however, it no longer works entirely to our benefit. By creating the illusion that there is a self-determining 'I' in each of us, it causes us to punish those who appear to behave badly, even when punishment clearly has no practical benefit. It also encourages us to see mechanical breakdowns of the brain as weaknesses of some non-material 'self' rather than as illnesses of the body. These distorted views were probably useful once because they would have driven antisocial and ailing people away from the tribe. Now they just cause pain. At an emotional level we may continue to believe that we are more than machines, but that need not stop us from accepting the opposite on a rational level and adapting our customs to reflect that knowledge – the brain, as we have seen, has no problem 'knowing yet not knowing'. Individually, it is the 'deep' knowledge ingrained in our emotional brain that invariably wins out, but in our dealings with others it is surely better that the rational brain is put in charge.

It seems unlikely to me that we will continue to punish people for misconduct when the crossed wires that spark their behaviour become as clear to see as a broken bone. Rather, I hope (and expect) we will use our knowledge of the brain to develop treatments for sick brains that will be infinitely more effective than the long-winded, hit-and-miss psychological therapies we use today. Restraint could then be used only when such treatment fails, or for those who would rather lose their liberty than their old habits.

I also hope that the ability to modulate brains will be used more widely to enhance those mental qualities that give sweetness and meaning to our lives, and to eradicate those that are destructive. Such an idea reeks of hubris today, and for a while it will be spoken of in the apocalyptic style that greets nearly every new thing that is made possible by science. Soon enough, though, the cries of 'Danger!' will give way to acceptance. Future generations will take for granted that we are programmable machines just as we take for granted the fact that the earth is round. Far from diminishing human existence, I believe that this acceptance will make our lives immeasurably better.

The findings outlined in this book give only the sketchiest impression of the landscape of the mind – the task of creating a detailed picture is one for the new millennium and beyond. Yet I believe one thing is already clear: there is no ghost in this place, no monsters in the depths, no lands ruled by dragons. What today's mind voyagers are discovering is instead a biological system of awe-inspiring complexity. There is no need for us to satisfy our sense of wonder by conjuring phantoms – the world within our heads is more marvellous than anything we can dream up.

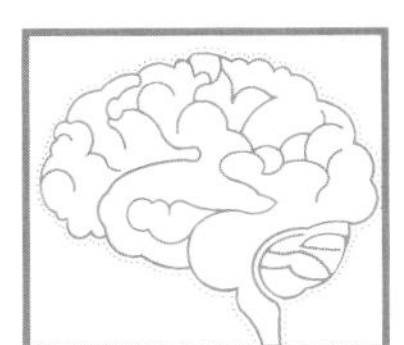

References

Chapter One

1 George Coombe, *Elements of Phrenology* (MA, Marsh, Capen and Lyon, 1834)
2 Itzhak Fried, 'Electrical current stimulates laughter', *Nature*, 391:6668 (1998), 650
3 Carlyle Jacobsen and John Fulton, Yale University, lecture at the Second International Congress of Neurology, London, 1935
4 Edward Shorter, *A History of Psychiatry: From the Era of the Asylum to the Age of Prozac* (New York, John Wiley, 1997), 228
5 Ibid.
6 Ian Cotton, 'Dr Persinger's God machine', *Independent on Sunday*, 2 July 1995
7 Daniel Pendick, 'The New Phrenologists', *New Scientist*, 155:2091 (1997), 34–47
8 J.M. Tanner, *Foetus into Man: Physical Growth from Conception to Maturity* (Ware, Castlemead Publications, 1989), 113
9 British Nutrition Foundation, Institute of Food Research report, January 1998
10 Joseph LeDoux, *The Emotional Brain* (New York, Simon and Schuster, 1996)
11 Ibid.
12 Tanner, *Foetus into Man*
13 J.M. Harlow, *'Recovery from the passage of an iron bar through the head'*, Publications of the Massachusetts Medical Society, 1868, cited in Antonio R. Damasio, *Descartes' Error: Emotion, Reason and the Human Brain* (London, Picador, 1995)
14 Private correspondence between the author and Professor Eric Wassermann
15 W. Penfield and P. Perot, 'The brain's record of auditory and visual experience', *Brain*, 86 (1963), 595–696
16 Steven Rose, *The Making of Memory: From Molecules to Mind* (London, Bantam Press, 1993)
17 The Edwin Smith Surgical papyrus, discovered in Luxor, 1862

Chapter Two

1 Sally P. Springer and Georg Deutsch (eds.), *Left Brain/Right Brain*, 4th edn (New York, W.H. Freeman, 1993)
2 Marthe J. Farah, *Visual Agnosia: Disorders of Object Recognition and What They Tell Us About Normal Vision* (Cambridge, MA, MIT Press, 1991)
3 R.C. Gur et al, 'Differences in the distribution of gray and white matter in human cerebral hemispheres', *Science*, 207:4436 (1980), 1226–8
4 M. Hobbs, 'A randomized controlled trial of psychological debriefing for victims of road traffic accidents', *British Medical Journal*, 313:7070 (1996), 1438–9
5 Michael S. Gazzaniga, *Nature's Mind: The Biological Roots of Thinking, Emotions, Sexuality, Language and Intelligence* (Harmondsworth, Penguin Books, 1992)
6 Richard Nisbett and Timothy Wilson, 'Telling more than we can know – verbal reports on mental processes', *Psychological Review*, 84 (1977), 231–59
7 R.W. Sperry, 'Hemisphere disconnection and unity in conscious awareness', *American Psychologist*, 23 (1968), 723–33

8 Alan J. Parkin, *Explorations in Cognitive Neuropsychology* (Oxford, Blackwell, 1996)
9 Alan J. Parkin, 'The alien hand' in Peter W. Halligan and John C. Marshall (eds.), *Method in Madness: Case Studies in Cognitive Neuropsychiatry* (Hove, Psychology Press, 1996)
10 Parkin in Halligan and Marshall
11 R. Leiguardia, S. Starkstein, M. Nogues, M. Berthier and R. Arbelaiz, 1993, cited in Halligan and Marshall
12 Joseph LeDoux, D. H. Wilson and Michael Gazzaniga, 'A divided mind', *Annals of Neurology*, 2 (1977), 417–21
13 R.W. Sperry, 'Brain bisection and consciousness', in John C. Eccles (ed.), *How the Self Controls Its Brain* (New York, Springer-Verlag, 1966)
14 R.W. Sperry, 'Hemispheric specialization: scope and limits' in F.O. Schmitt and Frederic G. Worden (eds.), *Neuroscience: Third Study Program* (Cambridge, MA, MIT Press, 1974)
15 Matthew 5:25
16 P.G. Hepper et al, 'Handedness in the human foetus', *Neuropsychologia*, 29:11 (1991), 1107–11
17 Luigi Gedda, Director, Gregor Mendel Institute, Rome, quoted by Lawrence Wright in 'Double mystery', the *New Yorker*, 7 August 1995
18 S. Coren and D.F. Helpern, 'Left-handedness: a marker for decreased survival fitness', *Psychological Bulletin*, 109 (1991), 90–106

Chapter Three

1 Peter Chadwick, *Schizophrenia – the Positive Perspective* (London, Routledge, 1997)
2 Richard Restak, *Brainscapes: An Introduction to What Neuroscience Has Learned About the Structure, Function and Ability* (New York, Hyperion, 1995), 106
3 Tourette newsgroup posting on the Internet, January 1998
4 Obsessive–compulsive disorder (OCD) newsgroup posting on the Internet, January 1998
5 Restak, *Brainscapes*, 107
6 OCD newsgroup posting on the Internet, January 1998
7 Frederick Toates, *Obsessional Thoughts and Behaviour* (Place, Thorsons, 1990)
8 Phyllida Brown, 'Over and over', *New Scientist*, 155:2093 (1997), 27
9 Kenneth Blum et al, 'Reward deficiency syndrome', *American Scientist*, 84 (1996)
10 John J. Ratey and Catherine Johnson (eds.), *Shadow Syndromes* (London, Bantam Press, 1997)
11 World Health Organization report on obesity, 1997
12 A. Rothenberger et al, 'What happens to electrical brain activity when anorectic adolescents gain weight', *European Archives of Psychiatry and Clinical Neurosurgery*, 240:3 (1991), 144–7
13 Kaye Weltzin, 'Serotonin activity in anorexia and bulimia', *Journal of Clinical Psychiatry*, 52, supplement (1991), 41–8
14 Wallin G. van der Ster et al, 'Selective dieting patterns among anorectics and

bulimics', *European Eating Disorders Review*, 2 (1994), 221–32

15 Y. Oomura, S. Aou, Y. Koyama and H. Yoshimatsu, 'Central control of sexual behaviour', *Brain Research Bulletin*, 20 (1988), 863–870

16 Jerome Goodman, Columbia University, quoted by Michael S. Gazzaniga, *Nature's Mind: The Biological Roots of Thinking, Emotions, Sexuality, Language and Intelligence* (Harmondsworth, Penguin Books, 1992), 155

17 Simon LeVay, *The Sexual Brain* (Cambridge, MA, MIT Press, 1994), 102

18 Ibid.

19 A. Nystrand, 'New discoveries on sex differences in the brain', National Institute of Ageing, NIH Bethesda, *Lakartidningen* 93:21 (1996), 2071–3 20J. Tiihonen et al, 'Increase in cerebral blood flow in man during orgasm', *Neuroscience Letters*, 170 (1994), 241–3

21 P.J. Reading and R.G. Will, 'Unwelcome orgasms', *The Lancet*, 350:9093 (1997)

22 Louis R. Franzini and John Grossberg (eds.), *Eccentric and Bizarre Behaviours* (New York, John Wiley, 1995)

23 F.R. Farnham, 'Pathology of love', *The Lancet*, 350:9079 (1997), 710

24 Susan Greenfield (ed.), *The Human Mind Explained: The Control Centre of the Living Machine* (London, Cassell, 1996)

Chapter Four

1 Antonio R. Damasio, *Descartes' Error: Emotion, Reason and the Human Brain* (London, Picador, 1995)

2 G. Hohman et al, 'Some effects of spinal cord lesions on feelings', *Psychophysiology*, 3 (1966), 143–56

3 Paul Ekman, quoted in *New Scientist* 'Emotions' supplement, 150:2027 (1996), 11

4 Paul Frank and Paul Ekman, 'Behavioural markers and recognizability of smile of enjoyment', *Journal of Personality and Social Psychology*, 64:1 (1993), 83–93

5 B. Kolb et al, 'Developmental changes in the recognition and comprehension of facial expression: implications for frontal lobe function', *Brain and Cognition*, 20 (1992), 74–84

6 J.S. Morris et al, 'A differential neural response in the human amygdala to fearful and happy facial expressions', letter to *Nature*, 383:6603 (1997), 812–15

7 M.L. Phillips et al, 'A specific neural substrate for perceiving facial expressions of disgust', letter to *Nature*, 389:6550 (1997), 495–7

8 Prosopagnosia newsgroup posting on the Internet, January 1998

9 Ekman, *New Scientist*

10 Ibid.

11 U. Hess et al, 'The facilitative effect of facial expression on the self-generation of emotion', *International Journal of Psychophysiology*, 12:3 (1992), 251–65

12 Interview on *Woman's Hour*, BBC Radio 4, 25 February 1998

13 Ibid.

14 A. Raine et al, 'Brain abnormalities in murderers indicated by positron emission tomography', *Biological Psychiatry*, 42 (1997), 495–508

15 S. Mineka et al, 'Observational condition-

ing of snake fear in rhesus monkeys', *Journal of Abnormal Psychology*, 93 (1984), 355–72
16 Joseph LeDoux, *The Emotional Brain* (New York, Simon and Schuster, 1996)
17 Ibid.
18 Riccardo Brambilla et al, 'A role for the Ras signalling pathway in synaptic transmission and long-term memory', *Nature*, 390:6657 (1997), 281
19 Andrew Young and Kate Leafhead, 'Cotard's delusion' in Peter W. Halligan and John C. Marshall (eds.), *Method in Madness: Case Studies in Cognitive Neuropsychiatry* (Hove, Psychology Press, 1996)
20 H. Forstl and B. Beats, 'Charles Bonnet's description of Cotard's delusion and reduplicative paramnesia in an elderly patient (1788)', *British Journal of Psychiatry*, 160 (1992), 416–18
21 Michael Posner and Marcus E. Raichle (eds.), *Images of Mind* (New York, W.H. Freeman, 1994)
22 Wayne Drevets et al, 'Subgenual prefrontal cortex abnormalities in mood disorder', letter to *Nature*, 386:6527 (1997), 824–7
23 Prosopagnosic newsgroup posting on the Internet, December 1997

Chapter Five

1 Richard Cytowic, *Synaesthesia: A Union of the Senses* (New York, Springer-Verlag, 1989)
2 C. Pantev et al, 'Increased auditory cortical representation in musicians', *Nature*, 392:6678 (1998), 81
3 A. M. Sillito in Richard L. Gregory (ed.), *The Oxford Companion to the Mind* (Oxford University Press, 1987)
4 E. Paulesu et al, 'The physiology of coloured hearing', *Brain*, 118 (1995), 661–76
5 Antonio R. Damasio, 'Neuropsychology: towards a neuropathology of emotion and mood', *Nature*, 386:6527 (1997), 769
6 Richard Cytowik, 'Synaesthesia: phenomenology and neurophysiology', *Psyche*, 2:10 (1995)
7 K. Patterson and J. Hodges in Alan D. Baddeley et al (eds.), *Handbook of Memory Disorders* (Chichester, John Wiley, 1996)
8 Glyn W. Humphreys and M. Jane Riddoch, *The Fractionation of Visual Agnosia in Visual Object Processing* (London, Lawrence Erlbaum Associates, 1997)
9 L.D. Kartsounis and Tim Shallice, 'Modality specific semantic knowledge loss for unique items', *Cortex*, 32:1 (1996), 109–19
10 G. Gainotti et al, *Cognitive Neuropsychology*, 13:3 (1996), 357–89
11 Prosopagnosia website
12 J.E. McNeil and E.K. Warrington, 'Prosopagnosia: a reclassification', *Quarterly Journal of Experimental Psychology*, A, 43:2 (1991), 267–87
13 Tim Shallice, *From Neuropsychology to Mental Structure* (Cambridge University Press, 1989), 390
14 H. Ellis and T. Szulecka, 'The disguised lover: a case of Fregoli delusion', in Peter W. Halligan and John C. Marshall (eds.), *Method in Madness: Case Studies in Cognitive Neuropsychiatry*

(Hove, Psychology Press, 1996)
15 G. Blount, 'Dangerousness of patients with Capgras syndrome', *Nebraska Medical Journal*, 71 (1986), 207
16 Fischer and Frederikson, 'Extraversion, neuroanatomy and brain function – a PET study of personality', *Personality and Individual Differences*, 23, 345–52
17 Joel Katz, 'Phantom limb pain', *The Lancet*, 350:9088 (1997), 1338
18 R. Melzack, 'Phantom limbs', *Scientific American* supplement, 7:1 (1997), 84
19 Steven Rose, *The Making of Memory: From Molecules to Mind* (London, Bantam Press, 1993), 103
20 Morton Schatzman, *The Story of Ruth* (London, Gerald Duckworth, 1980)
21 Oliver Sacks, *The Man Who Mistook His Wife for a Hat* (New York, Summit Books, 1985), 135
22 C.D. Frith, 'Functional imaging and cognitive abnormalities', *The Lancet*, 346:8975 (1995), 615–20
23 Louis R. Franzini and John Grossberg (eds.), *Eccentric and Bizarre Behaviours* (New York, John Wiley, 1995)
24 Paul Sieveking, 'Then I saw her face', *Sunday Times*, 2 November 1997

Chapter Six

1 Oliver Sacks, *Seeing Voices* (London, Picador, 1991), 40
2 Robert Finn, 'Different minds', *Discover*, June 1991
3 Ursula Bellugi, 'Williams syndrome and the brain', *Scientific American*, December 1997, 42–7
4 Uta Frith, 'Autism', *Scientific American*, special issue, April 1997, 92
5 Philip Cohen, 'Hunting the language gene', *New Scientist*, 157:2119 (1998)
6 Paul Fletcher et al, 'Other minds in the brain – a functional imaging study', *Cognition*, 57 (1995), 109–28
7 Simon Baron-Cohen, 'Is there a language of the eyes?', *Visual Cognition*, 4:3 (1997), 311–31
8 Lorna Wing, 'The autistic spectrum', *The Lancet*, 350:9093 (1997), 1762
9 John J. Ratey and Catherine Johnson (eds.), *Shadow Syndromes* (London, Bantam Press, 1997), 230
10 L.W. Olsho, 'Infant frequency discrimination', *Infant Behaviour and Development*, 7 (1984), 27–35
11 Jaak Panksepp, reported in *New Scientist* 'Emotions' supplement, 150:2027 (1996)
12 A.C. North et al, 'In-store music affects product choice', *Nature*, 390:6656 (1997), 132
13 G.A. Ojemann, 'Subcortical language' in H.A. Whitaker (ed.), *Studies in Neurolinguistics,* vol. 1 (New York, Academic Press, 1976)
14 (1) G.A. Donnan et al, 'Identification of brain region for co-ordinating speech articulation', *The Lancet*, 349:9047 (1997), 221; (2) J.R. Binder et al, 'Human brain areas identified by fMRI', *Journal of Neuroscience*, 17:1 (1997), 353–62
15 D.V.M. Bishop, 'Listening out for subtle deficits', *Nature*, 387:6629 (1997), 129
16 J. Graham Beaumont (ed.), *The Blackwell Dictionary of Neuropsychology* (Oxford, Blackwell, 1997)

17 Alan J. Parkin, *Explorations in Cognitive Neuropsychology* (Oxford, Blackwell, 1996), 133
18 H. Goodglass and D. Kaplan, 'The Boston diagnostic aphasia examination' (Philadelphia, Lea and Febiger, 1983)
19 M. Kinsbourne and E.K. Warrington, 'Jargon aphasia', *Neuropsychologia*, 1 (1963), 27–37
20 Christine Temple, *The Brain* (Harmondsworth, Penguin Books, 1993), 90
21 Stephen M. Kosslyn and Oliver Koenig, *Wet Mind: The New Cognitive Neuroscience* (New York, Free Press, 1992)
22 Ralph, Sage and Ellis, 'Word meaning blindness: a new form of acquired dyslexia', *Cognitive Neurology* 13:5 (1996), 617–39
23 E. Paulesu, et al, 'Is developmental dyslexia a disconnection syndrome?', *Brain*, 119 (1996), 143–7
24 Margaret Donaldson, *Human Minds: An Exploration* (Harmondsworth, Penguin Books, 1992)
25 Russ Rymer, *Genie: A Scientific Tragedy* (Harmondsworth, Penguin Books, 1993)
26 Oliver Sacks, *Seeing Voices*
27 K.H.S. Kim et al,'Distinct cortical areas associated with native and second languages', letter to *Nature*, 388:6538 (1997), 171

Chapter Seven

1 Farana Vargha-Khadem et al, 'Differential effects of early hippocampal pathology on episodic and semantic memory', *Science*, 277:5324 (1997), 376–80
2 N. McNaughton et al, 'Reactivation of hippocampal ensemble memories during sleep', *Science*, 19 July 1994, 676–9
3 R.J. Dolan, E. Paulesu and P. Fletcher in Richard Frackowiak et al (eds.), *Human Brain Function* (New York, Academic Press, 1998)
4 E. Maguire et al, 'Recalling routes around London: activation of the right hippocampus in taxi drivers', *Journal of Neuroscience*, 17 (1997), 7103–10
5 Elizabeth F. Loftus, 'Creating false memories', *Scientific American*, September 1997, 51–5
6 Elizabeth F. Loftus, *The Myth of Repressed Memory* (New York, St Martin's Press, 1994)
7 David Schacter et al, 'Recognition memory for recently spoken words', *Neuron*, 17:2 (1996), 267–74
8 Alan D. Baddeley et al (eds.), *Handbook of Memory Disorders* (Chichester, John Wiley, 1996)
9 D. Stuss and D. Benson, 'Neuropsychological studies of the frontal lobes', *Psychological Bulletin*, 95, 3–28
10 Yadin Dudai, *The Neurobiology of Memory: Concepts, Findings, Trends* (Oxford University Press, 1989)
11 Philip J. Hilts, *Memory's Ghost: The Nature of Memory and The Strange Tale of Mr M* (New York, Simon and Schuster, 1996)
12 Colin Blakemore, *The Mind Machine* (London, BBC Books, 1988)
13 R. Lane, letter in *Journal of Neurology, Neurosurgery and Psychiatry*, 63:2 (1997)
14 Baddeley et al, *Handbook of Memory Disorders*
15 Ibid.

16 L.M. Williams, 'Recovered memories of abuse in women with documented sexual victimization histories', *Journal of Traumatic Stress*, 8, 649–73
17 J.D. Bremner et al, 'MRI-based measurement of hippocampal volume in post-traumatic stress disorder', *Biological Psychiatry*, 41 (1997), 23–32
18 F.W. Putnam, 'Dissociative phenomena', *American Psychiatric Press Review of Psychiatry*, 10, 145–60
19 Aleksandr Luria, *The Mind of a Mnemonist* (London, Jonathan Cape, 1969)

Chapter Eight

1 Lawrence Weiskrantz, *Blindsight: A Case Study and Its Implications* (Oxford, Clarendon Press, 1986)
2 Daniel C. Dennett, *Consciousness Explained* (Harmondsworth, Penguin Books, 1993), 331
3 Marshall Miles, 'A description of various aspects of anencephaly', Lafayette College, USA, http://www/lafayette.edu/~loerc/miles.html
4 J.L. Barbur et al, 'Conscious visual perception without V1', *Brain*, 116 (1993), 1293–302
5 C.D. Frith, *Schizophrenia* (Hove, Psychology Press, 1997)
6 Peter W. Halligan et al, 'The functional anatomy of a hysterical paralysis', *Cognition*, 64:1 (1997), B1–8
7 G. Rees, 'Too much for our brains to handle', *New Scientist*, 158:2128 (1998), 11
8 Michael Posner and Marcus E. Raichle (eds.), *Images of Mind* (New York, W.H. Freeman, 1994)
9 D. Stuss in *The Blackwell Dictionary of Neuropsychology* (Oxford, Blackwell, 1997), 350
10 Wayne Drevets et al, 'Subgenual prefrontal cortex abnormalities in mood disorders', letter to *Nature*, 386:6527 (1997), 824–7
11 E. Bisiach and C. Luzzatti, 'Unilateral neglect of representational space', *Cortex*, 14, 129–33
12 S.S. Ackerly and Arthur Benton, 'Report of a case of bilateral frontal lobe defect', *Publication of Association for Research in Neurology and Dental Disease*, 27, 479–504, cited in Antonio R. Damasio, *Descartes' Error: Emotion, Reason and the Human Brain* (London, Picador, 1995)
13 R.A. O'Connell, 'SPECT imaging study of the brain in acute mania and schizophrenia', *Journal of Neuroimaging*, 2 (1995), 101–4; also Ian Daly, 'Mania', *The Lancet*, 349:9059 (1997), 1157–9
14 Itzhak Fried, 'Syndrome E', *The Lancet*, 350:9094 (1997), 1845–7
15 François L'Hermitte, 'Human autonomy and the frontal lobes', *Annals of Neurology*, 19 (1986), 335–43
16 *Daily Telegraph* report, 8 November 1997

BIBLIOGRAPHY

Books

Baddeley, Alan D. et al (eds.). *Handbook of Memory Disorders* (Chichester, John Wiley, 1996)

The Blackwell Dictionary of Neuropsychology (Oxford, Blackwell, 1997)

Blakemore, Colin. *The Mind Machine* (London, BBC Books, 1988)

Calvin, William H. *How Brains Think* (London, Weidenfeld and Nicolson, 1997)

Chadwick, Peter. *Schizophrenia – the Positive Perspective* (London, Routledge, 1997)

Coombe, George *Elements of Phrenology* (MA, Marsh, Capen and Lyon, 1834)

Cytowic, Richard *Synaesthesia: A Union of the Senses* (New York, Springer-Verlag, 1989)

Damasio, Antonio R. *Descartes' Error: Emotion, Reason and the Human Brain* (London, Picador, 1995)

Dennett, Daniel C. *Consciousness Explained* (Harmondsworth, Penguin, 1993)

Donaldson, Margaret *Human Minds: An Exploration* (Harmondsworth, Penguin, 1992)

Dudai, Yadin *The Neurobiology of Memory: Concepts, Findings, Trends* (Oxford University Press, 1989)

Eccles, John C. (ed.) *The Evolution of the Brain: Creation of the Self* (London and New York, Routledge, 1989)

Farah, Marthe J. *Visual Agnosia: Disorders of Object Recognition* and *What They Tell Us About Normal Vision* (Cambridge, MA, MIT Press, 1991)

Frackowiak, Richard et al (eds.) *Human Brain Function* (New York, Academic Press, 1998)

Franzini, Louis R. and Grossberg, John (eds.), *Eccentric and Bizarre Behaviours* (New York, John Wiley, 1995)

Freeman, Walter J. *Societies of Brains: A Study in the Neuroscience of Love and Hate* (Hillsdale, NJ, Lawrence Erlbaum Associates, 1995)

Frith, C.D. *Schizophrenia* (Hove, Psychology Press, 1997)

Frith, Uta *Autism: Explaining the Enigma* (Oxford, Basil Blackwell, 1989)

Gazzaniga, Michael S. *Nature's Mind: The Biological Roots of Thinking, Emotions, Sexuality, Language and Intelligence* (Harmondsworth, Penguin, 1992)

Goodglass, H. and Kaplan, D. *The Boston Diagnostic Aphasia Examination* (Philadelphia, Lea and Febiger, 1983)

Greenfield, Susan (ed.) *The Human Mind Explained: The Control Centre of the Living Machine* (London, Cassell, 1996)

Gregory, Richard L. (ed.) *The Oxford Companion to the Mind* (Oxford University Press, 1987)

Gregory, Richard L. *Eye and Brain, 4th ed.* (Oxford, Oxford University Press)

Halligan, Peter W. and Marshall, John C. (eds.). *Method in Madness: Case Studies in Cognitive Neuropsychiatry* (Hove, Psychology Press, 1996)

Hilts, Philip J. *Memory's Ghost: The Nature of Memory and The Strange Tale of Mr M* (New

York, Simon and Schuster, 1996)
Humphreys, Glyn W. and Riddoch, M. Jane *The Fractionation of Visual Agnosia in Visual Object Processing* (London, Lawrence Erlbaum Associates, 1997)

Kosslyn, Stephen M. and Koenig, Oliver *Wet Mind: The New Cognitive Neuroscience* (New York, Free Press, 1992)

LeDoux, Joseph *The Emotional Brain* (New York, Simon and Schuster, 1996)
LeVay, Simon *The Sexual Brain* (Cambridge, MA, MIT Press, 1994)
Loftus, Elizabeth F. *The Myth of Repressed Memory* (New York, St Martin's Press, 1994)
Luria, Aleksandr *The Mind of a Mnemonist* (London, Jonathan Cape, 1969)

Mithen, Steven *The Pre-History of the Mind* (London, Thames and Hudson, 1996)
Morton, William H. *The Cerebral Code* (Cambridge, MA, MIT Press, 1996)

Nabokov, Vladimir *Speak, Memory* (London, Weidenfeld and Nicolson, 1967)

Ojemann, G.A. *Subcortical language* in H.A. Whitaker (ed.), *Studies in Neurolinguistics, vol. 1* (New York, Academic Press, 1976)

Parkin, Alan J. *Explorations in Cognitive Neuropsychology* (Oxford, Blackwell, 1996)
Penrose, Roger *The Emperor's New Mind* (Oxford, Oxford University Press, 1989)
Piatelli-Palmarini M *Inevitable Illusions* (New York, John Wiley, 1994)
Posner, Michael and Raichle, Marcus E. (eds.) *Images of Mind* (New York, W.H. Freeman, 1994)

Ratey, John J. and Johnson, Catherine (eds.) *Shadow Syndromes* (London, Bantam Press, 1997)
Redfield Jamison, Kay *Touched with Fire: Manic-depressive Illness and the Artistic Temperament* (New York, Free Press, 1995)
Restak, Richard *Brainscapes: An Introduction to What Neuroscience Has Learned About the Structure, Function and Ability* (New York, Hyperion, 1995)
Rose, Steven *The Making of Memory: From Molecules to Mind* (London, Bantam Press, 1993)
Rymer, Russ *Genie: A Scientific Tragedy* (Harmondsworth, Penguin, 1993)

Sacks, Oliver *An Anthropologist on Mars* (London, Picador, 1995)
Sacks, Oliver *The Man Who Mistook His Wife for a Hat* (New York, Summit Books, 1985).
Sacks, Oliver *Seeing Voices* (London, Picador, 1991)
Schatzman, Morton *The Story of Ruth* (London, Gerald Duckworth, 1980)
Schmitt, F.O. and Worden, Frederic G. (eds.) *Neuroscience: Third Study Program* (Cambridge, MA, MIT Press, 1974)
Shallice, Tim *From Neuropsychology to Mental Structure* (Cambridge University Press, 1989)
Shorter, Edward *A History of Psychiatry: From the Era of the Asylum to the Age of Prozac* (New York, John Wiley, 1997)
Silk, Kenneth R. *Biological and Neurobehavioural Studies of Borderline Personality Disorder* (Washington, American Psychiatric Press, 1994)
Springer, Sally P. and Deutsch, Georg (eds.) *Left Brain/Right Brain*, 4th edn (New York, W.H. Freeman, 1993)

Tanner, J.M. *Foetus into Man: Physical Growth from Conception to Maturity* (Ware, Castlemead Publications, 1989)
Temple, Christine *The Brain* (Harmondsworth, Penguin, 1993)
Toates, Frederick *Obsessional Thoughts and Behaviour* (Place, Thorsons, 1990)

Weiskrantz, Lawrence *Blindsight: A Case Study and Its Implications* (Oxford, Clarendon Press, 1986)

Articles and Reports

Barbur, J.L. et al 'Conscious visual perception without V1', *Brain*, 116 (1993), 1293–302
Baron-Cohen, Simon 'Is there a language of the eyes?', *Visual Cognition*, 4:3 (1997), 311–31
Bellugi, Ursula 'Williams syndrome and the brain', *Scientific American*, December 1997, 42–7
Binder, J.R. et al 'Human brain areas identified by fMRI', *Journal of Neuroscience*, 17:1 (1997), 353–62
Bishop, D.V.M. 'Listening out for subtle deficits', *Nature*, 387:6629 (1997), 129
Bisiach, E. and Luzzatti, C. 'Unilateral neglect of representational space', *Cortex*, 14, 129–33
Blount, G. 'Dangerousness of patients with Capgras syndrome', *Nebraska Medical Journal*, 71 (1986), 207
Blum, Kenneth et al 'Reward deficiency syndrome', *American Scientist*, 84 (1996)
Brambilla, Riccardo et al 'A role for the Ras signalling pathway in synaptic transmission and long-term memory', *Nature*, 390:6657 (1997), 281
Bremner, J.D. et al 'MRI-based measurement of hippocampal volume in post-traumatic stress disorder', *Biological Psychiatry*, 41 (1997), 23–32
British Nutrition Foundation, Institute of Food Research report, January 1998
Brown, Phyllida. 'Over and over', *New Scientist*, 155:2093 (1997), 27

Cohen, Philip 'Hunting the language gene', *New Scientist*, 157:2119 (1998)
Coren, S. and Helpern, D.F. 'Left-handedness: a marker for decreased survival fitness', *Psychological Bulletin*, 109 (1991), 90–106
Cotton, Ian 'Dr Persinger's God machine',

Independent on Sunday, 2 July 1995
Cytowik, Richard. 'Synaesthesia: phenomenology and neurophysiology', *Psyche*, 2:10 (1995)

Daily Telegraph report, 8 November 1997
Damasio, Antonio R. 'Neuropsychology: towards a neuropathology of emotion and mood', *Nature*, 386:6527 (1997), 769
Donnan, G.A. et al 'Identification of brain region for co-ordinating speech articulation', *The Lancet*, 349:9047 (1997), 221
Drevets, Wayne et al 'Subgenual prefrontal cortex abnormalities in mood disorder', letter to *Nature*, 386:6527 (1997), 824-7

Farnham, F.R. 'Pathology of love', *The Lancet*, 350:9079 (1997), 710
Finn, Robert 'Different minds', *Discovery*, June 1991
Fischer and Frederikson 'Extraversion, neuroanatomy and brain function – a PET study of personality', *Personality and Individual Differences*, 23, 345–52
Fletcher, Paul et al 'Other minds in the brain – a functional imaging study', *Cognition*, 57 (1995), 109–28
Forstl, H. and Beats, B. 'Charles Bonnet's description of Cotard's delusion and reduplicative paramnesia in an elderly patient (1788)', *British Journal of Psychiatry*, 160 (1992) 416–18
Frank, Paul and Ekman, Paul 'Behavioural markers and recognizability of smile of enjoyment', *Journal of Personality and Social Pychology*, 64:1 (1993), 83–93
Fried, Itzhak 'Electrical current stimulates laughter', *Nature*, 391:6668 (1998), 650
Fried, Itzhak 'Syndrome E', *The Lancet*, 350:9094 (1997), 1845–7
Frith, C.D. et al 'Functional imaging and cognitive abnormalities', *The Lancet*, 346:8975 (1995), 615–20
Frith, Uta 'Autism', *Scientific American*, special issue, April 1997, 92

Gainotti, G. et al *Cognitive Neuropsychology*, 13:3 (1996), 357–89
Gur R.C. et al 'Differences in the distribution of gray and white matter in human cerebral hemispheres', *Science*, 207:4436 (1980), 1226–8

Hepper, P.G. et al 'Handedness in the human foetus', *Neuropsychologia*, 29:11 (1991), 1107–11
Hess, U. et al 'The facilitative effect of facial expression on the self-generation of emotion', *International Journal of Psychophysiology*, 12:3 (1992), 251–65
Hobbs M. 'A randomized controlled trial of psychological debriefing for victims of road traffic accidents', *British Medical Journal*, 313:7070 (1996), 1438–9
Hohman, G. et al 'Some effects of spinal cord lesions on feelings', *Psychophysiology*, 3 (1966), 143–56

Kartsounis, L.D. and Shallice, Tim 'Modality specific semantic knowledge loss for unique items', *Cortex*, 32:1 (1996), 109–19
Katz, Joel 'Phantom limb pain', *The Lancet*, 350:9088 (1997), 1338
Kim, K.H.S. et al 'Distinct cortical areas associated with native and second languages', letter to *Nature*, 388:6538 (1997), 171
Kinsbourne, M. and Warrington, E.K. 'Jargon aphasia', *Neuropsychologia*, 1 (1963), 27–37
Kolb, B. et al 'Developmental changes in the recognition and comprehension of facial expression: implications for frontal lobe func-

tion', *Brain and Cognition*, 20 (1992), 74–84

L'Hermitte, François 'Human autonomy and the frontal lobes', *Annals of Neurology*, 19 (1986), 335–43

Lane, R. Letter in *Journal of Neurology, Neurosurgery and Psychiatry*, 63:2 (1997)

LeDoux, Joseph, Wilson, D. H. and Gazzaniga, Michael *'A divided mind'*, Annals of Neurology, 2 (1977), 417–21

Loftus, Elizabeth F. 'Creating false memories', Scientific American, September 1997, 51–5

Maguire, E. et al 'Recalling routes around London: activation of the right hippocampus in taxi drivers', *Journal of Neuroscience*, 17 (1997), 7103–10

McNaughton, N. et al 'Reactivation of hippocampal ensemble memories during sleep', *Science*, 19 July 1994, 676–9

McNeil, J.E. and Warrington, E.K. 'Prosopagnosia: a reclassification', *Quarterly Journal of Experimental Psychology*, A, 43:2 (1991), 267–87

Melzack, R. 'Phantom limbs', *Scientific American* supplement, 7:1 (1997), 84

Miles, Marshall 'A description of various aspects of anencephaly', Lafayette College, USA, http://www/lafayette.edu/~loerc/miles.html

Mineka, S. et al 'Observational conditioning of snake fear in rhesus monkeys', *Journal of Abnormal Psychology*, 93 (1984), 355–72

Morris, J.S. et al 'A differential neural response in the human amygdala to fearful and happy facial expressions', letter to *Nature*, 383:6603 (1997), 812–15

New Scientist 'Emotions' supplement, 27 April 1996

Nisbett, Richard and Wilson, Timothy 'Telling more than we can know – verbal reports on mental processes', *Psychological Review*, 84 (1977), 231–59

North, A.C. et al 'In-store music affects product choice', *Nature*, 390:6656 (1997), 132

Nystrand, A. 'New discoveries on sex differences in the brain', National Institute of Ageing, NIH Bethesda, *Lakartidningen* 93:21 (1996), 2071–3

O'Connell, R.A. 'SPECT imaging study of the brain in acute mania and schizophrenia', *Journal of Neuroimaging*, 2 (1995), 101–4; also Ian Daly, 'Mania', *The Lancet*, 349:9059 (1997), 1157–9

Olsho, L.W. 'Infant frequency discrimination', *Infant Behaviour and Development*, 7 (1984), 27–35

Oomura, Y., Aou, S., Koyama, Y. and Yoshimatsu, H. 'Central control of sexual behaviour', *Brain Research Bulletin*, 20 (1988), 863–870

Pantev, C. et al 'Increased auditory cortical representation in musicians', *Nature*, 392:6678 (1998), 81

Paulesu, E. et al 'The physiology of coloured hearing', *Brain*, 118 (1995), 661–76

Paulesu, E., Frith, U. et al 'Is developmental dyslexia a disconnection syndrome?', *Brain*, 119 (1996), 143–7

Pendick, Daniel 'The New Phrenologists', *New Scientist*, 2091 (1997), 34–47

Penfield, W. and Perot, P. *'The brain's record of auditory and visual experience'*, *Brain*, 86 (1963), 595–696

Phillips, M.L. et al 'A specific neural substrate for perceiving facial expressions of disgust', letter to *Nature*, 389:6550 (1997), 495–7

Putnam, F.W. 'Dissociative phenomena',

American Psychiatric Press Review of Psychiatry, 10, 145–60

Raine, A. et al 'Brain abnormalities in murderers indicated by positron emission tomography', *Biological Psychiatry*, 42 (1997), 495–508

Ralph, Sage and Ellis 'Word meaning blindness: a new form of acquired dyslexia', *Cognitive Neurology* 13:5 (1996), 617–39

Reading, P.J. and Will, R.G. 'Unwelcome orgasms', *The Lancet*, 350:9093 (1997)

Rees, G. 'Too much for our brains to handle', *New Scientist*, 158:2128 (1998), 11

Rothenberger, A. et al 'What happens to electrical brain activity when anorectic adolescents gain weight', *European Archives of Psychiatry and Clinical Neurosurgery*, 240:3 (1991), 144–7

Sieveking, Paul 'Then I saw her face...', *Sunday Times*, 2 November 1997

Sperry, R.W. 'Hemisphere disconnection and unity in conscious awareness', *American Psychologist*, 23 (1968), 723–33

Stuss, D. and Benson, D. 'Neuropsychological studies of the frontal lobes', *Psychological Bulletin*, 95, 3–28

Tiihonen, J. et al 'Increase in cerebral blood flow in man during orgasm', *Neuroscience Letters*, 170 (1994), 241–3

Van der Ster, Wallin G. et al 'Selective dieting patterns among anorectics and bulimics', *European Eating Disorders Review*, 2 (1994), 221-32

Vargha-Khadem, Farana 'Differential effects of early hippocampal pathology on episodic and semantic memory', *Science*, 277:5324 (1997), 376–80

Weltzin, Kaye 'Serotonin activity in anorexia and bulimia', *Journal of Clinical Psychiatry*, 52, supplement (1991), 41–8

Williams, L.M. 'Recovered memories of abuse in women with documented sexual victimization histories', *Journal of Traumatic Stress*, 8, 649–73

Wing, Lorna 'The autistic spectrum', *The Lancet*, 350:9093 (1997), 1762

Wright, Lawrence 'Double mystery', *New Yorker*, 7 August 1995

INDEX